Management and Industrial Engineering

Series Editor

J. Paulo Davim, Department of Mechanical Engineering, University of Aveiro, Aveiro, Portugal

This series fosters information exchange and discussion on management and industrial engineering and related aspects, namely global management, organizational development and change, strategic management, lean production, performance management, production management, quality engineering, maintenance management, productivity improvement, materials management, human resource management, workforce behavior, innovation and change, technological and organizational flexibility, self-directed work teams, knowledge management, organizational learning, learning organizations, entrepreneurship, sustainable management, etc. The series provides discussion and the exchange of information on principles, strategies, models, techniques, methodologies and applications of management and industrial engineering in the field of the different types of organizational activities. It aims to communicate the latest developments and thinking in what concerns the latest research activity relating to new organizational challenges and changes world-wide. Contributions to this book series are welcome on all subjects related with management and industrial engineering. To submit a proposal or request further information, please contact Professor J. Paulo Davim, Book Series Editor, pdavim@ua.pt

Diego Carou · Antonio Sartal · J. Paulo Davim
Editors

Applying Drones to Current Societal and Industrial Challenges

Springer

Editors
Diego Carou
Department of Design in Engineering
University of Vigo
Ourense, Spain

J. Paulo Davim
Department of Mechanical Engineering
University of Aveiro
Aveiro, Portugal

Antonio Sartal
Department of Business Management
and Marketing
University of Vigo
Vigo, Spain

ISSN 2365-0532 ISSN 2365-0540 (electronic)
Management and Industrial Engineering
ISBN 978-3-031-55573-2 ISBN 978-3-031-55571-8 (eBook)
https://doi.org/10.1007/978-3-031-55571-8

This Springer imprint is published by the registered company Springer Nature Switzerland AG
The registered company address is: Gewerbestrasse 11, 6330 Cham, Switzerland

Paper in this product is recyclable.

Preface

Although drone technology cannot be considered a new technology, in the last years it is emerging as a key technology with a wide range of applications. In this sense, drones are not used anymore by a specific economic sector. Instead, drones are used by both public bodies and companies, by individual users and by researchers.

The present book aims to present state-of-the-art applications of drones to solve some of the challenges of society. Thus, the book deals more with applications than design and technology, though some insights on these topics can be also found.

The book is organized into 12 chapters. The book starts with an introduction to present the topic and review the main concepts related to the book. After that, the following ten chapters deal with specific applications. Thus, chapters "Unmanned Aerial Vehicles (UAVs) in Firefighting" and "Monitoring of the Evolution of the Fire-front of Forest Fires Using Unmanned Aerial Vehicles (UAVs) in Night Conditions" analyse the use of drones in firefighting. Chapter "Key Technologies for UAVs in Search and Rescue (SAR) and Surveillance Operations" is devoted to studying the application of drones in search, rescue, and surveillance operations. Chapter "C-UAS in the Protection of Critical Infrastructures" focuses attention to the protection of large critical infrastructure. Logistics is a clear application for drones. In this sense, the chapter "Transport of Light Parts for Logistics Supply in Industrial Manufacturing Plants by Means of UAV" deals with the use of drones in industrial manufacturing plants for moving light parts. Chapter "Drones for Space Exploration" discusses potential applications in space exploration. In the chapter "Drone Journalism: Where the Human Eye Cannot Reach—Narratives and Journalistic Uses", the use of drones in journalism is presented. Challenges in the application of drones, particularly for commercial purposes are discussed in the chapter "Challenges of Civil Application of 'Drones' for Commercial Purposes in the Logistics and Transportation Sector". Then, the chapter "Autonomous Drones and Smart Containers in Medicine" studies the potential application of drones in delivering critical products in healthcare. Finally, chapter "Drones in Aerospace Engineering Education: The Case of UVigo Aerotech" presents the application of drones as a tool to improve engineering education. After presenting these ten specific applications, the book is completed with a chapter that summarizes the main conclusions of the work and presents some future implications.

The editors would like to thank the work of all the contributors from the Universidad de Sevilla, Universidad Nacional de Educación a Distancia, Universidad Politécnica de Madrid, Istituto di Fisica Applicata "Nello Carrara", IFAC (Consiglio Nazionale delle Ricerche, CNR), Smart Medical Theatre Laboratory, Azienda Toscana Nord Ovest, Universidade de Vigo, and Gradiant Research Center. Especially would like to thank the involvement of the Escola de Enxeñaría Aeronáutica e do Espazo of the Universidade de Vigo and UVigo Aerotech.

Ourense, Spain

Diego Carou
Antonio Sartal
J. Paulo Davim

Contents

Brief Introduction to Unmanned Aerial Systems

Pedro Orgeira-Crespo and Uxía García-Luis

Abstract People have gotten used to seeing what is generally known as a drone in many aspects of life in the last two decades. Present in a great variety of events, it is now considered as a key asset for security, industry, sports, or as a hobby. However, the idea of a manned or unmanned aircraft is quite older than the public might think. Although the television industry brought the operations of Unmanned Aerial Vehicles (UAVs) during the First Gulf War in the early 1990s to screens, the remote-controlled and autonomous aerial vehicles date to several years and run parallel to the history of aviation. This introduction reviews the path that Unmanned Aerial Systems (UASs) have undergone throughout the years, from the very early and simple pioneers to the latest developments, with many applications in different areas of life. A clear distinction between the numerous terms is also delved to provide an overview of the several elements that comprise a whole aircraft system, and all companion systems that make the mission possible. Finally, this chapter provides a classification of these aerial systems, specifically focusing on their structure, and how they fly.

1 Unmanned Aerial Systems

1.1 Introduction

In recent years, Unmanned Aerial Vehicles (UAVs), also known as drones, have become ubiquitous in various sectors, revolutionizing the way of approaching tasks in fields ranging from entertainment to critical infrastructure inspection. The widespread adoption of drones has been fueled by advancements in technology, leading to increased affordability, improved capabilities, and enhanced safety features. This introduction offers a comprehensive overview of the fundamental concepts commonly referred to as drones in modern society, present in many sports

P. Orgeira-Crespo (✉) · U. García-Luis
Department of Mechanical Engineering, Aerospace Engineering School, University of Vigo, Heat Engines and Machines, and Fluids, Aerospace Area, Campus Ourense, 32004 Ourense, Spain
e-mail: porgeira@uvigo.es

D. Carou et al. (eds.), *Applying Drones to Current Societal and Industrial Challenges*, Management and Industrial Engineering, https://doi.org/10.1007/978-3-031-55571-8_1

events, military and security operations, infrastructure inspection, and recreational activities.

Drones have seamlessly integrated into the world of sports, providing captivating aerial perspectives that enhance the viewer experience. From capturing dynamic shots during high-profile events, such as the Olympics, to delivering real-time footage in sports like racing and extreme sports, drones have emerged as valuable tools for sports broadcasters and enthusiasts alike. Their ability to navigate challenging environments and capture high-quality footage has redefined the way people consume sports content [1].

The military and security sectors have extensively adopted drones for reconnaissance, surveillance, and tactical operations. UAVs offer a cost-effective and versatile solution for gathering intelligence in real time without exposing human personnel to potential risks. Armed with advanced sensors and imaging technology, military drones play a crucial role in monitoring hostile territories, tracking targets, and executing precise airstrikes. The integration of artificial intelligence further enhances their autonomy, making them invaluable assets in modern defense strategies [2].

In the realm of infrastructure, drones have proven instrumental in inspecting complex and large structures, such as power lines, bridges, and pipelines. With their sensors and camera payloads, drones can access hard-to-reach or hazardous areas, providing detailed visual inspections. This not only improves the efficiency of maintenance processes but also minimizes risks to human inspectors. The application of drones in infrastructure inspection contributes to the overall safety and reliability of critical systems [3].

Beyond professional and industrial applications, drones have become a popular tool for recreational activities. Enthusiasts and hobbyists use consumer-grade drones for aerial photography, videography, and racing. The accessibility of user-friendly models has democratized drone technology, allowing individuals to explore the skies and capture breathtaking moments. However, the surge in recreational drone use has also prompted discussions around safety, regulations, and privacy concerns [4].

The widespread integration of drones into various aspects of modern life highlights their transformative impact on society. The versatility, cost-effectiveness, and technological advancements of UAVs have led to their adoption in the mentioned sports, military and security operations, infrastructure inspection, recreational pursuits, and many more. As drone technology continues to evolve, it is imperative to address associated challenges, including regulatory frameworks, privacy considerations, and safety protocols, to ensure the responsible and beneficial use of this revolutionary technology [5]. However, it is not that long before people can see drones flying in the sky. Or not?

The First Gulf War, which took place from 1990 to 1991, marked a turning point in the public's perception of drones and unmanned aircraft. During this conflict, televised coverage brought the concept of UAVs into the living rooms of millions worldwide, introducing many to the term "drone" for the first time. This war acted as a catalyst in shaping public perception of drones, notably through television, which played a central role in introducing the concept of unmanned aircraft to a global audience. Viewers were exposed to images of unmanned aircraft, such as the

Pioneer and the Predator, conducting surveillance and reconnaissance missions in the vast deserts of the Middle East [6]. As viewers witnessed the capabilities of these machines in real-time, the term "drone" became emblematic of a new era in military technology. This seminal event laid the groundwork for the integration of unmanned technology into the public consciousness [7].

For many public, the term drone became synonymous with these unmanned aerial vehicles as they witnessed their deployment during the conflict. The captivating footage of drones in action relayed through television broadcasts, presented a new and futuristic dimension to military operations [8]. The public's exposure to these cutting-edge technologies sparked curiosity and, in some cases, apprehension, as people grappled with the idea of aircraft operating without a human pilot on board. While television coverage provided valuable insights into the capabilities of drones, it also gave rise to certain misconceptions. The anonymity and detachment associated with unmanned aircraft, viewed through the lens of television, led to questions about the ethics and implications of remote warfare. The public began to wrestle with the idea of military actions carried out from a distance, challenging traditional notions of combat and its human toll [9]. Anyway, the televised coverage of drones in the First Gulf War laid the groundwork for subsequent advancements in unmanned technology. As public awareness grew, so did the interest in the potential applications and ethical considerations surrounding drones [10]. This crucial moment in history set the stage for the integration of UAVs into military strategies and civilian life, paving the way for their adoption in various sectors, including surveillance [11], reconnaissance [11], and, eventually, commercial [12] and recreational use [13].

1.2 UAV, UAS, Drones

In popular speech, the terms UAV, UAS, and drones can be seen as equivalents. The field of Unmanned Aerial Systems (UAS) is rife with terminology that often interchanges, leading to confusion among both professionals and the public [14].

The term UAV refers specifically to the aircraft itself—an aerial platform that operates without a human pilot on board. There are UAVs in different shapes and sizes, ranging from small (even launched by hand) systems to large, high-altitude, long-endurance vehicles. These vehicles can be utilized for a wide array of applications, including reconnaissance, surveillance, data collection, and, in some cases, combat.

While UAV focuses solely on the aircraft, the term UAS encompasses the entire unmanned system, comprising not only the aerial vehicle but also the control station on the ground, communication links, and any supporting infrastructure. UAS emphasizes the holistic nature of unmanned aerial operations, acknowledging that effective deployment involves a network of interconnected components working together seamlessly [7].

The term drone is perhaps the most colloquially used and widely recognized by the public. It is an umbrella term that can encompass both UAVs and UAS but is often used more broadly to describe any unmanned vehicle, whether aerial, ground, or maritime. In popular culture, the term drone has become synonymous with small consumer-grade quadcopters and hexacopters, often used for recreational purposes and aerial photography.

Beyond UAV, UAS, and drones, the lexicon of unmanned aviation includes several other terms reflecting different aspects of the technology. Remotely Piloted Aircraft Systems (RPAS) is a term often used interchangeably with UAS, emphasizing the remote-control aspect of the aircraft. Unmanned Combat Aerial Vehicle (UCAV) specifically refers to UAVs designed for combat purposes, capable of carrying and deploying weapons [7].

The key differences among these terms lie in their scope and emphasis. UAV isolates the aircraft, UAS encompasses the entire system, and drone serves as a versatile, catch-all term. Understanding these distinctions is crucial for effective communication within the industry and for the public to grasp the intricacies of unmanned aviation.

At its core, a standard Unmanned Aerial System comprises several essential components that collectively contribute to its functionality. These fundamental elements typically encompass air vehicles, one or more Ground Control Stations (GCS), although sometimes referred to as Mission Planning and Control Stations (MPCS), payload, and a data link. The air vehicles form the airborne segment of the system, ranging from drones to more sophisticated unmanned aircraft. The GCS or MPCS serves as the centralized hub for managing and orchestrating the UAS operations, overseeing tasks such as flight control and mission planning [14].

Moreover, beyond the foundational components, many UAS systems boast additional features to enhance their operational capabilities. These supplementary elements may encompass launch and recovery subsystems, ensuring the safe deployment and retrieval of the air vehicles. Furthermore, air-vehicle carriers facilitate the transportation and deployment of unmanned aircraft. Additionally, specialized ground handling and maintenance equipment are often integrated into UAS systems to streamline maintenance procedures and ensure the sustained reliability of the overall system. This comprehensive infrastructure underscores the complexity and versatility of modern UAS designs, showcasing the amalgamation of advanced technologies and logistical support systems [9].

In essence, the integration of these diverse components into a cohesive UAS system represents a synergistic approach aimed at optimizing aerial capabilities. This systematic arrangement enables efficient communication, data exchange, and precise control throughout their operational life cycle. As technology continues to advance, UAS systems evolve to incorporate even more sophisticated features, further expanding their applications across various domains, including military, civilian, and scientific endeavors.

At the heart of a UAS, the aircraft represents the aerial component that encompasses essential elements such as the frame, electric or reciprocating propulsion unit, flight systems, and power system. This airborne platform serves as the carrier for various mission-specific equipment and subsystems. The air vehicle's design and configuration vary widely, accommodating different mission requirements and operational needs. It does not matter if the propulsion is by a ducted fan, if the wings are fixed or are rotary wings, the air vehicle forms the foundation for the entire UAS, dictating its maneuverability, range, and payload capacity [2].

Within the air vehicle, the air data terminal assumes a critical role as part of the communications system. Mounted within the UAS, it facilitates the link of information between the air vehicle and the ground control station or other relevant components. This airborne communication system is pivotal for real-time data transmission, enabling seamless control and coordination during their operational phases. The air data terminal is integral to maintaining a robust and reliable link between the UAS and its ground-based control infrastructure [7].

Furthermore, the payload, housed onboard the air vehicle, constitutes an independent subsystem designed to execute specific mission objectives. This modular and interchangeable component distinguishes itself as a versatile element that can be easily swapped between different air vehicles. The payload's versatility allows for customization based on mission requirements, making it adaptable for diverse applications. Its functionalities may include surveillance equipment, sensors, or other mission-specific instruments, emphasizing the flexibility of UAS systems to cater to a range of operational scenarios. The capability to interchange payloads underscores the UAS's adaptability and capacity to fulfill multifaceted roles across various sectors, from military reconnaissance to scientific research. Finally, it is noteworthy that the term UAV extends to include lighter-than-air vehicles, broadening the spectrum of aerial platforms eligible for unmanned applications [14].

The Mission Planning and Control Station stands as the central operational hub of the Unmanned Aerial Vehicle system. Within this station, critical functions take place, including the processing and display of the images from the camera, the commands from the pilot, and telemetry information originating from the airborne vehicle. Typically, these data streams are transmitted through a ground terminal, constituting the terrestrial segment of the overall data link. MPCS aimed to be a comprehensive facility, incorporating diverse features essential for the effective management and control of UAV operations [15].

The shelter housing the MPCS is equipped with a range of facilities, ensuring its multifunctional role in the UAV system. Among its components are a mission planning facility, providing a space for meticulous pre-mission strategizing, and control and display consoles that serve as interfaces for operators to manage and monitor the UAV in real-time. Furthermore, the shelter encompasses video and telemetry instrumentation, contributing to a comprehensive understanding of the vehicle's operational status. A dedicated computer and signal-processing group manages the vast data streams, ensuring efficient data handling and analysis. The ground data terminal, responsible for relaying information between the MPCS and the UAV, is also housed within this facility. It is complemented by an environmental subsystem as well as the communication electronic equipment to further enhance the MPCS's functionality, emphasizing the importance of this central station in orchestrating the seamless functioning of the entire UAV system [15].

In essence, the MPCS serves as the key element of the UAV system, orchestrating the flow of critical information and ensuring precise control over the airborne vehicle. Its multifaceted design reflects the need for a comprehensive and integrated facility to handle the complexities of UAV operations. The inclusion of mission planning capabilities, control consoles, and advanced instrumentation underscores the role of the MPCS in enabling operators to execute missions with accuracy, efficiency, and adaptability, solidifying its significance in the overall success of the missions [16].

The MPCS assumes a decisive role, not only as the operations center and control system of the UAV system but also as the base post for the personnel involved in mission planning. This dual functionality underscores the MPCS's comprehensive nature within the UAV framework. In the context of mission planning, the MPCS serves as the central command post where individuals receive mission assignments from supported headquarters. These assignments can span various domains, including the weaponry control system, combat information gathering, or overall command and control. The MPCS thus becomes a crucial nexus for integrating mission planning with the broader operational objectives set by higher echelons of command [15].

The MPCS plays a key role in the reporting structure within the UAV system. Individuals stationed at the MPCS are responsible for reporting acquired data and information to the appropriate units. This reporting mechanism ensures that the valuable intelligence and insights gathered during missions are disseminated to relevant entities, such as the mission commander or other operational units. The MPCS, functioning as a strategic command post, enhances communication and coordination, contributing to the overall effectiveness of the UAV system [16].

Within the MPCS, dedicated positions are allocated not only for air vehicle operators but also for mission payload operators. These personnel are tasked with crucial responsibilities related to monitoring and executing missions. Air vehicle operators oversee the monitoring of the UAV itself, ensuring its safe and efficient operation in real-time. On the other hand, mission payload operators are responsible for overseeing the functions of the mission-specific payload onboard, ensuring that it operates according to mission parameters. The allocation of specific roles within the MPCS reflects the need for a specialized and organized approach to managing UAV operations, with operators focused on distinct aspects of the mission to optimize efficiency and mission success [17].

In summary, the dual role of the MPCS as both a command post for mission planning and an operational control center underscores its significance within the UAV system. The integration of command and reporting functions, coupled with specialized operator positions, exemplifies the comprehensive and organized nature of the MPCS, essential for the seamless execution of the missions.

Unmanned Aerial Systems exhibit a spectrum of ground control station designs, varying from portable configurations to complex installations situated remotely from the airborne vehicle [18]. In certain compact UAS, the GCS is ingeniously included within a portable case, facilitating mobility [19]. These cases, often backpack-sized, can be easily transported and set up on the ground. Comprising essentials like a remote control and a display, the station may also integrate embedded microprocessors or operate on a ruggedized laptop. This minimalist setup underscores the agility and versatility of small UAS, suitable for scenarios demanding rapid deployment and maneuverability [20].

Conversely, at the other end of the spectrum, some GCSs are housed in dedicated static structures located away from the UAS's operational airspace (even in different countries) [21]. These sophisticated setups leverage satellite relays to establish and maintain seamless communication with the airborne vehicle. In such instances, the control consoles of the operator may find their place in a dedicated internal room within a large building. The station's infrastructure is extended to include satellite dishes positioned on the roof, ensuring robust and reliable connectivity across vast distances. This configuration enables operators to remotely manage missions with precision, making it an ideal choice for scenarios requiring extensive coverage and global reach.

The diversity in ground control station designs highlights the adaptability of UAS technology to varying operational needs. The range, from portable cases for small UAS to remotely located, satellite-linked stations for large-scale operations, underscores the scalability and flexibility inherent in UAS deployment. This diversity in ground control station setups reflects the evolving landscape of unmanned systems, catering to a broad spectrum of applications across military, civilian, and research domains.

2 Historical Perspective

As with many technical advances throughout history, the history of UAS has had important milestones related to war events that humanity has been involved with. Since the beginning of time, human beings have felt the curiosity and need to fly, which has led humanity to build flying devices; clearly, the history of UAS runs parallel to the history of aircraft, as it is going to be elaborated in this section.

Among the earliest and most captivating endeavors in this pursuit was the invention and ascent of hot air balloons. The story of human flight with hot air balloons begins in 1783 with the Montgolfier brothers, Joseph and Étienne, in Annonay, France. Inspired by earlier experiments with paper and fabric bags billowing with hot air, the brothers sought to scale up their ideas for a public demonstration. On June 4, 1783, in front of a crowd, the Montgolfier brothers launched their unmanned hot-air balloon, heralding the beginning of aerial exploration [22].

Following the success of the unmanned flight, the Montgolfier brothers set their sights on achieving the first manned ascent. On November 21, 1783, in Paris, a large crowd gathered at the Bois de Boulogne to witness history in the making. Pilâtre de Rozier, a scientist, and Marquis d'Arlandes, a military officer, boarded the gondola suspended beneath the hot air balloon [23]. As the balloon ascended into the sky, propelled by the heated air inside, the sight of humans defying gravity captivated onlookers. The balloon reached an altitude of approximately 500 feet and covered a distance of about five and a half miles, marking the first successful human flight in a hot air balloon.

News of the successful manned ascent spread rapidly across Europe, sparking a craze known as "ballooning fever." Aeronauts and adventurers from various countries sought to replicate the Montgolfier brothers' achievement, leading to a series of ballooning endeavors and spectacles. The newfound ability to navigate the skies opened up possibilities for scientific exploration, military reconnaissance, and, eventually, recreational pursuits.

The new creation by the human being was to have its role in history at the hand of military action: on August 22, 1849, the Austrian army launched balloons filled with explosives into the city of Venice. When they reached the vertical of the city, they were made to fall using an electrical device. The balloons were launched from the ship Vulcano [22].

On November 8, 1898, the renowned Serbian inventor Nikola Tesla was granted a patent in the United States for a groundbreaking invention titled "Method and Apparatus for Controlling the Movement Mechanism of Ships or Vessels" commonly referred to as the Teleautomaton [24]. Tesla's patent outlined an innovative system for remotely controlling the movement of vehicles, laying the foundation for future developments in the field of remote guidance and automation. The primary mechanism of the Teleautomaton involved a combination of radio waves and electrical signals to transmit commands from a remote-control station to the vehicle, be it a ship or another mode of transportation. Tesla's invention was a visionary leap, foreseeing the potential for wireless communication and automation in transportation long before such concepts became integral to modern technology [25].

At the core of Tesla's Teleautomaton was a sophisticated system that utilized radio frequency transmission to send signals wirelessly between the control station and the vehicle. The control station was equipped with a transmitter capable of generating specific radio frequency signals corresponding to desired movements such as steering, propulsion, and braking. These signals were then received by a corresponding receiver on the vehicle, which would interpret the commands and actuate the various mechanical components accordingly [24]. The brilliance of Tesla's invention lay not only in its technical aspects but also in its prescient recognition of the transformative power of remote control and automation, foreshadowing the future integration of similar principles in the development of unmanned vehicles and autonomous systems. Tesla's Teleautomaton was a testament to his visionary thinking and a crucial milestone in the evolution of transportation technology.

Another focal event can be traced back to the era of airship flights. An airship, essentially a self-propelled and maneuverable aircraft, operates on the principles of aerostatic lift, achieved by utilizing gas-filled tanks with a lower density than the surrounding atmosphere. While early airships predominantly used hydrogen, the modern preference shifted to helium due to its non-flammable nature, albeit with a slight decrease in lift efficiency. Unlike the aerodynamic lift generated by the rapid movement of wing foils on airplanes or helicopter blades, airships marked the advent of sustained, controlled flight [22]. The zenith of their utilization occurred roughly between 1900 and the 1930s, gradually waning as airplanes surpassed their capabilities and after enduring several significant accidents [26].

The lineage of airships extends back to the late eighteenth century, with the Blanchard airship crossing the English Channel in 1785 [27]. Subsequent developments included Giffard's addition of a steam engine in 1852 and various instances catalyzed by war, such as the Franco-Prussian conflict of the 1870s [22]. By 1883, Tissandier integrated an electric machine into an airship, a precursor to the renowned German zeppelin [26]. Notably, on March 5, 1912, Italian forces became the first to deploy military-grade airships for reconnaissance west of Tripoli, operating behind Turkish lines [22]. However, it wasn't until World War I that airships were officially introduced as weapons, showcasing their strategic importance in military operations [27]. After its golden age, a series of accidents ended the history of airships. the most spectacular and remembered accident was the fire of the hydrogen that provided lift in the LZ 129 Hindenburg on May 6, 1937, which evaporated confidence in airships and marked the final triumph of airplanes, which cost less to build, but more expensive propulsion. Of the 97 people on board, 36 died (13 passengers and 23 crew) [28].

A pivotal moment in the annals of aviation history unfolded with the historic flight of the Wright brothers. On December 17, 1903, at the windswept expanse of Kitty Hawk, Orville and Wilbur Wright conducted their inaugural flight aboard the Flyer I [29]. This pioneering aircraft propelled into the air using an external catapult, undertook a brief yet momentous journey, serving as a testament to the Wright brothers' groundbreaking innovations in aircraft design and control systems. Crucially, the Wright brothers' transformative contribution to aviation lay in their revolutionary approach to controlling mechanisms. Unlike previous aircraft, which grappled with inherent difficulties in maneuverability due to a lack of consideration for wing tilting, the Wright brothers implemented a systematic and meticulous methodology. Their ingenious solution involved the utilization of a swaying motion to control the aircraft during flight, a concept that profoundly enhanced the maneuverability and stability of heavier-than-air crafts [30]. The Wright brothers' methodical work not only facilitated their successful flight but also laid the foundational principles that would govern the design and control of subsequent generations of aircraft. In essence, their pioneering contributions marked a paradigm shift in aviation, propelling the development of more advanced and controllable forms of aerial transportation [31].

Continuing with the historical evolution of aviation, the next milestone has, if possible, even more related to the seed for the history of UAS. In 1915, Tesla proposed to the United Kingdom the concept of a fleet of unmanned aerial vehicles. The idea materialized with the Aerial Target Project, designed for defense against Zeppelins, and as a guided bomb [24].

In 1917, Captain Archibald Low of the British Army embarked on a pioneering initiative with the development of the "Aerial Target" (AT), a radio-controlled drone that left an indelible mark on the evolution of unmanned aerial systems [32]. Conceived in the latter part of 1916, the Aerial Target served a dual purpose: it was envisaged as a defensive tool against Zeppelins, with ground control directing its movements, and as a flying bomb, capable of being controlled from a manned aircraft accompanying it. The construction of the Aerial Target took place at the Royal Aircraft Factory in Putnam under Captain Low's guidance.

Captain Low, a visionary in the realm of wireless radio control, drew inspiration for the Aerial Target from his role in the Army's wireless radio-control unit. Although his initial attempts to construct the aircraft from spare parts were unsuccessful, Captain Low sought the expertise of the Royal Aircraft Factory to bring his vision to fruition. While he designed the wireless device for this innovative unmanned aerial system, the aircraft itself featured a small monoplane powered by a 35-horsepower air-cooled 2-cylinder engine. The distinctive design incorporated a vertical radio antenna wiring that ran along the fuselage and formed a rope-like structure through the wings. Lateral control was achieved through warping, with an extensive dihedral configuration maintaining lateral stability. Six units were manufactured and underwent testing in the summer of 1917. Despite these advancements, military leaders at the time failed to foresee the potential of this groundbreaking concept, ultimately leading to the unfortunate abandonment of the Aerial Target project [33].

On September 12, 1917, a significant milestone in the development of remote-controlled weaponry was achieved with the inaugural flight of the Flying Bomb [9]. Serving as an early iteration of a remotely guided missile, the Flying Bomb represented a pioneering step toward the creation of contemporary cruise missiles. Distinct from conventional bomber planes designed for dropping bombs and returning to base for reuse, the Flying Bomb diverged in its modus operandi. This aerial vehicle, whether manned or unmanned, carried a substantial explosive warhead and, in stark contrast to bombers, intentionally crashed into its target, resulting in its destruction during the attack. The term "flying bomb" has become most closely associated with two prominent weapons from World War II: the German V-1 [34] and the Japanese Ohka [33]. The former, a pilotless vehicle, marked the inception of cruise missiles deployed in combat, while the latter involved a pilot undertaking a kamikaze mission [9].

The Flying Bomb system, credited to Hewit and Sperry, was a sophisticated apparatus comprising various components designed for precision and autonomy. Among its integral elements were a gyroscopic stabilizer, a directive gyroscope, an aneroid barometer for altitude regulation, servo motors responsible for controlling the rudder and aileron, and a distance change device. This comprehensive suite of instruments allowed for the integration of the Flying Bomb system onto an aircraft, which could then be either catapulted into the air or launched from water. Once airborne, the system followed a meticulously predetermined sequence: ascending to a specified altitude, traversing a pre-set course, and, upon covering a predetermined distance, either releasing its payload of bombs or executing a self-launch maneuver upon reaching the ground [33].

While the military initially doubted the system's precision, particularly in targeting ships, it boasted a remarkable range of 50 to 100 miles (160 km) [33]. This extensive operational radius rendered the Flying Bomb system suitable for a range of missions beyond naval targets. The adaptable nature of the system, capable of being launched from both aircraft and water, hinted at its potential versatility in diverse military scenarios. Despite initial reservations about its accuracy, the Flying Bomb system's impressive range positioned it as a strategic asset for missions where precision targeting might be less critical, opening avenues for its application in varied military operations.

In the year 1918, Charles Kettering made a groundbreaking contribution to the field of unmanned aviation with the development of the Kettering Bug, an unmanned biplane that never saw active deployment due to the conclusion of World War I [35]. The biplane featured an innovative clockwork mechanism that powered its flight, and notably, it possessed the capability to fold its wings when it reached the vertical position above its target. This experimental unmanned aerial torpedo, although unrealized in wartime, emerged as a significant precursor to the contemporary cruise missiles that can be recognized today. The Kettering Bug showcased impressive operational characteristics, with the ability to engage ground targets within a radius of up to 120 km from its departure, all while maintaining a speed of 80 km per hour [33].

Despite the Kettering Bug's ambitious design and capabilities, its expensive nature led to its discontinuation. However, the Bug's unique features inspired Dr. Henry W. Walden to conceptualize a rocket system that would allow a pilot to retain control after launch, facilitated by the use of radio waves [36]. These visionary designs laid the foundation for the development of modern missiles, emphasizing the evolution from clockwork-powered experimental unmanned aerial torpedoes to more sophisticated and adaptable missile systems guided by advanced technologies. The Kettering Bug represents a pivotal moment in the historical trajectory of unmanned aviation, steering the course toward the sophisticated missile systems integral to contemporary military operations.

In 1931, a significant leap in unmanned aviation was achieved as the first successful UAV, known as the Queen Bee, underwent a successful test [37]. Derived from a Fairey biplane seaplane, the Queen Bee represented a pilotless, radio-controlled adaptation of the Tiger Moth, specifically designed for the training of anti-aircraft gunners. The nomenclature drone as a generic term for pilotless aircraft is believed to have originated from the Queen Bee's name and its intended purpose [28]. The DH.82, the model from which the Queen Bee was developed, featured a wooden fuselage based on the DH.60 Gipsy Moth, incorporating structural modifications related to cockpit strut placement, and wings reminiscent of the Tiger Moth II. The Queen Bees maintained a standard forward cockpit for test or transfer flights, while the rear cockpit housed a radio-controlled system, employing pneumatically controlled servo systems to operate the controls [36].

A collaborative effort between de Havilland and Scottish Aviation produced a total of 400 Queen Bee examples, with an additional 70 built by the latter; approximately 300 of these aircraft were in active service at the onset of World War II [33]. Piloted by radio control from a boat, the Queen Bee played a crucial role in naval operations, serving as a test target until 1943. This pioneering UAV laid the groundwork for subsequent developments in the field, demonstrating the feasibility and practical applications of radio-controlled pilotless aircraft for military training and target testing.

In a significant development in 1936, the U.S. Navy officially introduced the term drone [33], emphasizing the unmanned nature of these radio-controlled aircraft [31]. This designation was specifically assigned to radio-controlled aircraft that were part of a nascent and low-priority research program initiated by the Navy. The adoption of the term drone represented a departure from traditional aviation language, signifying a shift towards acknowledging the distinct characteristics and applications of these unmanned flying machines. The low-priority research program indicated that, at the time, the full extent of the applications and strategic importance of drones was not fully grasped. However, this moment marked the inception of a trajectory that would lead to the evolution and widespread integration of drones in military and civilian contexts.

As the term drone gained traction within the U.S. Navy, it laid the groundwork for a broader understanding of unmanned aircraft and their diverse applications. While the initial research program may have been considered low priority, it set the stage for subsequent advancements and investments in drone technology, paving the way for the sophisticated unmanned aerial systems that play integral roles in contemporary military operations, surveillance, exploration, and various other fields. The U.S. Navy's introduction of the term drone in 1936 influenced the language and perception surrounding unmanned aerial vehicles [31].

In 1940, a watershed moment in the history of UAVs occurred with the commencement of large-scale production. Denny Industries, under contract with the U.S. Navy, manufactured a staggering 15,000 units of the OQ-2 radio plane [38]. This marked the initial foray into mass production before Northrop acquired Denny Industries in the 1950s. The OQ-2, a relatively uncomplicated aircraft, was propelled by a two-stroke twin-cylinder piston engine generating 6 horsepower, with counter-rotating propellers providing the necessary thrust. Bendix played a crucial role in crafting the radio control system. Launched exclusively by catapult, the OQ-2 was designed to be recovered by parachute after target practice, and its landing gear was exclusively integrated into Army versions for cushioning parachute landings. Intriguingly, the Navy variants, even in upgraded versions, eschewed the use of landing gear.

In the evolution of UAV technology during World War II, a significant breakthrough emerged with the Fox project in 1942 [39]. This initiative gave rise to the TG-2, a drone that achieved a historic first by being guided with the aid of onboard television imagery, successfully striking a target located 50 km away. The year 1944 witnessed the birth of the Interstate TDR, a drone controlled by another aircraft, typically an Avenger. Despite the production of 189 units, the program was eventually terminated, influenced by the efficacy of conventional weaponry in the ongoing conflict with Japan. Post-World War II, experimentation with unmanned radio plane targets continued, culminating in a notable 1950 endeavor where a derivative of the radio plane OQ-3 drone was deployed for laying military communications cables, showcasing the expanding utility of UAVs in various applications [37].

In 1951, the United States initiated the widespread production of the AQM-34, a key advancement in the realm of reconnaissance UAVs [40]. This aircraft, strategically designed for reconnaissance missions, was uniquely launched from a larger aircraft, serving as its mother plane. Following the completion of its mission, the

AQM-34 descended safely by parachute, showcasing innovative recovery methods. This aircraft played a crucial role in intelligence-gathering operations, proving to be an asset for the U.S. military for an impressive span of three decades. Its extended service life underscored the effectiveness of its design and its enduring utility in providing valuable surveillance data [41].

The deployment of the AQM-34 exemplified a shift in reconnaissance strategies, as the UAV's ability to be launched from a larger aircraft offered a versatile and strategic advantage. The reliance on parachute descent for recovery demonstrated an innovative approach to overcoming logistical challenges. Over its three decades of service, the AQM-34 not only contributed significantly to reconnaissance efforts but also paved the way for the development of subsequent generations of UAVs, shaping the trajectory of unmanned aerial systems in military applications [35].

On May 1, 1960, a watershed incident occurred when a U2 aircraft was downed while operating within Soviet airspace [42]. This event served as a catalyst, providing both military and economic impetus for the resurgence of unmanned aircraft into active service. The downing of the U2 triggered a reevaluation of reconnaissance strategies, compelling a shift toward the development and deployment of unmanned alternatives. The incident underscored the vulnerability of piloted reconnaissance missions and emphasized the need for more covert and technologically advanced solutions, thus reigniting the interest and investment in unmanned aerial systems [9]. In response, the military recognized the potential of unmanned systems to navigate and operate in hostile environments without risking human lives. This realization spurred a resurgence in the utilization of unmanned aircraft, not only for their tactical advantages but also for the economic benefits they offered in terms of cost-effectiveness and reduced risk [37]. The resurrection of unmanned aircraft marked a strategic shift in military thinking, laying the foundation for the widespread adoption of these systems in subsequent decades [33].

From 1961 to 1975, the deployment of UAVs as reconnaissance aircraft gained extensive prominence in conflict zones [25]. During this period, the AQM-34 played a pivotal role, undertaking over 3,400 missions, particularly in the tumultuous context of the Vietnam War. Notably, the Vietnam War, marred by its challenging conditions and contentious nature, saw UAVs become indispensable tools for gathering critical intelligence [22]. The AQM-34's operational significance was particularly pronounced, offering a reliable and versatile platform for reconnaissance missions during the complex and demanding environment of the Vietnam War.

The Vietnam War saw UAVs emerge as valuable assets for obtaining crucial information in a cost-effective and less risky manner. The use of UAVs in this conflict era demonstrated their adaptability and effectiveness, solidifying their role as integral components in modern military operations. The extensive deployment of the AQM-34 not only underscored the practical advantages of UAVs but also set the stage for the continued evolution and diversification of unmanned systems in subsequent military engagements [33].

In the year 1980, during the ongoing conflict between Israel and Syria, a notable strategic shift occurred with Israel incorporating a collaborative approach involving both UAVs and manned aircraft [43]. The context of this development lies within the broader historical backdrop of hostilities between Israel and Syria, marked by territorial disputes and geopolitical tensions in the Middle East. The conflict between Israel and Syria during this period was characterized by a series of skirmishes, border tensions, and regional power dynamics [44]. In this conflict, Israel strategically leveraged UAVs not only for aerial reconnaissance but also for a diverse array of electronic warfare functions [9]. These UAVs were not merely confined to traditional reconnaissance roles but were also utilized as electronic decoys, electronic jammers, and countermeasure assets. The integration of these vehicles allowed Israel to introduce a dynamic and technologically advanced dimension to its military operations, contributing significantly to its strategic advantage [33].

The critical role of UAVs was particularly evident in Israel's successful execution of electronic warfare tactics against Syria. By employing UAVs as electronic decoys and jammers, Israel disrupted and deceived Syrian defenses, creating openings for manned aircraft to operate with heightened effectiveness. The combination of UAVs and manned aircraft showcased the synergy between human and autonomous systems, providing Israel with a tactical edge that contributed to its success in the conflict. The utilization of UAVs in diverse roles underscored their adaptability and transformative impact on modern warfare, influencing subsequent military doctrines and strategies [45].

In 1990, the Gulf War emerged as a watershed moment, showcasing the strategic deployment of UAVs in an open and televised fashion [46]. The Gulf War, a conflict between the United States-led coalition forces and Iraq, stemmed from Iraq's invasion of Kuwait in 1990. This geopolitical crisis prompted a robust response from the international community, culminating in Operation Desert Storm, a military campaign led by the U.S. to liberate Kuwait from Iraqi occupation [47].

During the Gulf War, UAVs played a crucial role in reshaping the landscape of modern warfare [22]. The strategic use of UAVs was prominently featured on television, providing the public with unprecedented visibility into the military's reliance on unmanned systems for both detailed reconnaissance and signaling of strategic targets [25], providing crucial intelligence on enemy positions and movements, and contributing to the precision and efficiency of military operations. The televised coverage allowed viewers worldwide to witness firsthand the impact and effectiveness of UAVs in a real-time military setting. This marked a significant departure from previous conflicts, as the Gulf War served as a public showcase of the evolving capabilities and contributions of UAVs to contemporary military strategies [31].

3 Clasification of UAV

3.1 Size

In recent years, a burgeoning interest has emerged in broadening the size spectrum of UAVs [48], encompassing devices ranging from insect-sized units [49] to medium-sized air transport vehicles [50]. The impetus for developing smaller UAVs is driven by the desire for portability, allowing soldiers or border guards to carry, launch, and control model-airplane-sized UAVs easily [51]. These compact vehicles empower individuals to gain visual access beyond immediate obstacles, whether it be a hill or buildings obstructing their line of sight. As miniaturization advances further, reaching the scale of small birds or even insects, the goal is to enable UAVs to navigate confined spaces, fly inside buildings, or inconspicuously perch on windowsills or roof gutters. This realm of small aircraft presents a unique niche, uncontested by manned vehicles, capitalizing on micro-miniaturized sensors and electronics to offer humans the capability to explore and observe the world from an airborne vehicle that can be held in the palm of their hand. These diminutive UAVs can access locations beyond human reach, providing unparalleled perspectives for reconnaissance and surveillance [52].

Concurrently, the motivation for larger UAVs centers on achieving extended endurance at high altitudes, allowing them to cover vast distances and linger over specific areas for prolonged periods [53, 54]. These larger vehicles are equipped with an extensive array of sensors, enhancing their ability to conduct thorough searches or surveillance over designated regions [55]. In military applications, larger UAVs are increasingly sought for their capacity to transport sizable weapons payloads across considerable distances, delivering them precisely to target areas. The discourse surrounding unmanned systems has expanded to include discussions about employing these larger UAVs in roles traditionally reserved for manned aircraft, such as transportation of heavy payloads, war aircraft, and even transportation of people. While the outcome of these discussions remains uncertain, the trajectory indicates an eventual integration of unmanned systems across the entire size spectrum, reflecting the diverse roles these autonomous vehicles are poised to play in the future [22]. Whether navigating confined spaces or fulfilling long-endurance, high-altitude missions, unmanned systems are poised to permeate various sectors, offering a versatile and transformative impact on aerial capabilities [9].

The smallest UAVs are those known as very small, which encompasses a spectrum extending from "micro" dimensions up to dimensions roughly in the range of half a meter [52]. This category encompasses two predominant types of small UAVs: the first type mimics natural flight through the use of flapping wings, resembling the flight patterns of insects or birds; the second type adopts a more conventional aircraft configuration, often employing rotary wings, especially within the range of micro-UAV. The selection between flapping and rotary wings is often driven by specific operational considerations, such as the need to land and perch on small surfaces, facilitating continued reconnaissance without the high energy demanding hovering. Additionally, the use of flapping wings contributes to covert operations, enabling them to fly near subjects under surveillance or perch conspicuously without revealing their identity as sensor platforms [33].

When delving into the smaller end of this size spectrum, especially concerning UAVs with flapping wings, unique aerodynamic challenges come to the forefront. These challenges stem from the intricate aerodynamics that enable small UAVs to achieve flight [37]. Nevertheless, it is essential to underscore that, across all cases, the fundamental aerodynamic principles and equations remain applicable. A comprehensive understanding of these foundational principles is imperative before navigating the intricacies associated with very small-size or flapping-wing configurations. This foundational knowledge serves as a critical precursor to addressing the specialized considerations and conditions inherent to the realm of very small UAVs or those employing bird-like flapping wings [55].

Typically referred to as small UAVs, this category encompasses unmanned aerial vehicles with 50 cm or more in one of its dimensions, and extending up to dimensions reaching at least one meter, and no more than two [38]. A prevalent configuration within this size range mirrors that of fixed-wing model airplanes, often requiring manual launching by the operator, akin to throwing a toy glider into the air. This launching method enhances the accessibility and ease of deployment for operators. Additionally, a substantial proportion of vehicles in this size class adopt a rotary-wing configuration, contributing to the diversity of designs and operational capabilities within the small category. These rotary-wing UAVs offer distinctive advantages, including enhanced maneuverability and the ability to navigate confined spaces with greater agility, making them suitable for various applications ranging from surveillance to reconnaissance [39].

The versatility of small UAVs is evident in their widespread use across different configurations and launch mechanisms [56]. The prevalence of fixed-wing models, reminiscent of traditional model airplanes, highlights the user-friendly aspect of these UAVs, requiring minimal infrastructure for deployment. Simultaneously, the inclusion of rotary-wing designs speaks to the adaptability of these small vehicles, as they can excel in scenarios where vertical takeoff and landing capabilities are crucial. Whether launched into the sky like a glider or leveraging rotary-wing mechanisms for enhanced aerial control, these small UAVs contribute significantly to diverse applications, spanning hobbyist interests [57] to professional surveillance [58] and reconnaissance endeavors [59].

In general terms, a UAV assumes the classification of "medium" when its size surpasses the dimensions suitable for one-person portability yet remains smaller than a light aircraft [52]. Characteristically, these medium-sized vehicles exhibit wingspans ranging from approximately five to ten meters, striking a balance between operational capabilities and manageable dimensions; the payload capacity of medium UAVs is substantial, with the ability to carry loads ranging from 100 to over 200 kg [49]. This payload capacity facilitates the integration of various mission-specific equipment, sensors, or technology, making medium UAVs versatile tools across a broad spectrum of applications. Furthermore, this class encompasses a significant number of rotary-wing configurations, adding an extra layer of adaptability and maneuverability to their operational repertoire [55].

Within the realm of medium-sized UAVs, the emphasis lies in achieving a size that maximizes operational capabilities while retaining practical transportability [52]. The wingspan, a key parameter, ensures a balance between stability during flight and adaptability for deployment. The substantial payload capacity, coupled with the option for rotary-wing designs, makes medium UAVs invaluable for a diverse range of missions. Whether used for reconnaissance, surveillance, or various industrial applications, medium-sized UAVs serve as pivotal assets that bridge the operational gap between the more agile smaller and larger manned aircraft. This classification captures the essence of UAVs designed for versatility and proficiency, adapting to the dynamic needs of contemporary aerial missions.

Ultimately, a UAV assumes the classification of "large" when it possesses the capability to conduct extended-range missions, covering substantial distances from its base, and lingering in the airspace for prolonged durations to fulfill surveillance objectives. These large UAVs are characterized not only by their expansive operational range but also by their ability to carry out persistent monitoring and reconnaissance over designated areas. The size aligns with their intended function of covering vast territories, making them well-suited for applications requiring extensive coverage and prolonged aerial observation.

Moreover, the designation as a large UAV is often associated with the capacity to carry significant quantities of weaponry [48]. This feature underscores the dual role of large UAVs, not only as surveillance platforms but also as potential carriers for various mission payloads. Their size and payload capacity make them formidable assets for military applications, allowing them to carry an array of weapons over long distances, contributing to their strategic value in modern military operations. In essence, the classification of large highlights the comprehensive capabilities of these UAVs, encompassing both extended-range surveillance and the potential for carrying substantial payloads, demonstrating their versatility across a spectrum of mission profiles [52].

3.2 Lift Generation Method

As in any type of aircraft, the fundamental principles of flight revolve around the balance of forces that act upon the UAV. One critical force is a lift, an upward force generated by the wings or rotors that opposes the downward force of gravity, commonly referred to as weight [60]. Lift is crucial for enabling flight, and it is generated through the shape and design of the wings or rotors [50]. Wings exploit the principles of aerodynamics, utilizing airfoil shapes that create pressure differences above and below the wing surface, resulting in an upward force [60]. In the case of rotary-wing UAVs, such as helicopters or quadcopters, the lift is produced by the rotating blades, generating aerodynamic lift to counteract the gravitational pull [61]. Thrust and drag are two additional forces that play key roles in flight [60]. Thrust is the forward force produced by the propulsion system, typically provided by engines or electric motors that propel the aircraft through the air. The UAV generates thrust to overcome drag (the aerodynamic resistance encountered as the vehicle moves through the air). Drag is influenced by the UAV's shape, size, and the airspeed at which it is traveling [61]. Minimizing drag is crucial for optimizing fuel efficiency and achieving higher speeds. UAVs must carefully balance thrust and drag to achieve stable and controlled flight. The propulsion system must provide sufficient thrust to overcome drag and allow the UAV to maintain its intended speed and trajectory [62].

Understanding and managing these aerodynamic forces—lift, weight, thrust, and drag—are essential for designing and operating UAVs effectively. Engineers need to consider these forces in the design, ensuring that the aircraft can generate enough lift to counteract its weight, and enough thrust to overcome drag. Precise control and balance of these forces enable UAVs to achieve the desired flight characteristics, whether it be efficient cruising, hovering, or executing dynamic maneuvers, contributing to the versatility and effectiveness of UAV technology.

Consequently, there is another classification of UAV according to how lift is generated.

Fixed-wing UAVs have their two wings attached to the fuselage of the aircraft in a fixed position and cannot perform any type of movement during flight, only using hyper lift devices such as flaps or slats. The way of generating lift of these aircraft is through their wings, due to the difference in pressures that exists between the under and the upper parts of the airfoil, due to the incidence of an air current against the airfoil in the lower part produces greater pressure than in the upper part and this difference in pressures is what produces a positive net force from the lower to the upper producing lift [60].

Thanks to aerodynamic profiles calculated for pre-established flight conditions (Reynolds number, cruising speed, operating height, etc.), the design process is focused on achieving an efficient lift generation, i.e., a high lift/drag ratio. The higher the Lift/Drag ratio of an airfoil, the higher the efficiency of the airfoil and the more suitable it is for those flying conditions [63].

Fixed-wing UAVs can be further subdivided into four subcategories depending on where the main wing is located concerning the fuselage [64]. Each wing shape has its advantages and disadvantages when the aircraft is in flight:

High Wing UAV: this is perhaps the most convenient way to locate the wing concerning the fuselage, as this positioning confers great stability to the aircraft [65].

Mid-Wing UAV: intermediate configuration between high and low wing, giving you the benefits and disadvantages of each configuration. Aircraft with this configuration have acceptable maneuverability and stability [66].

Low Wing UAV: in this configuration, the wing is located at the bottom of the fuselage. Many Class III UAVs tend to select this configuration because they are large aircraft with high flight inertia and this wing configuration allows them to perform agile maneuvers in flight. These aircraft take advantage of the so-called

pendulum effect to perform rapid maneuvers due to the inherent instability of this configuration [67].
Blended Wing Body: the wing is practically part of the fuselage of the aircraft, it is the most aerodynamically efficient configuration, and many of the new manufacturers of class I aircraft are developing new models with this configuration [68].

The second type of aircraft according to the way they generate lift are rotary-wing UAVs, which produce lift through one or more rotors. The wings of a rotary-wing aircraft are also called propellers since a propeller is nothing more than a rotating wing that produces lift due to a rotation at high revolutions concerning an axis of rotation. The rotation at high revolutions allows these aircraft to generate enough lift to take off and perform the necessary maneuvers to perform their mission and land [69].

This category of aircraft has its advantages and disadvantages, on the one hand, they provide great control and maneuverability, but on the other hand, they are mechanically more complex than fixed-wing aircraft [60]. As they are intrinsically more complex aircraft due to their configuration, a greater number of components are usually needed, and the possibility of failure is greater. These aircraft are indicated for missions in which great maneuverability and positioning precision are needed, many of them are used in photogrammetry, inspection of high voltage lines, and infrastructure control, and in recent years they have also begun to be used in closed interior spaces, without GNSS connectivity.

As with fixed-wing aircraft, this category can be divided into several subcategories: helicopter-type rotary aircraft, twin-engine rotary aircraft, and multirotors (tricopter, quadcopter, or hexacopter). Each subcategory has its strengths and weaknesses as follows:

- Helicopters: these unmanned aircraft feature the conventional configuration of helicopters, with a main rotor and a tail rotor to balance the momentum of the main rotor.
- Bicopters: they can act in a coaxial or tandem manner. It is one of the least used configurations so far, although there are efforts to develop this configuration, focusing on the need to increase flight time.
- Multirotors—Tricopter: multirotors have three or more rotors. Tricopters are one of the most unstable configurations and this is the main reason for their low use at present, as they must have very complex electronics and very sophisticated control algorithms to control these inherently unstable aircraft.
- Multirotors—Quadcopters: this is the most widely used type of multi-rotator aircraft configuration today and the public associates them with 'drones'. The most common is a multirotor-quadcopter, although there are other types of multirotors depending on the number of rotors in the aircraft; hexacopters, octocopters, etc.

References

1. Fan, B., Li, Y., Zhang, R., & Fu, Q. (2020). Review on the technological development and application of UAV systems. *Chinese Journal of Electronics, 29*(2), 199–207.
2. Mohsan, S. A. H., Khan, M. A., Noor, F., Ullah, I., & Alsharif, M. H. (2022). Towards the unmanned aerial vehicles (UAVs): A comprehensive review. *Drones, 6*(6), 147.
3. Greenwood, W. W., Lynch, J. P., & Zekkos, D. (2019). Applications of UAVs in civil infrastructure. *Journal of Infrastructure Systems, 25*(2), 04019002.
4. Bassi, E. (2019). European drones regulation: Today's legal challenges. In *2019 international conference on unmanned aircraft systems (ICUAS)* (pp. 443–450). IEEE.
5. Stöcker, C., Bennett, R., Nex, F., Gerke, M., & Zevenbergen, J. (2017). Review of the current state of UAV regulations. *Remote Sensing, 9*(5), 459.
6. Bone, E., & Bolkcom, C. (2003). Unmanned aerial vehicles: Background and issues for congress. Congressional Research Service, Library of Congress.
7. Dalamagkidis, K., Valavanis, K. P., Piegl, L. A., Dalamagkidis, K., Valavanis, K. P., & Piegl, L. A. (2012). Aviation history and unmanned flight. *On Integrating Unmanned Aircraft Systems into the National Airspace System: Issues, Challenges, Operational Restrictions, Certification, and Recommendations*, 11–42.
8. Burridge, B. (2003). UAVs and the dawn of post-modern warfare: A perspective on recent operations. *The RUSI Journal, 148*(5), 18–23.
9. Cook, K. L. (2007). The silent force multiplier: The history and role of UAVs in warfare. In *2007 IEEE Aerospace Conference* (pp. 1–7). IEEE.
10. Finn, R. L., & Wright, D. (2012). Unmanned aircraft systems: Surveillance, ethics and privacy in civil applications. *Computer Law & Security Review, 28*(2), 184–194.
11. Li, X., & Savkin, A. V. (2021). Networked unmanned aerial vehicles for surveillance and monitoring: A survey. *Future Internet, 13*(7), 174.
12. Shakhatreh, H., Sawalmeh, A. H., Al-Fuqaha, A., Dou, Z., Almaita, E., Khalil, I., ... & Guizani, M. (2019). Unmanned aerial vehicles (UAVs): A survey on civil applications and key research challenges. *IEEE Access, 7*, 48572–48634.
13. Saulnier, A., & Thompson, S. N. (2016). Police UAV use: Institutional realities and public perceptions. *Policing: An International Journal of Police Strategies & Management, 39*(4), 680–693.
14. Huttunen, M. (2017). Unmanned, remotely piloted, or something else? analysing the terminological dogfight. *Air and Space Law, 42*(3), 349–368.
15. Vachtsevanos, G., Tang, L., Drozeski, G., & Gutierrez, L. (2005). From mission planning to flight control of unmanned aerial vehicles: Strategies and implementation tools. *Annual Reviews in Control, 29*(1), 101–115.
16. Atyabi, A., MahmoudZadeh, S., & Nefti-Meziani, S. (2018). Current advancements on autonomous mission planning and management systems: An AUV and UAV perspective. *Annual Reviews in Control, 46*, 196–215.
17. Stecz, W., & Gromada, K. (2020). UAV mission planning with SAR application. *Sensors, 20*(4), 1080.
18. Arnold, K. P. (2016). The uav ground control station: Types, components, safety, redundancy, and future applications. *International Journal of Unmanned Systems Engineering, 4*(1), 37–50.
19. Hong, Y., Fang, J., & Tao, Y. (2008). Ground control station development for autonomous UAV. In *Intelligent Robotics and Applications: First International Conference, ICIRA 2008 Wuhan, China, October 15–17, 2008 Proceedings, Part II 1* (pp. 36–44). Springer, Berlin, Heidelberg.
20. Kang, Y., & Yuan, M. (2009). Software design for mini-type ground control station of UAV. In *2009 9th International Conference on Electronic Measurement & Instruments* (pp. 4–737). IEEE.
21. Sadraey, M. H. (2017). Ground Control Station. *Unmanned aircraft design: A review of fundamentals* (pp. 141–156). Springer International Publishing.
22. Torenbeek, E., & Wittenberg, H. (2009). History of aviation. *Flight Physics: Essentials of Aeronautical Disciplines and Technology, with Historical Notes*, 1–46.

23. Gillispie, C. C. (2014). *The Montgolfier brothers and the invention of aviation 1783–1784: With a word on the importance of ballooning for the science of heat and the art of building railroads* (Vol. 684). Princeton University Press.
24. Petrescu, R. V., Aversa, R., Apicella, A., Abu-Lebdeh, T., & Petrescu, F. I. (2017). Nikola Tesla. *American Journal of Engineering and Applied Sciences, 10*(4), 868–877.
25. Chant, C. (2002). *A Century of Triumph: The History of Aviation*. Simon and Schuster.
26. Lace, W. W. (2008). *The Hindenburg disaster of 1937*. Infobase Publishing.
27. Kotar, S. L., & Gessler, J. E. (2011). *Ballooning: A History, 1782–1900*. McFarland.
28. Mackenzie, D. (2010). *ICAO: A history of the international civil aviation organization*. University of Toronto Press.
29. Anderson, J. D. (2004). *Inventing Flight: the Wright brothers & their predecessors*. JHU Press.
30. Padfield, G. D., & Lawrence, B. (2003). The birth of flight control: An engineering analysis of the Wright brothers' 1902 glider. *The Aeronautical Journal, 107*(1078), 697–718.
31. Haulman, D. L. (2003). *US unmanned aerial vehicles in combat, 1991–2003*. Defense Technical Information Center.
32. Miličević, Z. M., & Bojković, Z. B. (2021). From the early days of unmanned aerial vehicles (UAVS) to their integration into wireless networks. *Vojnotehnički glasnik/Military Technical Courier, 69*(4), 941–962.
33. Fahlstrom, P. G., Gleason, T. J., & Sadraey, M. H. (2022). *Introduction to UAV systems*. Wiley.
34. Evans, S. G., & Delaney, K. B. (2018). The V1 (Flying Bomb) attack on London (1944–1945); the applied geography of early cruise missile accuracy. *Applied Geography, 99*, 44–53.
35. Sullivan, J. M. (2006). Evolution or revolution? The rise of UAVs. *IEEE Technology and Society Magazine, 25*(3), 43–49.
36. Wolf, W. (2021). *Off Target: America's Guided Bombs, Missiles and Drones 1917–1950*. Fonthill Media.
37. Mátyás, P., & Máté, N. (2019). Brief history of UAV development. *Repüléstudományi Közlemények, 31*(1), 155–166.
38. Cai, G., Lum, K. Y., Chen, B. M., & Lee, T. H. (2010). A brief overview on miniature fixed-wing unmanned aerial vehicles. *IEEE ICCA, 2010*, 285–290.
39. Carrillo, L. R. G., López, A. E. D., Lozano, R., & Pégard, C. (2012). *Quad rotorcraft control: Vision-based hovering and navigation*. Springer Science & Business Media.
40. Gutiérrez, G., & Searcy, M. T. (2016). Introduction to the UAV special edition. *The SAA Archaeological Record, Special Issue Drones in Archaeology, 16*(2), 6–9.
41. Keane, J. F., & Carr, S. S. (2013). A brief history of early unmanned aircraft. *Johns Hopkins APL Technical Digest, 32*(3), 558–571.
42. Arthey, V. (2015). *Abel: the true story of the spy they traded for Gary Powers*. Biteback Publishing.
43. Ma'Oz, M. (1995). *Syria and Israel: From War to Peacemaking: From War to Peacemaking*. Clarendon Press.
44. Rabil, R. G. (2003). *Embattled Neighbors: Syria, Israel, and Lebanon*. Lynne Rienner Publishers.
45. Keegan, J. (2011). *A history of warfare*. Random House.
46. Allison, W. T. (2012). *The Gulf War, 1990–91*. Bloomsbury Publishing.
47. Doty, R. M., Winter, D. G., Peterson, B. E., & Kemmelmeier, M. (1997). Authoritarianism and American students' attitudes about the Gulf War, 1990–1996. *Personality and Social Psychology Bulletin, 23*(11), 1133–1143.
48. PS, R., & Jeyan, M. L. (2020). Mini Unmanned Aerial Systems (UAV)-A Review of the Parameters for Classification of a Mini UAV. *International Journal of Aviation, Aeronautics, and Aerospace, 7*(3), 5.
49. Abatti, J. M. (2005). *Small power: The role of micro and small UAVs in the future*. AIR WAR COLL MAXWELL AFB AL Center For Strategy and Technology.
50. Nagel, A., Levy, D. E., & Shepshelovich, M. (2006). Conceptual aerodynamic evaluation of MINI/MICRO UAV. In *44th AIAA Aerospace Sciences Meeting and Exhibit* (p. 1261).

51. Wang, H., Cheng, H., & Hao, H. (2020). The use of unmanned aerial vehicle in military operations. In *International Conference on Man-Machine-Environment System Engineering* (pp. 939–945). Singapore: Springer Singapore.
52. Arjomandi, M., Agostino, S., Mammone, M., Nelson, M., & Zhou, T. (2006). Classification of unmanned aerial vehicles. *Report for Mechanical Engineering class, University of Adelaide, Adelaide, Australia*, 1–48.
53. Nalepka, J., & Hinchman, J. (2005). Automated aerial refueling: extending the effectiveness of UAVs. In *AIAA modeling and simulation technologies conference and exhibit* (p. 6005).
54. Aboumrad, A., Haun, J., McGinnis, A., & Wu, N. (2020). An automatic platform for landing and charging of UAVs to extend UAV operations. In *2020 16th International Conference on Distributed Computing in Sensor Systems (DCOSS)* (pp. 343–347). IEEE.
55. Ladig, R., Paul, H., Miyazaki, R., & Shimonomura, K. (2021). Aerial manipulation using multirotor UAV: A review from the aspect of operating space and force. *Journal of Robotics and Mechatronics, 33*(2), 196–204.
56. Gokbel, E., & Ersoy, S. (2021). Launchable rotary wing UAV designs and launch mechanism designs for rotary wing UAV. *Journal of Mechatronics and Artificial Intelligence in Engineering, 2*(2), 102–113.
57. Cracknell, A. P. (2017). UAVs: Regulations and law enforcement. *International Journal of Remote Sensing, 38*(8–10), 3054–3067.
58. Ibrahim, A. W. N., Ching, P. W., Seet, G. G., Lau, W. M., & Czajewski, W. (2010). Moving objects detection and tracking framework for UAV-based surveillance. In *2010 Fourth Pacific-Rim Symposium on Image and Video Technology* (pp. 456–461). IEEE.
59. Mahmud, I., & Cho, Y. Z. (2018). Detection avoidance and priority-aware target tracking for UAV group reconnaissance operations. *Journal of Intelligent & Robotic Systems, 92*, 381–392.
60. Anderson, J. (2011). *EBOOK: Fundamentals of Aerodynamics (SI units)*. McGraw hill.
61. Johnson, W. (1986). Recent developments in rotary-wing aerodynamic theory. *AIAA journal, 24*(8), 1219–1244.
62. Lee, C., Kim, S., & Chu, B. (2021). A survey: Flight mechanism and mechanical structure of the UAV. *International Journal of Precision Engineering and Manufacturing, 22*(4), 719–743.
63. Kundu, A. K. (2010). *Aircraft design* (Vol. 27). Cambridge University Press.
64. Fielding, J. P. (2017). *Introduction to aircraft design* (Vol. 11). Cambridge University Press.
65. Rhodes, M. D., & Selberg, B. P. (1984). Benefits of dual wings over single wings for high-performance business airplanes. *Journal of Aircraft, 21*(2), 116–127.
66. Hileman, J., Spakovszky, Z., Drela, M., & Sargeant, M. (2006). Aerodynamic and Aeroacoustic Three-Dimensional Design for a" Silent" Aircraft. In *44th AIAA Aerospace Sciences Meeting and Exhibit* (p. 241).
67. Guo, S. (2007). Aeroelastic optimization of an aerobatic aircraft wing structure. *Aerospace science and Technology, 11*(5), 396–404.
68. Okonkwo, P., & Smith, H. (2016). Review of evolving trends in blended wing body aircraft design. *Progress in Aerospace Sciences, 82*, 1–23.
69. Filippone, A. (2006). *Flight performance of fixed and rotary wing aircraft*. Elsevier.

Unmanned Aerial Vehicles (UAVs) in Firefighting

Pedro Orgeira-Crespo, Guillermo Rey, Carlos Ulloa, and Fernando Aguado-Agelet

Abstract UAVs (Unmanned Aerial Vehicles) are present in life in different areas, from sports, leisure, scientific, and military. Forest fires pose a great problem for nature and society and require new observation and control mechanisms, a scenario in which UAVs can provide a fast, economical response without risking pilots' lives. Under the context of the FIREPOCTEP project, the design of a specific UAV for the observation of fires under night conditions was analyzed, designed, and tested. This design was carried out under specific premises for such a mission and concluded with an aircraft that provides surveillance of the firefront in forest fires under safe, reliable, and economical night conditions.

1 Introduction

In an increasingly technologically advanced world, the expansion of unmanned aerial systems represents a breakthrough in the history of aviation of equal magnitude as, for example, the first jet flight [1]. It is necessary to develop new aerial platforms to stay at the forefront both technologically and in the role of solutions to real problems of humanity. Aware of this, more and more countries are preparing for the full development of this sector, as is the case in Galicia (Spain) [2]. On the other hand, if UAS (Unmanned Aerial Systems) already have a large presence in the military [3] and civil recreational market [4], their growing role in different emergencies [5], natural disasters [6], etc., only increase the importance of these platforms that until recently were reserved for the great world powers [7].

P. Orgeira-Crespo (✉) · G. Rey · C. Ulloa
Aerospace Area, Department of Mechanical Engineering, Aerospace Engineering School, University of Vigo, Heat Engines and Machines, and Fluids, Campus Ourense, 32004 Ourense, Spain
e-mail: porgeira@uvigo.es

F. Aguado-Agelet
University of Vigo, Telecommunication Engineering School, 36310 Vigo, Spain

D. Carou et al. (eds.), *Applying Drones to Current Societal and Industrial Challenges*, Management and Industrial Engineering, https://doi.org/10.1007/978-3-031-55571-8_2

Fig. 1 List of entities participating in the FIREPOCTEP project

1.1 The FIREPOCTEP Project

The Firepoctep project is a project promoted by the Interreg VA Spain-Portugal Cooperation Program (POCTEP) 2014–2020 and 75% financed with ERDF Funds [8] (See all participants in Fig. 1). This initiative seeks to strengthen cross-border systems for the prevention and extinction of forest fires and improve resources for the generation of rural employment post-Covid-19, through three fundamental pillars:

- Educate the fixed and occasional rural population about the risk of large forest fires, good preventive practices, and self-protection. Train and equip personnel for cross-border operations.
- Identify Strategic Management Areas to minimize the risk and impact of large fires through comprehensive landscape management in the face of global climate change.
- Promote public and private investment through pilot experiences within the framework of the Circular Green Economy and the investigation of new market opportunities.

The first of these three objectives is part of the action of this project, to provide an aid mechanism for the observation of the fire front under night conditions, in forest fires.

1.2 Forest Fires: The Case of Galicia (Northwestern Spain)

Fires have been a big problem for civilization since its origins, as they cause damage to the environment, material losses, and, sometimes, human losses. Today, forest fires continue to be an environmental threat in Spain, which is one of the countries in Europe with the highest incidence in recent decades [9].

The classification of fires can vary depending on where you are in the world. However, the most common is to divide them into soil, surface, and corona fires, as seen in the following image [10].

Soil fires (also called underground or subsurface fires) are those in which only the humus layer of the soil, which is made up of decaying roots and dry vegetation, burns. These fires spread slowly by incandescent combustion (without flames) and without generating hardly any smoke, so they can go unnoticed for a long time [10].

In surface fires, the bed and understory that is on or near the ground surface, such as mosses, lichens, herbaceous vegetation, shrubs, and small trees, burn. This type is the easiest to extinguish and the one that causes the least damage to forests.

Crown fires (also called crown fires) are those in which the treetops and forest canopy, which includes the leaves, branches, and lichens of trees, burn. They are usually started due to a superficial fire and are the most intense and dangerous fires.

The most common thing in a forest fire is that the three types mentioned above occur. However, the proportion of each of these types can vary from one moment to another depending on the type of fuel, weather conditions, and topography of the place. In fact, any change in one of these factors can turn a soil fire into a surface fire (and vice versa) or a surface fire into a corona fire (and vice versa).

According to the World Wide Fund report [11], 95% of fires in Spain are due to human causes, of which 53% are intentional. The percentage of intentionality increases to 70% in the northwest of the peninsula, which makes Galicia the most affected community with more than 29% of the total fires. In Fig. 2, a map of the fires produced in 2020 by the Ministry for the Ecological Transition and the Demographic Challenge (MITECO) is observed. The map helps verify that the community with the most incidence is the Galicia and, specifically, the province of Ourense [12].

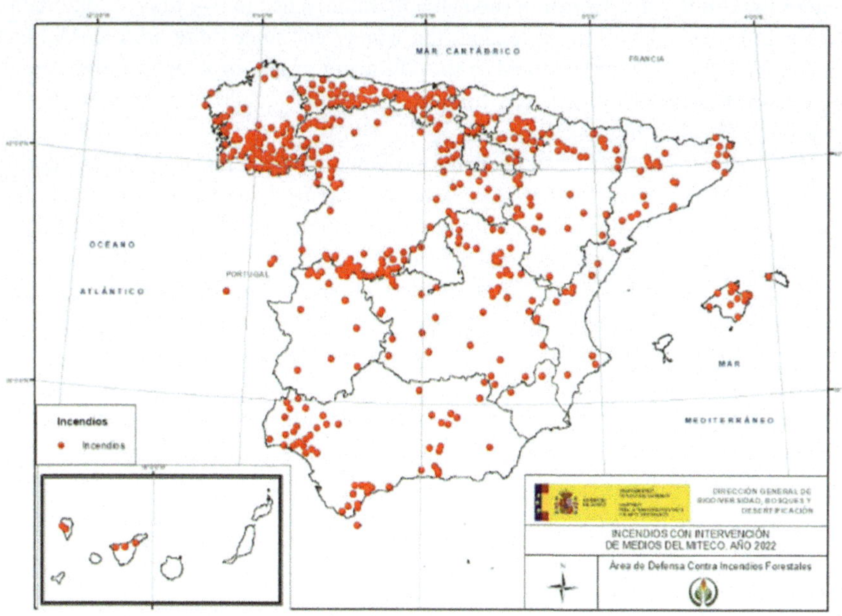

Fig. 2 Fires in Spain in 2022 (*Source* Government of Spain)

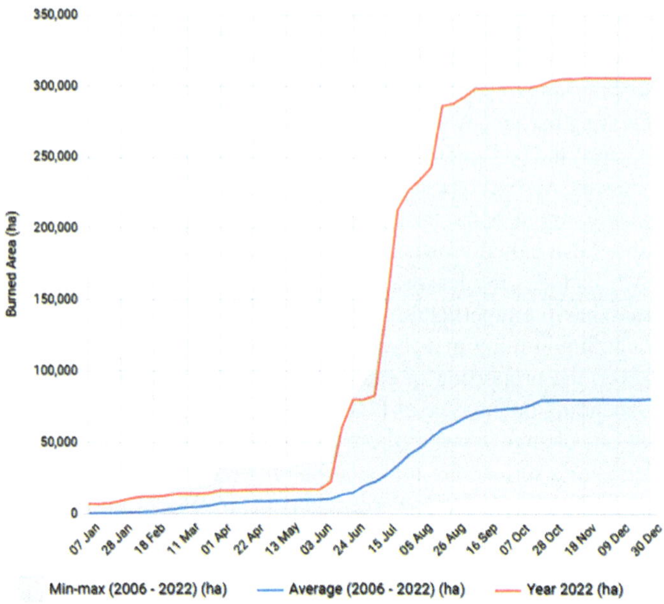

Fig. 3 Hectares burned in Spain in 2022 [13]

The European Forest Fire Information System (EFFIS) considers 2022 the worst year of the century in terms of burned area in Spain, since in less than seven months the figure of 2012 has been exceeded (the worst year since there are records) (see Fig. 3). This has been represented in the following graph, in which in mid-August the 280,000 hectares burned (red line) are exceeded, quadrupling the average of 2006–2021 (dashed line) [13].

The most affected area has been the province of Zamora, where fires such as Losacio or Sierra de la Culebra have devastated more than 60,000 hectares in less than a month. In fact, the Losacio fire has become the largest and fastest advancing in the history of Spain since records began [14].

In Galicia, there have been fires in the four provinces, being again the most punished Lugo and Ourense. The fire of A Pobra do Brollón (Lugo) has become the largest in the history of Galicia with some 11,000 hectares burned in the Sierra de O Courel [15]. It is also worth mentioning the fire of Verín (Ourense), started in 10 different and simultaneous foci. The Ministry of Rural Affairs has pointed out the strategic intentionality of this fire and has indicated its origin in a car moving through the mountain [16].

1.3 The Problem

In the fight against forest fires, there is a critical need, which is the observation of the fire front of the fire: the determination of the propagation of the real perimeter, and the advance always concerning the time of the position of the fire.

Although there are different models of propagation in the literature, which are used in large forest fires, in most of those that usually punish the territory, this mathematical simulation is not used before its extinction for practical reasons of time. On the contrary, at present the most used mechanism for the monitoring of this firefront is visual observation and comparison concerning a map. This mechanism is carried out (in Galicia), both by observation using the means of extinction on the ground, and using the ACO (coordination and observation aircraft, equipped with a camera operator and the presence of the coordinator of aerial means of the fire). Other means include the Superpuma (tracking helicopters) and the specific extinguishing aircraft (including in this section both the FOCA, the TANGO, ALFA, MIKE, and KILO).

The restriction common to all these types of aerial means for monitoring and observation is the same: they are aircraft that, with very few and unusual exceptions will not be able to operate under night conditions. This introduces an added problem and with a great impact on the means of extinction. This occurs not only because of the loss of highly valued information to contract the prediction of the possible evaluation of the fire with reality from a privileged point of view (the sky), but because this absence of aerial means tries to supply reality with an approach to the firefront by terrestrial means. Specifically by people who put their lives at risk by standing near the fire, at night, sometimes on forest tracks, in a terrain that they typically do not know.

The objective, therefore, is to provide, using another mechanism, this observation of the fire front, by aerial means that can fly in night conditions, to obtain on the map the evolution of the perimeter of the fire concerning time. In this context, UAVs can be highly useful, due to their ease and speed of deployment, and practically zero risk for the pilots who are operating them.

This chapter reviews the design of a UAV specifically defined to operate under these conditions and for this purpose, with special emphasis on providing it with exceptional flight autonomy, necessary to cover the long distances to determine the flight perimeter of a forest fire.

A fixed-wing configuration has been chosen since this gives the UAV:

- Increased aerodynamic efficiency: wing airfoils are designed to provide efficient lift when moving through the air. By adding a fixed wing to the drone, the amount of energy needed to maintain flight can be reduced compared to a pure quadcopter, where the rotors must generate all the thrust. This translates into greater energy efficiency and greater flight range.
- Glide flight: with a wing airfoil, the drone can take advantage of the lift generated by the wing to fly more efficiently in glide flight. This means the drone can shut down the engines or reduce power and glide through the air, taking advantage of

updrafts and minimizing aerodynamic drag. The glide flight significantly increases the autonomy of the drone, since constant energy consumption is not required to maintain the flight.

- Increased payload capacity: by adding a wing airfoil, some of the payload capacity that would normally go into batteries needed for the flight of a pure quadcopter can be released. This allows the drone to carry an additional payload, such as sensors, cameras, or observation equipment, without affecting the range too much.

It is important to note that the inclusion of a wing airfoil implies a compromise in terms of design and maneuverability. Fixed-wing drones are not as agile or able to change direction as quickly as quadcopters. However, for applications that require long flight times and large coverage areas, such as wildfire observation, the increased range provided by a wing airfoil can be very beneficial. After an analysis of the shapes and sizes of forest fires of the last ten years in the study areas, it has been determined that autonomy of 100 min is a goal for this project.

2 Aerodynamic Study

This section deals with the aerodynamic analysis of the different parts of the UAV, as well as a whole. The following prerequisites were based on the design, the result of the analysis of the different aircraft that, under the conditions of maximization of autonomy, can be found in Table 1.

In addition, it is necessary to set a series of initial parameters to be able to perform the analyses, such as the speed of flight, the angle of attack, or the height at which the drone will fly. The decision has been made to fly at 25 m/s with a maximum angle of attack of 12°, values considered typical for this type of UAV. Due to the proximity to possible fires, the flight height will depend on factors such as the material chosen, the operating temperature of the components, the wind speed, and the height of the flames. Figure 4 shows the variation of air temperature concerning the distance it is from the flames in a corona fire. Considering this graph, a flight height of 50 m will be assumed from now on, so it is possible to assume an air temperature of around 150 °C in the most extreme cases:

Table 1 Design requirements

Design requirement	Interval
MTOW	2.5–15.0 kg
Wingspan (b)	2.0–5.0 m
Wing surface (S)	0.7–1.9 m^2
Medium chord (c_m)	0.22–0.40 m
Fuselage length (L)	0.7–2.4 m

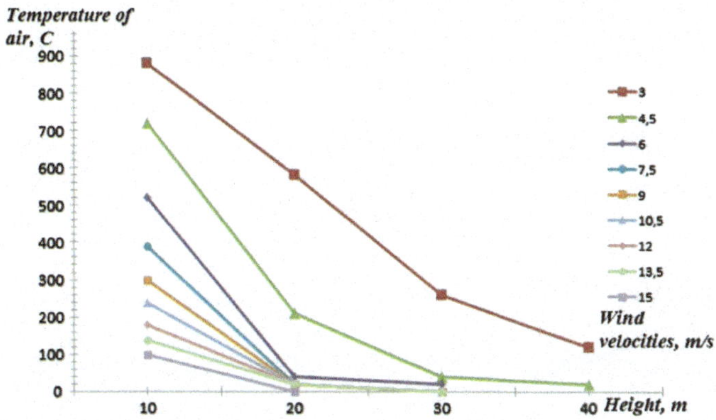

Fig. 4 Air temperature according to height in a corona fire [17]

2.1 *Comparison of Wing Airfoils*

To make an adequate selection of the wing airfoil of the drone, it has been decided to analyze many airfoils. Once the initial comparisons have been carried out, Table 2 shows the airfoils that stand out most for the conditions described above.

Next, in Fig. 5, the behavior of the four finalist airfoils is analyzed in a range of Re = 100,000 to Re = 900,000 with increases of 50,000.

In the first place, in the graph of Cl versus Cd (Fig. 6) the airfoil that stands out most is 9412. For the same drag coefficient value, airfoil 9412 has a higher lift coefficient.

In Fig. 7, it is again stated that the NACA 9412 is the airfoil that offers the highest lift coefficients for both positive and negative values of the angle of attack. In this

Table 2 Selection of initial airfoils under study

Block	Airfoil	Color
A	NACA 8412	Yellow
A	NACA 9412	Red
B	GOE 227	Pink
B	GOE 801	Blue

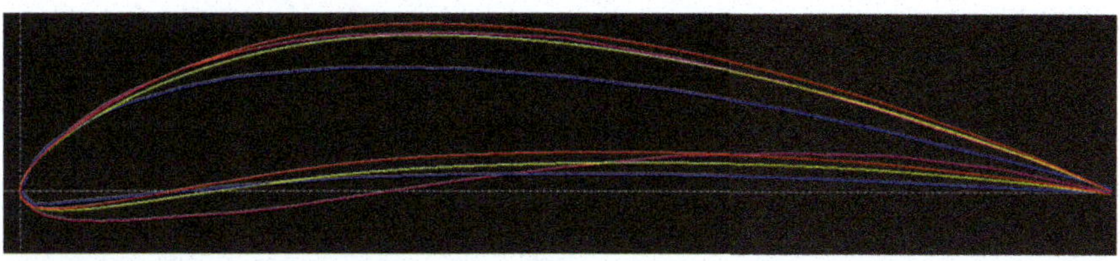

Fig. 5 Geometry of candidate airfoils

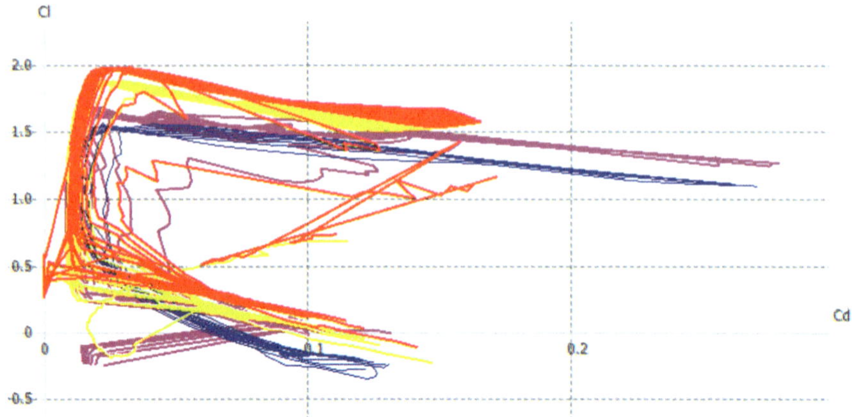

Fig. 6 Cl versus Cd graph

graph, you can also see that for angles of attack greater than 12° the lift begins to fall.

In the case of the graph of the coefficient of moments versus the angle of attack (Fig. 8), it is interesting to choose the curve that separates less than zero for better stability in flight. In this case, the GOE 801 airfoil stands out against the rest of the more similar airfoils.

Finally, as shown in Fig. 9, the airfoils with the best efficiency are NACAS 8412 and 9412. Furthermore, these two airfoils possess enhanced curvature, enabling them to achieve higher levels of efficiency and a wider range of flight angles.

After comparing airfoils and analyzing the polar graphs, the NACA 9412 airfoil has been chosen because it presents the best results overall (Fig. 10). Despite not

Fig. 7 Cl versus α graph

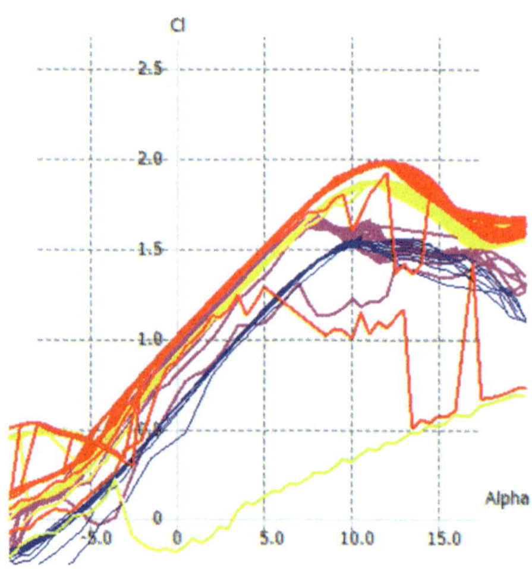

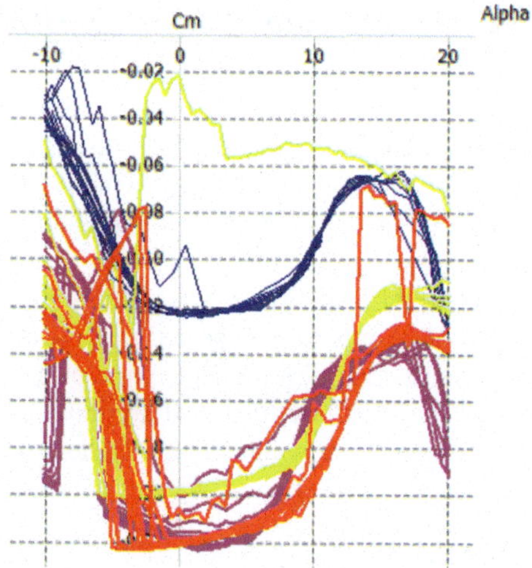

Fig. 8 Cm versus α graph

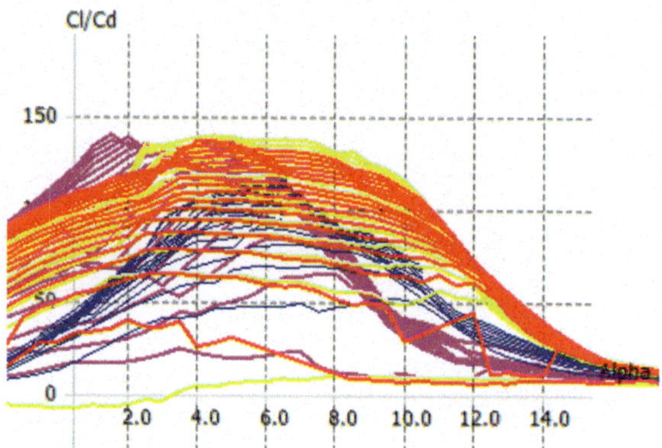

Fig. 9 Cl/Cd-α graph

offering one of the most powerful moment curves, it stands out the most in terms of efficiency and lift.

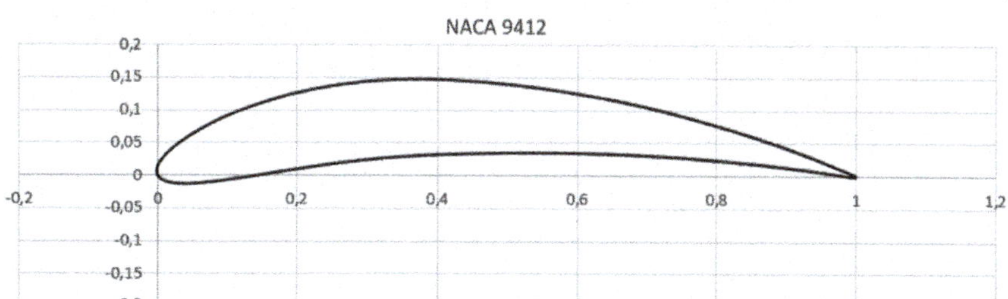

Fig. 10 NACA 9412 airfoil

2.2 *Comparison of 3d Wing Airfoils*

After selecting the airfoil that the drone will carry, a comparison of wings with different types of geometry will be made. For this comparison has been based on three initial wings of different wings (3.0, 3.3, and 4.1 m) to which modifications have been made, such as applying dihedral or aerodynamic torsion. Table 3 lists the analyzed wings and the color code.

Figure 11 displays the polar graphs obtained as results of the analysis at a constant speed of 25 m/s. It is expected that the wings of greater span will be the ones that generate the greatest lift. In the first place, in the graph of Cl versus Cd (Fig. 14) the wings that stand out the most are those of greater size. For the same value of drag coefficient, these wings have a higher coefficient of lift. The modifications in this case do not present great changes concerning the initial wings.

Table 3 Different wings under study

Wingspan (m)	Wing	Color
4.1	1	Red
4.1	1–1	Light blue
4.1	1–2	Orange
3.0	2	Yellow
3.0	2–1	Dark green
3.0	2–2	Light green
3.3	3	Pink
3.3	3–1	Dark blue

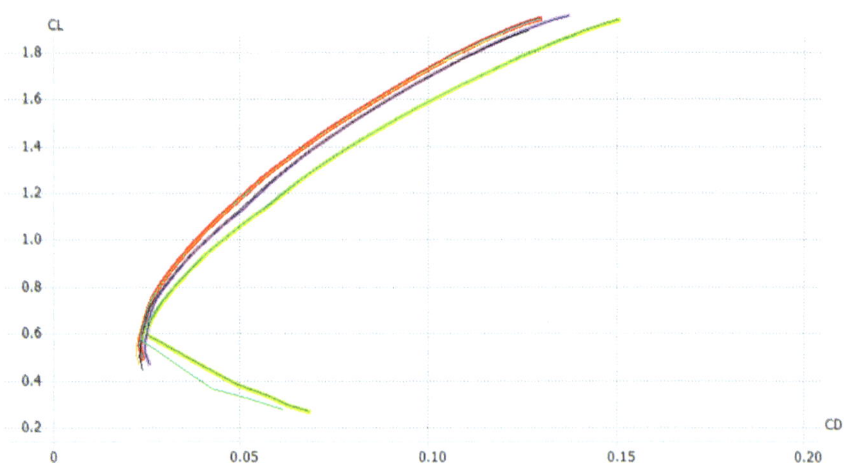

Fig. 11 CL-CD graph

In Fig. 12, it is again obtained that the 4 m wings are those that offer the highest lift coefficients for both positive and negative values of the angle of attack. In this case, the wings with aerodynamic torsion have a lower lift than the corresponding initial wings; therefore, the modification would not be interesting.

In the graph of the CD versus α (Fig. 13) the wings of greater span are those that offer less drag. Contrary to the previous graph, the wings with aerodynamic torsion have less drag than the corresponding initial wings, so the modification would be interesting.

Finally, as can be seen in Fig. 14, the wings with the best efficiency are again the largest. In this case, the wings with aerodynamic torsion have better efficiency than the corresponding initial wings, so the modification would be interesting.

Fig. 12 CL-α graph

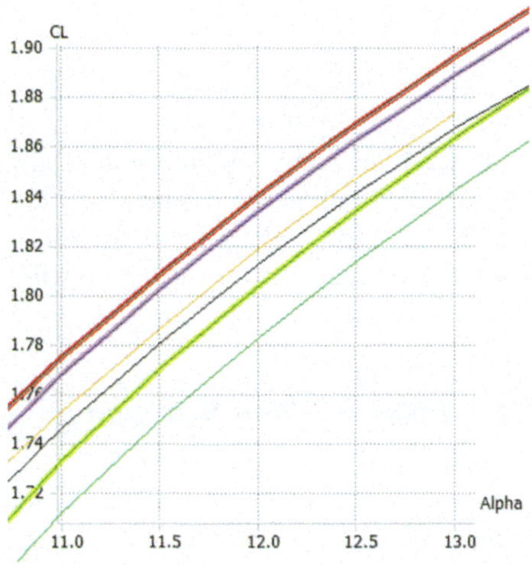

Fig. 13 CD versus α graph

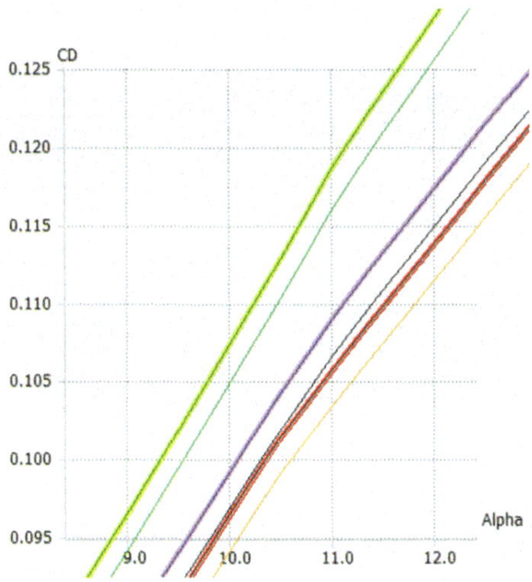

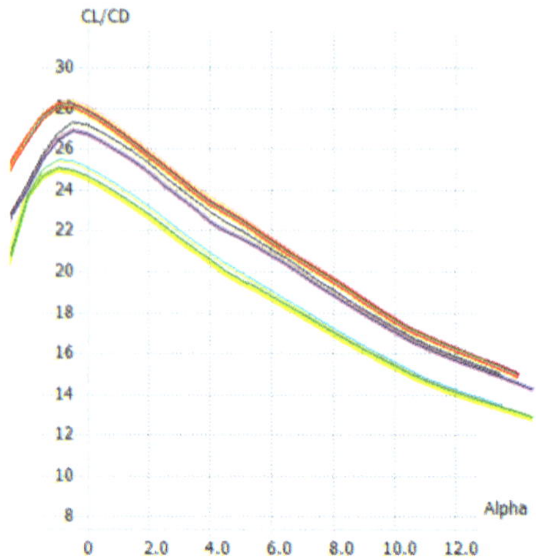

Fig. 14 CL/CD versus α graph

In short, the wings with dihedral behave practically the same as the initial wings, so it is ruled out to add a dihedral to the wings of the drone. On the other hand, aerodynamic torque improves behavior in some aspects. However, it considerably reduces the lift concerning the initial wings, so this modification is also ruled out.

2.3 *Final 3d Wing Selection*

After making the comparison of wings and analyzing the polar graphs, wing 1 with a wingspan of 4 m has been chosen because it is the one that offers the best results. The design can be seen in Fig. 15.

Then, in Table 4, the main characteristics of the designed wing are collected.

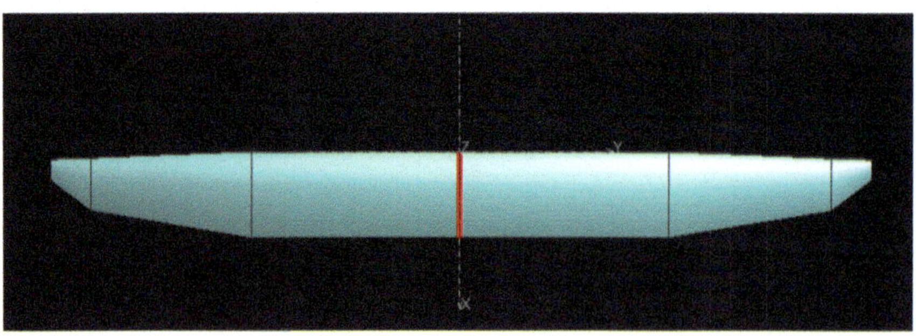

Fig. 15 Designed wing geometry

Table 4 Parameters of the designed wing

Feature	Value
Wingspan (b)	4.1 m
Wing surface (S)	1.45 m^2
Mean chord (c_m)	0.35 m
Aspect ratio (AR)	11.19
Taper ratio (TR)	4.12

2.4 Fuselage Design

The design of the fuselage has been made from the default model of Xflr5 to which small modifications have been applied until reaching the result. The following aspects have been considered when designing the fuselage:

- The purpose of the fuselage is to join the wings with the stabilizers while housing all the components.
- Its shape must be as aerodynamic as possible since it affects certain variables such as the parasitic drag coefficient.

Figure 16 shows the result of the fuselage and Table 5 shows the main features of the design.

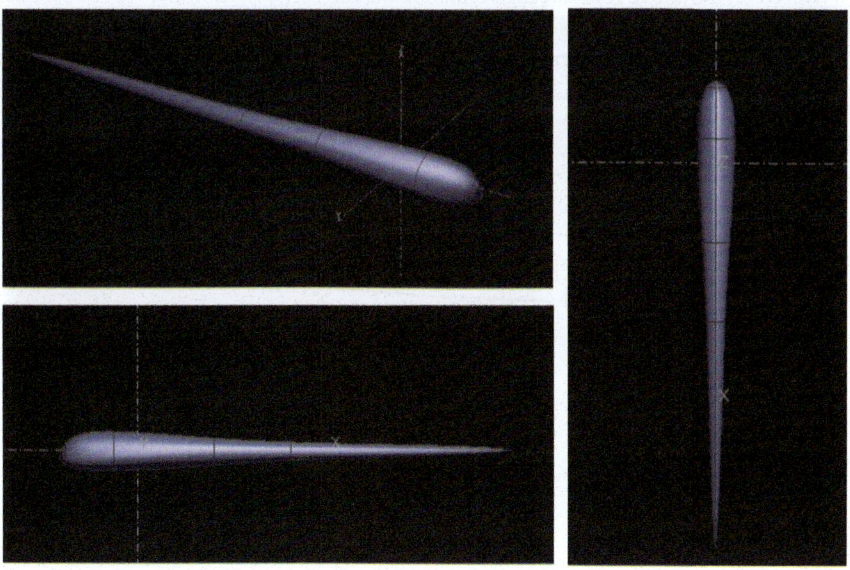

Fig. 16 Designed fuselage geometry

Table 5 Parameters of the designed fuselage

Feature	Value (m)
Length (L)	1.25
Maximum diameter (D_{max})	0.14

2.5 *Empennage*

To perform the queue design, an analysis of several symmetric airfoils has first been performed. In this case, only symmetrical airfoils have been analyzed, since the main objective of the tail surfaces is not to generate lift, but to generate moments to stabilize the drone. After comparing airfoils and analyzing the polar graphs in an analogous way to the analysis of the wing airfoil, the NACA 0012 airfoil has been chosen because it presents a higher efficiency value for low angles of attack (see Fig. 17).

Table 6 shows the main geometric characteristics of the airfoil.

For the horizontal stabilizer, a trapezoidal design with a wingspan of 1 m and a slight arrow has been chosen, as can be seen in Fig. 18.

Table 7 shows the main characteristics of the horizontal stabilizer.

After testing several models for the vertical stabilizer, a simple design of 42 cm in height and with a certain arrow has been chosen, as shown in Fig. 19.

Table 8 shows the main characteristics of the vertical stabilizer.

Based on the above references of real models, the vertical stabilizer will be set back a little more than the horizontal, so that the moving part of the rudder has complete freedom of movement. Figure 20 shows how the final design of the empennage looks.

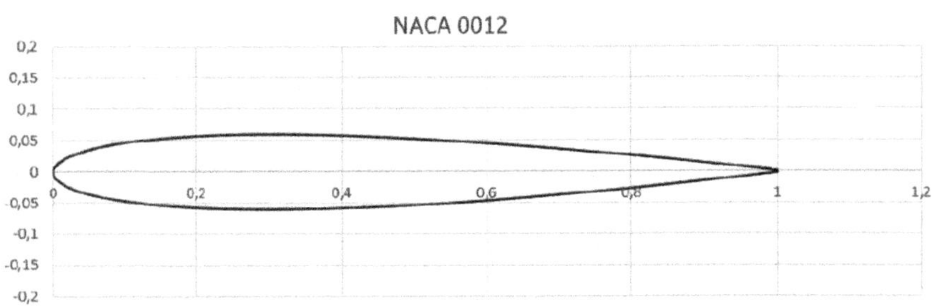

Fig. 17 NACA 0012 airfoil

Table 6 NACA 0012 airfoil parameters

Feature	Value (%)
Maximum curvature	0
Maximum thickness	12
Position maximum thickness	30

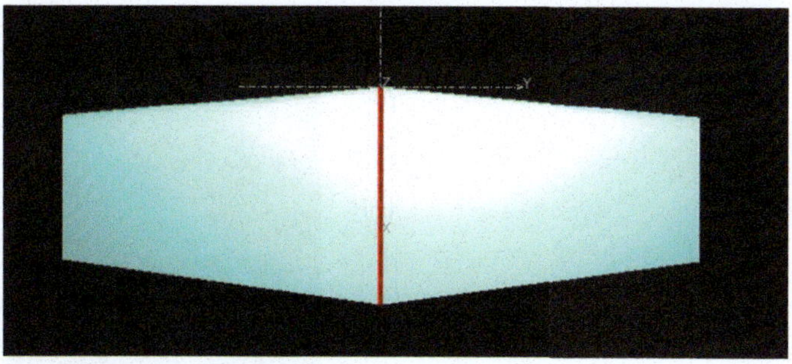

Fig. 18 Designed horizontal stabilizer geometry

Table 7 Parameters of the designed horizontal stabilizer

Feature	Value
Wingspan (be)	1.00 m
Wing surface (Se)	0.25 m^2
Mean chord (cme)	0.23 m
Aspect Ratio (AR)	3.70
Taper Ratio (TR)	1.45

Fig. 19 Geometry of the designed vertical stabilizer

Table 8 Parameters of the designed vertical stabilizer

Feature	Value
Wingspan (br)	0.42 m
Wing surface (Sr)	0.06 m^2
Mean chord (cmr)	0.28 m
Aspect Ratio (AR)	1.55
Taper Ratio (TR)	1.49

Fig. 20 Geometry of the designed empennage

2.6 *Final Aerodynamic Study of the Drone*

Figure 21 shows the complete design of the drone.

Finally, the analysis of the final drone has been carried out at a constant speed of 25 m/s. The graphs of the coefficients of lift, drag, and moment of the final configuration, as well as the relationship of the coefficients of lift and drag, are shown in Fig. 22.

The software also makes it possible to represent the forces that the aircraft will withstand during the cruise flight. Next, the forces for an angle of attack of 0° will be shown, however, it should be clarified that these forces will vary with the angle of attack.

First, as seen in Fig. 23, the induced drag (orange) and the parasitic drag (pink) have been superimposed to show the total drag that the aircraft would face. This drag will increase with the angle of attack.

In Fig. 24, the lift distribution generated by the aircraft is shown in green. In addition, the downwash is observed in red, which is the descending current of air deflected by the wing. Both distributions increase with angle of attack.

Fig. 21 Final drone design

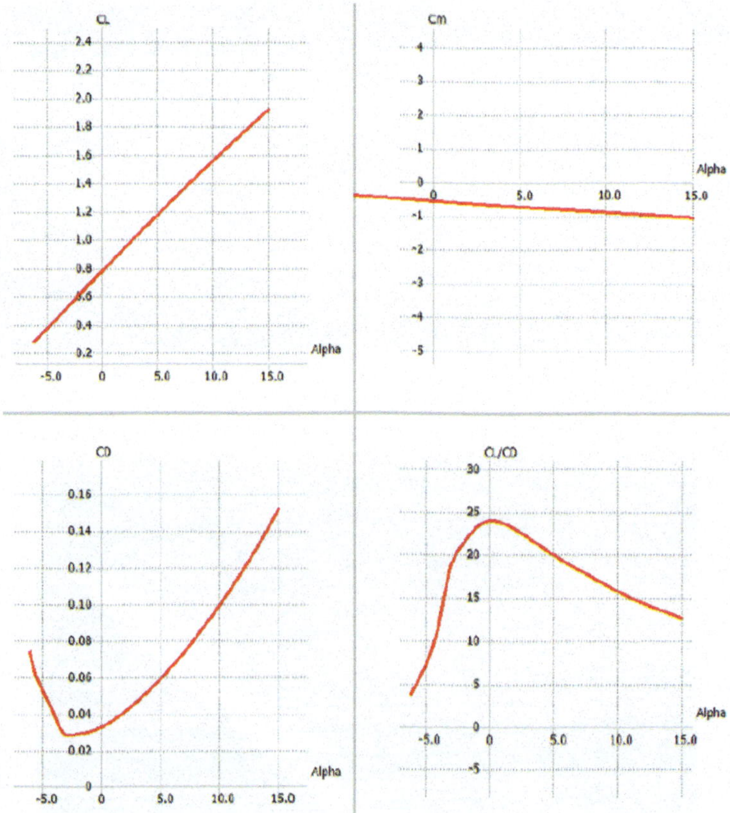

Fig. 22 Polar graphs of the drone's final design

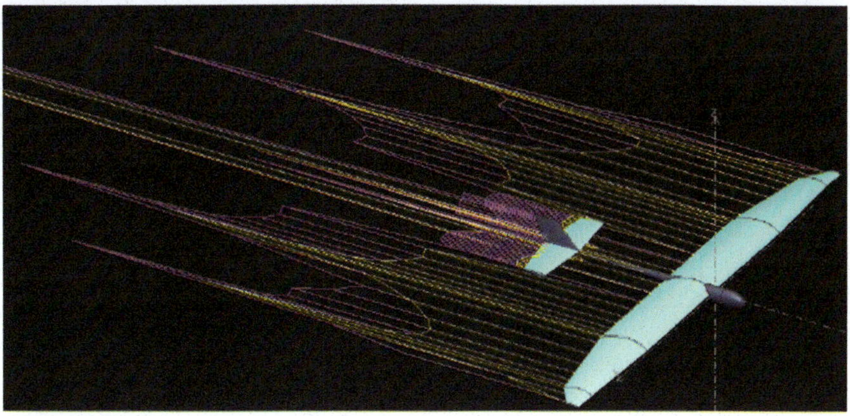

Fig. 23 Induced and parasitic drag of the final design

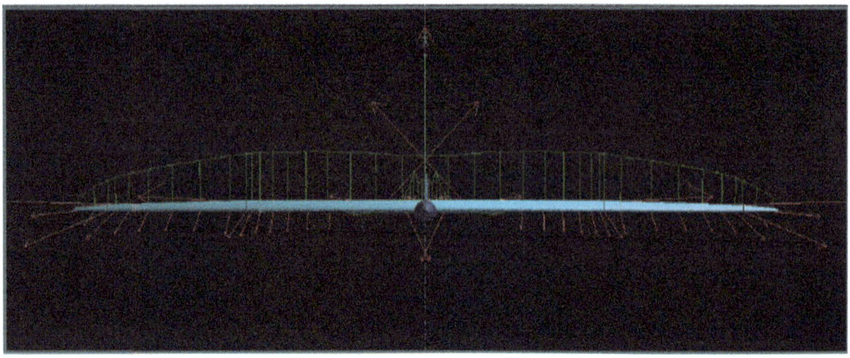

Fig. 24 Lift and downwash of the final design

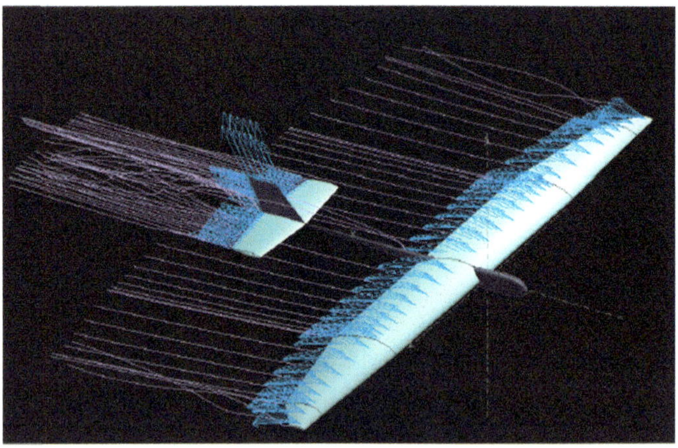

Fig. 25 Surface velocity and current trails of the final design

Finally, Fig. 25 represents in blue the velocity vectors of the flow when passing over the aerodynamic surfaces and the purple current trails that these surfaces will leave behind.

2.7 *Wing Design*

The structure of the wings, as well as that of the stabilizers, is based on the traditional method of ribs and stringers. The ribs are structures perpendicular to the stringers, which give the shape to the wings and provide rigidity to the torsion of these. The stringers are rebars that run the wings longitudinally. Its function is to support the wings and increase the drag to bending of these.

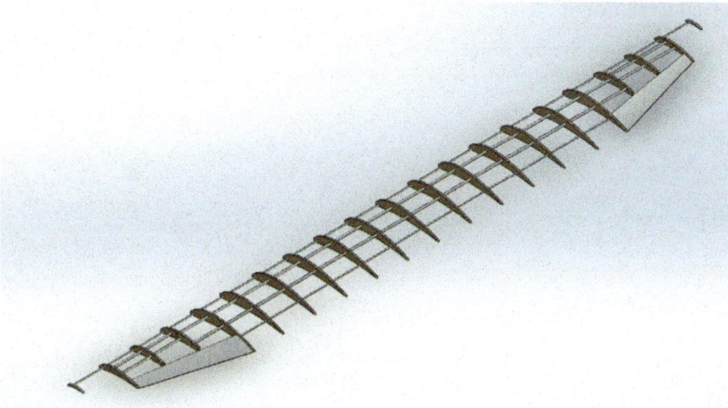

Fig. 26 Internal structure of the wings

Fig. 27 External structure of the wings

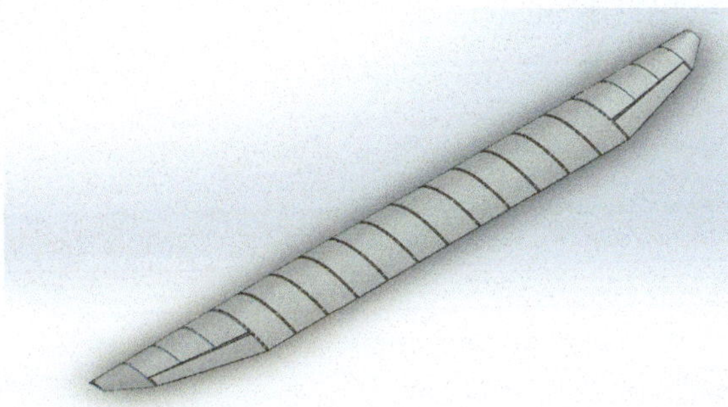

As can be seen in Fig. 26, the structure consists of 21 ribs of 0.01 m thickness, 2 central stringers, a rear spar, and another to reinforce the leading edge. In addition, 4 stringers of shorter length have had to be placed at the ends, due to the narrowing of these towards the tips. The central and pointed stringers have a diameter of 0.01 m, while those at the ends have a diameter of 0.005 m. All the stringers of this piece, as well as the rest of the pieces presented in this section, have a thickness of 1 mm.

Figure 27 also shows that the wings have two ailerons of 0.6 m in length. The mission of the ailerons is to carry out the turns by providing a roll motion to the aircraft. Each spoiler is mounted on a bar attached to the ribs, allowing them freedom of movement. Finally, the ribs are covered by an envelope that forms the outer structure of the wing (Fig. 27).

2.8 Fuselage Design

The internal structure of the fuselage (Fig. 28) consists of 7 ribs of 0.01 m thickness, 2 front stringers, and a longitudinal spar to increase the rigidity to flexion due to

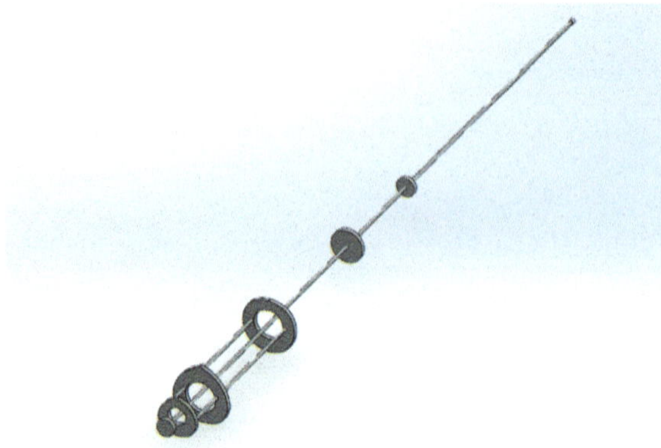

Fig. 28 Internal structure of the fuselage

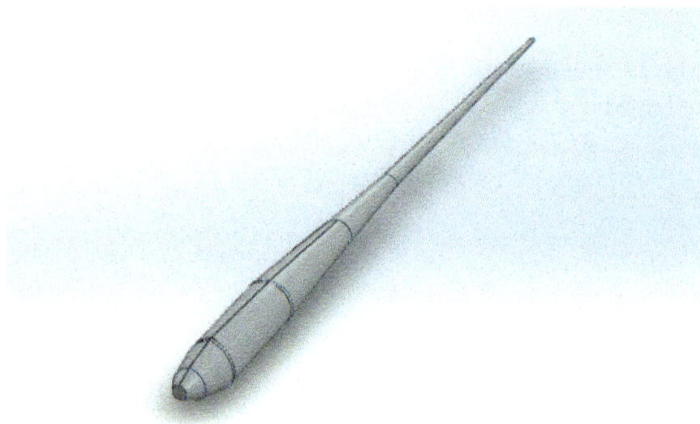

Fig. 29 External structure of the fuselage

the presence of the tail and the forces it generates. The stringers have a diameter of 0.005 m.

Figure 29 shows the lining of the ribs that forms the external structure of the fuselage.

2.9 *Horizontal Stabilizer Design*

The horizontal stabilizer has a structure analogous to that of the wings. As shown in Fig. 30, the internal structure of the horizontal stabilizer consists of 7 ribs of 0.01 m, a central spar, a rear spar, and another to reinforce the leading edge. The central spar has a diameter of 0.01 m, while the other two stringers have a diameter of 0.005 m.

It also has two ailerons of 0.3 m in length, which are mounted to the rear spar and fulfill the same function as the ailerons of the wings.

Finally, the ribs are covered by a wrap that forms the external structure of the stabilizer (Fig. 31).

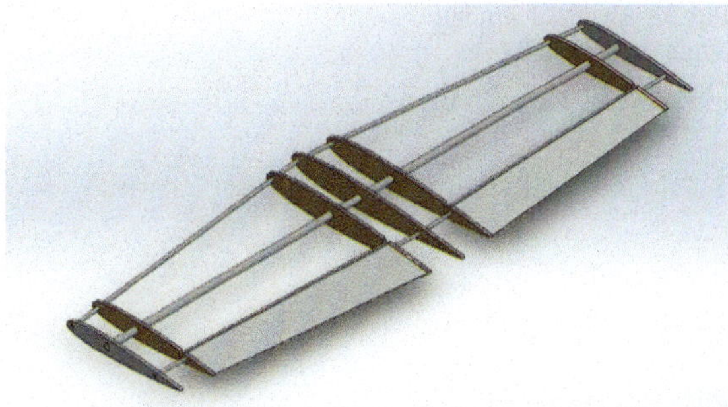

Fig. 30 Internal horizontal stabilizing structure

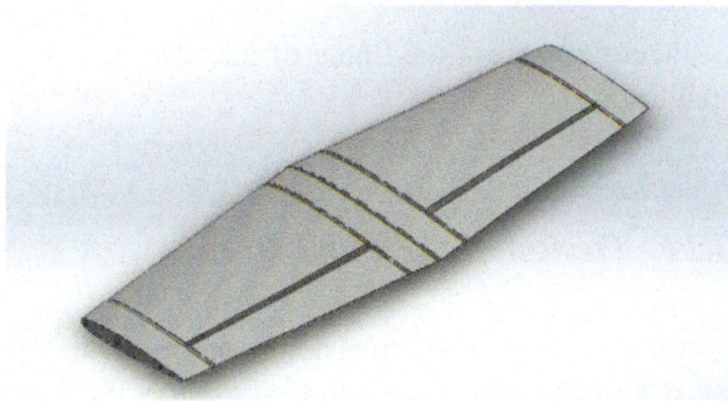

Fig. 31 External structure horizontal stabilizer

2.10 *Vertical Stabilizer Design*

The internal structure of the vertical stabilizer (Fig. 32) consists of two ribs, a central spar, a rear spar and another to reinforce the leading edge. The central spar has a diameter of 0.01 m, while the other two stringers have a diameter of 0.005 m.

In this case, a spoiler of the same length as the stabilizer has been designed, which is mounted on the rear spar, and its function is to provide yaw movement to the aircraft.

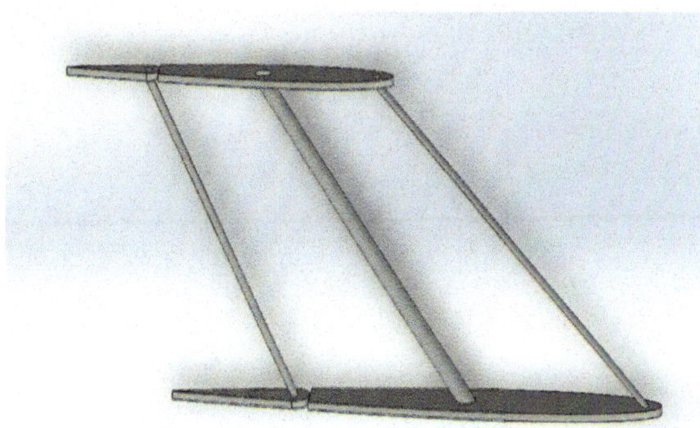

Fig. 32 Internal structure vertical stabilizer

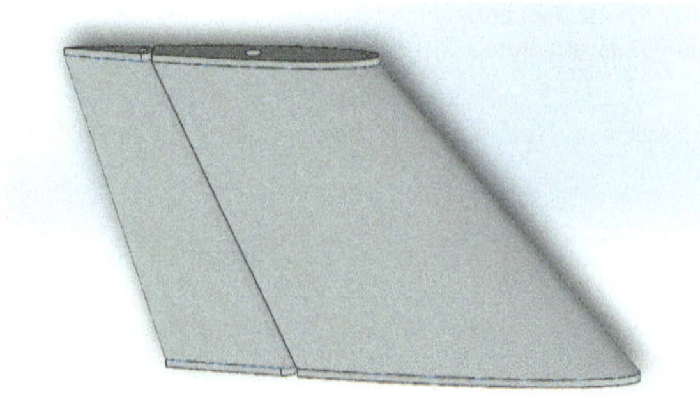

Fig. 33 External structure of the vertical stabilizer

Figure 33 shows the lining of the ribs that forms the external structure of the stabilizer.

2.11 Design of the Final Drone

As shown in Fig. 34, the final design of the drone resembles a glider. This is because, as in the case of these aircraft, the design of the drone has tried to reduce friction to achieve lower engine energy consumption and therefore more flight hours.

It should be noted that the design does not have landing gear, since the takeoff will be made from a drone launch ramp. For this, the Embention C400P launcher (Fig. 35) has been chosen, which is a pneumatic propulsion catapult that allows the launch of drones up to 35 kg.

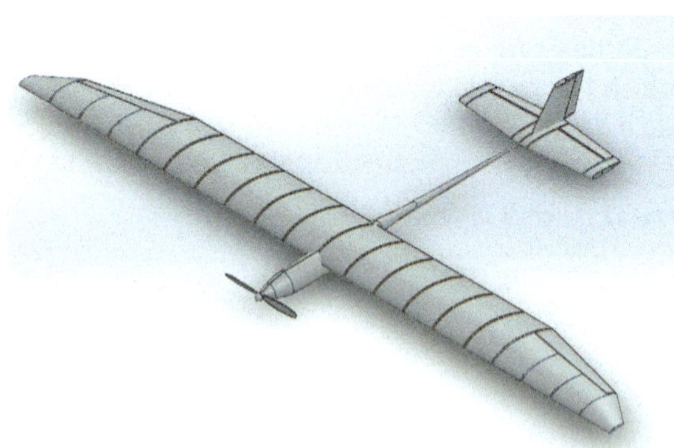

Fig. 34 Final assembly of the drone

Fig. 35 Embention C400P Launcher

2.12 *Calculation of the Power Required at Cruise*

First, it is necessary to know how much the drone will finally weigh to make a reliable estimate of the flight time. The weight of the drone is obtained by combining the weight of the structure with that of all standard components. As shown in Table 9, the final weight is 9244.8 g.

Table 9 Final weight of the UAV

Component	Weight (g)
General structure	3175
Camera FLIR AX8	120
Sensor Telaire ChipCap 2-SIP	5
Sensor CM1106H-NS	4
Pixhawk 4	17
RFD868X Telemetry Pack	15
Receptor FrSky R9 MM	0.8
GPS Module NEO-M8N	1.5
Servos Volz DA13	94
Motor MN701-S KV135	360
T-Motor propeller	65
Power inverter AIR 40A 6S	25
Tattu LiPo Battery 22,000 mAh	5000
Total Weight	9244.8

For power calculations, a speed (V) of 25 m/s, a propulsion system efficiency (ηprop) of 85% and a total mass (mTOT) of 9.25 kg (rounded upwards) are assumed.

As is known, in the cruise phase, in which the drone will mostly fly, lift equals weight and drag equals thrust. The system of equations that defines this balance of forces is as follows:

$$L = W \tag{1}$$

$$D = T \tag{2}$$

The power required by the propulsion system (Pprop) is calculated as:

$$P_{prop} = \frac{T \bullet V}{\eta_{prop}} \tag{3}$$

Using (2):

$$P_{prop} = \frac{D \bullet V}{\eta_{prop}} \tag{4}$$

Multiplying by the lift in the numerator and denominator gives:

$$P_{prop} = \frac{L \bullet V}{\eta_{prop} \bullet (\frac{L}{D})} \tag{5}$$

Using (1) yields the equation used in (4) to calculate the required power.

$$P_{prop} = \frac{W \bullet V}{\eta_{prop} \bullet (\frac{L}{D})} \tag{6}$$

First, the weight (W) is calculated with the total mass of the drone estimated:

$$W = m_{TOT} \bullet g = 9.25 \bullet 9.81 = 90.69N \tag{7}$$

To calculate the lift and drag forces it is first necessary to obtain the values of the corresponding dimensionless coefficients of the polar plots obtained in Xflr5 (Fig. 22). Going to these graphs, it can be assumed that for an angle of attack of 12° the coefficient of lift (CL) is 1.705 and the coefficient of drag (CD) is 0.118.

The density of the air at 50 m height (h) is obtained with the following equation considering ISA atmosphere conditions:

$$\frac{\rho}{\rho_{SL}} = \left(\frac{\alpha_T \bullet h}{\theta_{SL}} + 1\right)^{n-1} \tag{8}$$

Knowing that the constants of the previous equation at sea level are SL = 1.255 kg/m^3, ρSL = 288.15 K, α = 6.5 × 10–3 Kθ/m and n = 5.252, a density value of 1.219 kg/m^3 is obtained.

To calculate lift and drag forces, aerodynamic equations are applied:

$$L = \frac{1}{2}\rho V^2 S C_L = \frac{1}{2} \bullet 1.219 \bullet 25^2 \bullet 1.42 \bullet 1.705 = 922.29N \tag{9}$$

$$D = \frac{1}{2}\rho V^2 S C_D = \frac{1}{2} \bullet 1.219 \bullet 25^2 \bullet 1.42 \bullet 0.118 = 63.83N \tag{10}$$

Substituting the values into Eq. (6) gives a power required by the propulsion system of 184.6 W.

Additionally, it is necessary to calculate the power consumed by payload and avionics. For this, the formula (11) is used, which relates voltage and intensity to power.

$$P = V \bullet I \tag{11}$$

Applying the data from the component data sheets, 11.6 W is obtained, as shown in Table 10.

The total power shall be the sum of the power required by the propulsion system and that consumed by the components.

$$P = P_{prop} + P_{comp} = 184.6 + 11.6 = 196.2W \tag{12}$$

Finally, it is obtained that the power required for a cruise flight is approximately 196 W.

Table 10 Power consumed by components

Component	Voltage	Current	Power
FLIR AX8 Camera	10 V	0.2 A	2 W
Sensor Telaire ChipCap 2-SIP	5 V	0.5 A	2.5 W
Sensor CM1106H-NS	5 V	800 mA	4 W
Pixhawk 4	5 V	100 mA	0.5 W
RFD868X Telemetry Pack	3 V	21 mA	6.3 W
Receptor FrSky R9 MM	5 V	0.75 mA	3.75 W
GPS Module NEO-M8N	5 V	50 mA	0.25W
Servos Volz DA13[a]	5 V	0.3 A	0.45 W (×5)
Total Power			11.6 W

[a] It is assumed that servos only consume power 30% of the time since they are not in use all the time

3 Simulation of the Aerodynamic Flow on the Drone

This section will analyze the behavior of the aerodynamic flow around the drone, for which the Flow Simulation extension offered by SolidWorks will be used. First, the initial conditions of study of this simulation will be presented.

Figure 36 shows the computational domain defined for this simulation. As it is only intended to analyze the longitudinal aerodynamic flow around the surface, the domain has been adjusted to reduce the calculation time. Only the rear face has been moved a little further away to be able to visualize how the flow continues downstream.

As for the meshing, it has been based on the basic rectangular mesh as shown in Fig. 36. In addition, a refinement of the mesh has been made in certain parts of interest as shown in Fig. 37. With these parameters, a mesh composed of 268,032 cells has been obtained and the simulation has been completed in a total of 251 iterations.

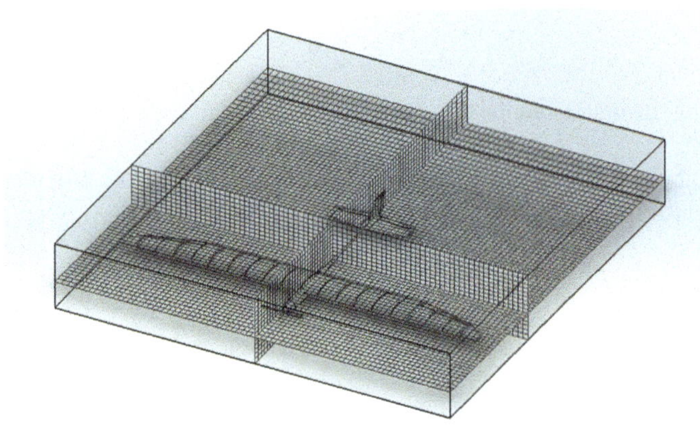

Fig. 36 Computational domain for simulation

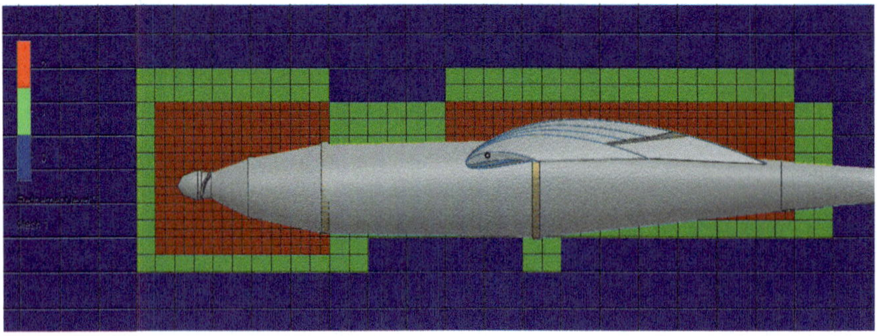

Fig. 37 Detailed view of mesh refinement

Below Fig. 38 shows an image of the flow around the surface of the drone obtained through simulation. The red arrows represent the direction of the velocity of the particles at each point at a given instant. That is, they represent the flow or current lines that the particles follow.

As expected, the surfaces that most disturb the flow of particles are the forward fuselage and the tail rudder, since they are the ones that directly face the flow. On the contrary, the wings and the horizontal stabilizer do not present great disturbances for the flow since they are more aerodynamic surfaces. Figure 39 shows a side view of the flow in which the streamlines can be better appreciated.

Finally, Fig. 40 shows the flow of several lines of particles that pass through the drone in different sections. The first line (a) passes through the axis of symmetry of the aircraft, the second (b) cuts the wing near the fuselage, the third (c) cuts the wing away from the fuselage, and the fourth (d) cuts the wing by the tip. Again,

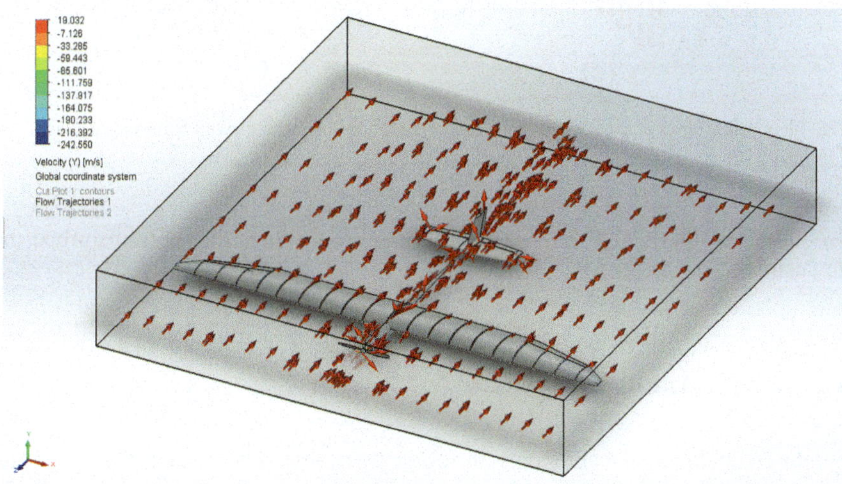

Fig. 38 General view of the aerodynamic flow around the drone

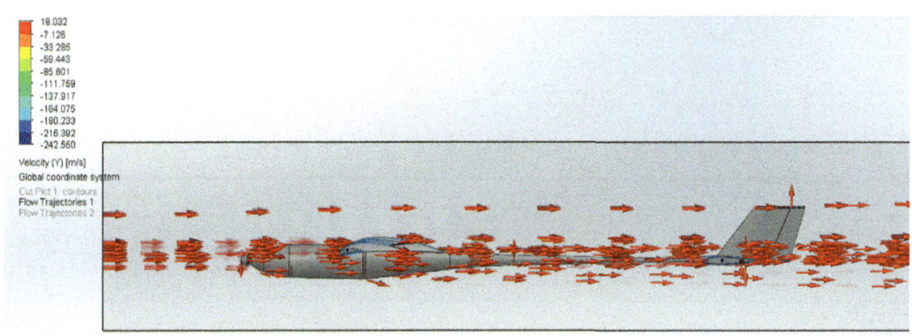

Fig. 39 Side view of aerodynamic flow around the drone

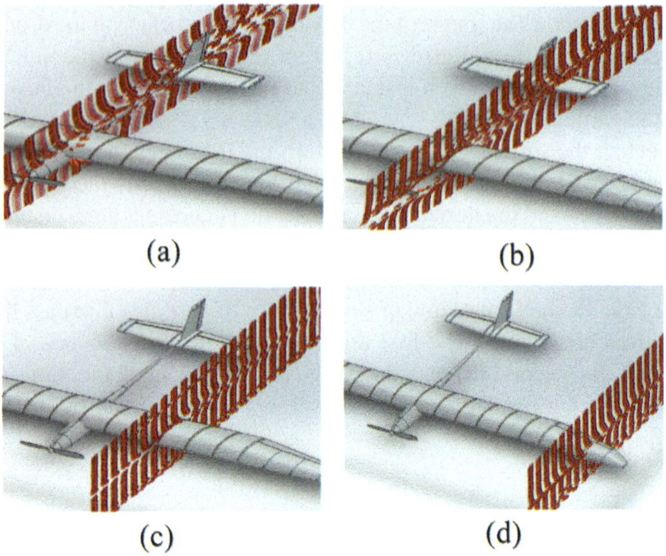

Fig. 40 Particle lines cutting by: **a** the center, **b** near the fuselage, **c** away from the fuselage, and **d** the wing tip

it is observed that the closer that one gets to the fuselage the more disturbed the downstream flow becomes.

3.1 *Flight Autonomy*

For the estimation of the autonomy, the eCalc software has been used. This result can only be taken as an indicative value of the autonomy since this cannot be calculated with a simple formula.

From the components selected above, the corresponding configuration parameters have been entered into eCalc, as shown in Fig. 41.

Fig. 41 Parameters entered in eCalc

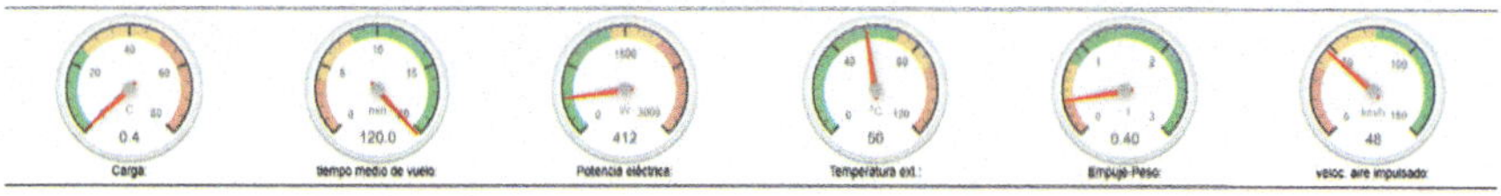

Fig. 42 Results obtained in eCalc

Once all aircraft parameters have been entered, the following results are obtained as shown in Fig. 42. The estimated average flight time is 120 min, which exceeds the minimum autonomy target set at the beginning of this project.

This autonomy was tested on a flying field, under flying conditions on an elliptical route in an authorized area (see Fig. 43).

In Table 11 it is possible to see the different flight times achieved in the various tests carried out. It can be observed that the time-of-flight prediction is fulfilled in practically all cases.

Fig. 43 Flight area

Table 11 Autonomy in minutes for every flight test

Flight #	Time (min)
1	131
2	119
3	127
4	124
5	128
6	121
7	127
8	125
9	129
10	126

4 Conclusions

In the present project, and under the context of FIREPOCTEP, it was sought to design, manufacture, and test a UAV with the capacity to provide the conditions under which fire front monitoring could be carried out in forest fires, with sufficient autonomy to be a reasonable alternative to more dangerous observation methods.

It began by reviewing the characteristics of the forest fires that usually hit the areas of interest of the project, such as the border area between Spain and Portugal, and particularly, in the area of Galicia. Analyzed this casuistry, a study was conducted based on wing airfoils aimed at an increase in autonomy, as well as 3d wings, so that the best result was obtained from the point of view of the simulation. In the same way, the design of the rest of the elements that make up the fixed-wing aircraft (fuselage, empennage) was carried out, and the standard elements (engines, propellers, etc.), as well as the on-board electronics, were selected to fulfill the observation mission. For this, the calculations corresponding to the flight maneuvers were carried out that determined the required power, and that helped to properly size all the standard elements that were incorporated into the UAV.

Special emphasis was placed on the structural aspect of the design, to confer the greatest lightness to the elements of the structure of the UAV, maintaining the necessary drag so that no deformations are suffered above what is bearable, and guaranteeing that the whole could maintain the necessary configuration for the flight.

Finally, the autonomy that the UAV was evaluated, both through software simulation and in different tests carried out on the flight field. The real autonomy has met expectations, even exceeding the values provided by the simulation, and above the initial objectives set to meet, the result of the study of the shapes and sizes of forest fires in the area under study.

References

1. Nonami, K. (2007). Prospect and recent research & development for civil use autonomous unmanned aircraft as UAV and MAV. *Journal of system Design and Dynamics, 1*(2), 120–128.
2. Picos, J., Bastos, G., Míguez, D., Alonso, L., & Armesto, J. (2020). Individual tree detection in a eucalyptus plantation using unmanned aerial vehicle (UAV)-LiDAR. *Remote Sensing, 12*(5), 885.
3. Glade, D. B. (2000). *Unmanned aerial vehicles: Implications for military operations* (No. 16). Center for Strategy and Technology, Air War College, Air University.
4. Ghamari, M., Rangel, P., Mehrubeoglu, M., Tewolde, G., & Sherratt, R. S. (2022). Unmanned aerial vehicle communications for civil applications: a review. *IEEE Access, 10,* 102492.
5. Alshamaila, Y., Papagiannidis, S., Alsawalqah, H., & Aljarah, I. (2023). Effective use of smart cities in crisis cases: A systematic review of the literature. *International Journal of Disaster Risk Reduction, 85,* 103521.
6. Kucharczyk, M., & Hugenholtz, C. H. (2021). Remote sensing of natural hazard-related disasters with small drones: Global trends, biases, and research opportunities. *Remote Sensing of Environment, 264,* 112577.
7. Mangewa, L. J., Ndakidemi, P. A., & Munishi, L. K. (2019). Integrating UAV technology in an ecological monitoring system for community wildlife management areas in Tanzania. *Sustainability, 11*(21), 6116.
8. *FIREPOCTEP*. (n.d.). https://firepoctep.eu/
9. Puente Puente, J. M. (2018). Educación ambiental e incendios forestales en España.
10. Northwest Fire Science Consortium. (2020). What are? TYPES OF FIRE. Accessed 23 March 2024 from https://www.nwfirescience.org/sites/default/files/publications/Types%20of%20Fire.pdf
11. Hernández, L. (2021). Paisajes cortafuegos. Propuesta de WWF España y ANP (WWF Portugal) para adaptar el territorio ibérico a los incendios. *WWF España.*
12. Gobierno de España. (2020). Los Incendios Forestales en España 1 enero—31 diciembre 2020. Avance Informativo. Ministerio para la transición Ecológica y el Reto.
13. RTVE. (8/4/2022). "El incendio intencionado de Verín, perimetrado y sin peligro para los vecinos tras arrasar 600 hectáreas". Accessed 23 March 2024 from https://www.rtve.es/noticias/20220804/incendio-intencionado-verinourenseperimetrado/2394024.shtml
14. Magnet. 7/22/2022 El incendio de Zamora es el peor de la historia de España. Accessed 23 March 2024 from https://magnet.xataka.com/que-pasa-cuando/incendio-zamora-peor-historia-espana-29-veces-destruccion-que-palma
15. Seijo, P., & Montes Coego, A. (7/29/2022). El alcalde de A Pobra do Brollón confiesa, tras los incendios, que aún "sueña" con ellos. Accessed 23 March 2024 from https://cadenaser.com/galicia/2022/07/29/el-alcalde-de-a-pobra-do-brollon-confiesa-tras-los-incendios-que-aun-suena-con-ellos-radio-lugo
16. RTVE. (8/4/2022). El incendio intencionado de Verín, perimetrado y sin peligro para los vecinos tras arrasar 600 hectáreas. Accessed March 23 2024 from https://www.rtve.es/noticias/20220804/incendio-intencionado-verin-ourense-perimetrado/2394024.shtml
17. Soprunenko, E. E., Perminov, V., Reyno, V. V., & Loboda, E. L. (2015). Simulation of impact assesment of crown forest fires on boudary layer of atmosphere using software PHOENICS. In *21st International Symposium Atmospheric and Ocean Optics: Atmospheric Physics* (Vol. 9680, pp. 1235–1240). SPIE.

Monitoring of the Evolution of the Firefront of Forest Fires Using Unmanned Aerial Vehicles (UAVs) in Night Conditions

Pedro Orgeira-Crespo, Carlos Ulloa, Guillermo Rey-González, Francisco Javier Rodríguez-Martínez, Juan Carlos González-Moreno, Alma Gómez-Rodríguez, Lorena Otero-Cerdeira, and Antonio Adrián González-Pedrouzo

Abstract Unmanned aerial vehicles (UAVs) are progressively reshaping the landscape of air power, introducing innovative methods for conducting surveillance and tracking operations, particularly in wildfire situations. This helps mitigate the risk of personal injuries. These aerial vehicles have become standard equipment, generating a global network of aircraft and specialized units for field operation. This chapter details the contributions carried out in the Firepoctep Interreg Spain Portugal cross-border cooperation project regarding hardware and software for the design and construction of a new drone that is suitable for flight in hostile environments, such as

P. Orgeira-Crespo · C. Ulloa · G. Rey-González
Department of Mechanical Engineering, Machines and Thermal and Fluid Engines and Motors, University of Vigo, Campus As Lagoas, Ourense, Spain
e-mail: porgeira@uvigo.es

C. Ulloa
e-mail: carlos.ulloa@uvigo.es

G. Rey-González
e-mail: guillermo.rey@uvigo.es

F. J. Rodríguez-Martínez (✉) · J. C. González-Moreno · A. Gómez-Rodríguez · L. Otero-Cerdeira · A. A. González-Pedrouzo
Computer Science Department, University of Vigo, Campus As Lagoas, Ourense, Spain
e-mail: franjrm@uvigo.es

J. C. González-Moreno
e-mail: jcmoreno@uvigo.es

A. Gómez-Rodríguez
e-mail: alma@uvigo.es

L. Otero-Cerdeira
e-mail: locerdeira@uvigo.es

A. A. González-Pedrouzo
e-mail: agpe-drouzo17@esei.uvigo.es

D. Carou et al. (eds.), *Applying Drones to Current Societal and Industrial Challenges*, Management and Industrial Engineering, https://doi.org/10.1007/978-3-031-55571-8_3

forest fires and the software developed to improve piloting conditions in low visibility circumstances (including night flights).

Keywords UAV · Drone · Forest firefighting · Onboard software

1 Introduction

The growing interest in unmanned aerial vehicles (UAVs), commonly known as drones, has allowed reduced development costs and allowed large sectors of the population to access this technology, acquiring low-cost commercial UAVs to perform simple missions [1–4]. The reduction in costs associated with UAV is largely due to the mobile technology or smartphones sector, which in recent years has developed sensors (accelerometers, gyroscopes, or magnetometers) of reduced dimensions and at increasingly lower prices, allowing benefit from the parallel development of this industry, embarking all this technology in commercial UAV at very competitive prices.

The utilization of drones in combating forest fires has emerged as a revolutionary tool for early detection, surveillance, and effective management of these devastating natural events. In recent years, drones have showcased their ability to address key challenges in wildfire prevention and control, offering an unparalleled aerial perspective and real-time data that enable response teams to make informed and strategic decisions [5]. These unmanned aircraft have transformed the way society approaches forest fires, providing a unique combination of versatility, agility, and efficiency in safeguarding forests, wildlife, and at-risk human communities. The integration of drones in wildfire management presents transformative opportunities, yet challenges must be navigated for their effective deployment. Regulatory complexities in shared airspace, limited flight times, and weather and fire condition constraints pose operational hurdles while data processing demands, communication infrastructure gaps, and the need for seamless coordination with ground crews are open challenges [4].

This chapter addresses the results obtained in the Firepoctep [6] project in two areas: new hardware designs for a specialized UAV tailored for monitoring wildfires and a new piloting aid system software for night conditions.

At the hardware level, the conception and manufacture of composite materials—carbon fiber of a fixed-wing UAV for monitoring forest fires—has been developed. The lamination patterns and the orientation of the laminates are some of the most relevant factors when manufacturing CFRP carbon fiber parts. The manufacture of parts is made in carbon fiber with aeronautical specifications from polyurethane molds previously designed for a certain stacking of layers. The results obtained from the XFLR5 simulation software tools are validated by obtaining empirical data from the wind tunnel: the test indicates that the lift generated by the left wing, obtained using additive manufacturing, is sufficient to counteract the weight of the UAV and verifying the flight capacity of the aircraft.

Additionally, at the software level, a UAV piloting aid system has been developed for night fire conditions. This software is based on overlaying cartographic images, projected from satellite information (position) and inertial information (attitude), onto images captured by onboard thermal images. Furthermore, this system introduces new characteristics tailored for extreme conditions, such as night flights, as well as facilitating the coordinated operation of multiple drones within the same wildfire scenario.

Researchers of the Department of Mechanical Engineering, Machines and Thermal and Fluid Engines and Motors designed the hardware, and researchers of the Computer Science Department developed the software.

The chapter will be divided into two distinct sections, focusing on both hardware and software aspects.

2 Hardware: Introduction

Composite materials are one of the biggest trends in the development of both piloted and unpiloted aircraft, and their percentage of total weight has only been increasing in recent years. Commercial aircraft, such as the Boeing 787 Dreamliner or the Airbus 350, are currently manufactured with a composite content of more than 50% [7].

One of the areas of greatest research and interest by manufacturers of UAVs with purposes other than those purely oriented to leisure is in the field of composite materials since they allow a reduction in the weight of the aircraft with a consequent improvement of its technical capabilities (flight time, operating height, and range). These materials allow for obtaining more rigid structures, supporting greater aerodynamic loads. In addition, they allow the manufacturing of lighter, stronger, and smaller models, essential characteristics in the aerospace sector. Thus, currently, much of the efforts to improve the capabilities of UAVs are focused on their research, specifically on the optimization of their application to the structures of unmanned aerial vehicles, being an area of great potential for improvement. In recent years, great progress has been made in the composite materials sector, which focuses on the improvement of each of its components: on the one hand, the improvement of synthetic fibers (for example, obtaining an improved aramid fiber such as Twaron instead of Kevlar), and, on the other hand, new epoxy resins or polymer matrices [8].

The parts or components of the military UAV developed will be manufactured in composite material—carbon fiber prepreg with epoxy resin, although to obtain the final pieces the molds with which the pieces will later be obtained in carbon fiber must be designed and manufactured. The molds can be made of different materials, such as carbon fiber or aluminum, although the project intends that they will be manufactured in polyurethane.

The final objective is to obtain the structural parts that give rise to the flight in carbon fiber. Thus, a study of the properties of the selected composite materials must be carried out, and all those parameters that will influence the characteristics of the final pieces (the lamination pattern, the properties of the synthetic fibers and the polymer matrix, and the thickness and the orientation of each layer). This study is included within a structural analysis based on composite materials that is carried out, thanks to the composite element calculation tool available in the ANSYS ACP (Ansys Composites Pre-Post) module.

Fixed-wing UAVs have their two wings attached to the fuselage of the aircraft in a fixed position and cannot perform any type of movement during flight, only by hyper-sustaining devices such as flaps or slats. The form of generation of lift for these aircraft is through their wings, due to the difference in pressures that exists between the intrados and the extrados of the airfoil. Due to the incidence of an air current against the airfoil in the intrados, there is greater pressure than in the extrados and this pressure difference is what produces a positive net force from the intrados to the extrados producing *Lift* or *Lift*.

Thanks to aerodynamic airfoils previously calculated for pre-established flight conditions (Reynolds number, cruising speed, operating height, etc.), an efficient lift generation, that is, a high Lift/Drag ratio can be achieved. The higher the Lift/Drag ratio of an airfoil, the higher the efficiency of the airfoil and the more suitable it will be for those flight conditions.

The absence of crew on board the UAV has allowed the development of multiple configurations and designs, such as VTOL unmanned aircraft, or designs as unique as bicopters or tricopters. Many of these models are based on traditional aviation such as fixed-wing UAVs inspired by conventional aircraft with fuselage/wings rigidly attached to the corresponding control surfaces both in the main wing (ailerons), as in the empennage (vertical rudder and depth rudder).

The object of study in the project focuses on a time-of-flight aircraft with requirements of 1.5 h, medium range, low cruising speed, and type of firefront follow-up mission. While these are the key values when parameterizing the aircraft, the elements that make up the UAV are like those of larger aircraft. The components that are part of fixed-wing UAV are the following:

1. Supporting surfaces or wings: In general, they are the element that produces the greatest amount of lift, although other elements such as the empennage or the fuselage itself also produce lift, although it is negligible concerning the lift produced by the wings.
2. Control surfaces: Depth rudder or elevator, vertical rudder, and ailerons to control pitch, yaw, and roll. Also included in this category are hyper-sustaining surfaces such as flaps, slats, and aerobrakes.
3. Fuselage: The main function of the fuselage is to store onboard sensors and wiring while providing a rigid and acceptable aerodynamic structure.
4. Propulsion system: Usually with a propeller, turbofan, or jet type. In the study system, the propulsion system focuses on propellers with electric brushless motors, rechargeable batteries, and a speed controller (ESC).
5. Autopilot and Communications System: Aircraft processing and control system. It is the most important element of the system because it is responsible for receiving and sending signals to the other elements of the system.

6. Payloads: Commonly onboard sensors, such as video capture systems, sensors of different gases, humidity, and temperature.
7. Additional Systems: This category includes the rest of the elements necessary for the air system to function properly, such as servomotors, wiring, and the possible integration of a landing gear.

3 Hardware: UAV Morphology Design

- Lifting surfaces:

The design of the supporting surfaces or lift generators (wings, depth rudder, fin or vertical tail rudder, etc.) for conventional aircraft such as an Airbus A380 or a Boeing 777 is a complex task in which many factors intervene due to the large structural loads that have to be supported, appearing. Effects such as aeroelasticity or vibrations emerge due to the large structural loads [7]. However, for the development of the wings of a small category UAV, many of these effects can be neglected. The objective when studying the supporting surfaces for fire monitoring aircraft is to obtain aerodynamically efficient sections in combination with a planform (view of the wing in plan) that allows to obtain high aspect ratios (AR) while obtaining small moments of flexion.

When working with a wing in three dimensions, the effect of induced drag appears, which is caused by the vortices that occur at the tips of the wings and is the product of the lift itself generated by the aircraft. Wings with a large Aspect Ratio (AR) produce less induced drag and improve the flight efficiency of the aircraft. The lift force generated by the wing is a function of several parameters: the airfoil itself, the geometry of the wing, the density of the air at the operating height of the aircraft (ISA table), the speed of the aircraft concerning the air, and the angle of attack, according to the formula shown below

$$L = \frac{1}{2}\rho v^2 S C_L \tag{1}$$

When selecting the aerodynamic airfoil and the shape of the aircraft, several factors must be assessed, mainly the balance between a good aspect ratio and weight. In general, a support surface with a high aspect ratio will be more aerodynamically efficient due to the reduction in induced drag. However, as a wing is nothing more than a cantilever beam supported to the structure, the higher the aspect ratio, the greater the bending moments and the heavier the internal structure. Aspect ratios between 8 and 12 are suitable for this type of aircraft. If the wings of the UAV are short, it will be necessary to insert a certain taper into the wings to reduce the induced drag, because this effect becomes increasingly important, the more reducing the wingspan.

A problem in the design of UAVs and aircraft, in general, is to obtain sufficient lift at low speeds during landing and takeoff. As noted in the lift formula, the lift is proportional to speed squared, so it is common to oversize the wing of the UAV more than necessary when flying at cruising speed to get enough lift at low speeds. Another important aspect of the wing is that the entry into loss occurs near the root rather than the tip because near the tip are usually located the control surfaces (ailerons) and in case of entering stall the aircraft would become uncontrollable.

- Fuselage:

After the wings or support surfaces, the next most important element of an unmanned aerial vehicle is the fuselage, because in this space all the onboard components and avionics of the aircraft will be subsequently housed: autopilot, payloads, sensors, communication modules, the propulsion system, etc. In most fixed-wing UAVs, the fuselage is an elongated structure in the longitudinal direction of the aircraft and in many cases manufactured in a modular way before possible future modifications, for example, the incorporation of a new payload that modifies the longitudinal Center of Gravity (CoG) [9].

The fuselages can be manufactured from different materials, always considering that the objective is to obtain light, rigid, and resistant structures. The most common materials used in the construction of low-cost fuselages are balsa or plywood, polyethylene, polypropylene, and rigid foam, and for higher cost fuselages composite materials based on synthetic fibers and epoxy polymer matrices are used, the most used fibers being carbon fiber (CFRP) for its high strength/weight ratio. Another high-cost material also used in the construction of UAVs, specifically for the fuselage, is nylon used for manufacturing parts using the Selective Laser Sintering (SLS) technique to print complete monocoques of fuselages. The advantage of this technique concerning parts made of composite materials is that it is not necessary to use molds or manufacturing tools.

- Empennage:

The next decision is regarding the type of empennage of the aircraft, from a simple T (Fig. 1: A) *to an inverted T* (Fig. 1: B), V (Fig. 1: F), or V—reversed, but there are other less conventional empennage configurations such as the U (Fig. 1: D) or H (Fig. 1: E). The different empennage configurations that can be integrated into an aircraft have their advantages and disadvantages. In the conceptual design phase, ideally, all possible empennage configurations must be considered to assess which is the ideal configuration for the system studied. In turn, estimates and approximations are usually made based on other similar models on the market; this last point has been repeated throughout the study because if several models of similar characteristics make use of a certain feature, it is because the problems associated with other configurations have been finally solved by this configuration [9].

UAV empennages can be manufactured with rigid foam inside, with CFRP longerons going through the rigid foam, and this structure is coated with GFRP or CFRP. The empennage can be directly coupled to the fuselage of the aircraft or make use of booms to join the empennage with the fuselage. The advantage of the booms is that they allow manufacturing smaller tails because the moments are

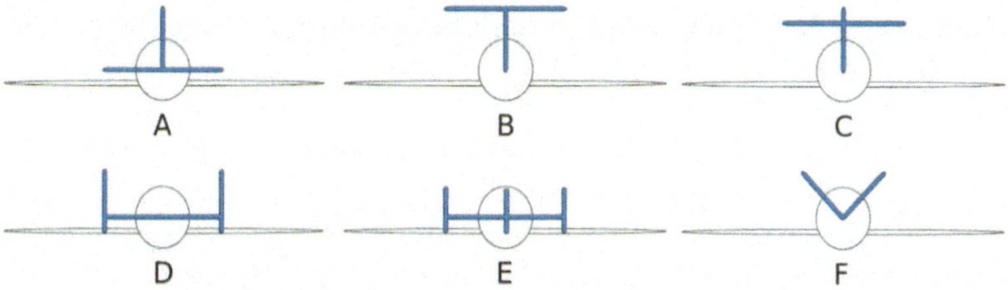

Fig. 1 Empennage types

increased concerning the CoG by delaying the position of the empennage, but as a counterpart, they can limit the angle of ascent during the moments after takeoff.

- Control surfaces:

Conventional aircraft configurations make use of control surfaces near the wingtips called ailerons, to control the warping of the aircraft. The ailerons are in continuous movement during flight making corrections to stabilize the moments around the longitudinal axis and maintain the balance position. Given that these devices are in continuous motion during flight, it's crucial to account for their aerodynamic drag. However, it is also important that they have sufficient actuation capacity for warping, especially during landing and takeoff, situations in which the aircraft operates at low speed and can stall.

- Materials and manufacturing methods:

The main characteristics of a material so that it can be incorporated into an aircraft are high mechanical strength, good elasticity, good fatigue resistance, corrosion resistance, and the price per kilogram of the material. A very important factor is the mechanical strength/density ratio or specific resistance; the higher this ratio the better resistance the material will have. Many of the new synthetic fibers, such as twaron, vectran, dynema, and zylon (PBO), have very high specific strength values.

The aerospace industry is a sector in which research into new materials plays a priority role, and both in new commercial aircraft and in unmanned aerial vehicles, there is still a long way to go for structural improvement, thanks to the great advances that have been made in recent years in new materials [10].

The incorporation of new composite materials into the most modern aircraft is carried out progressively, because before incorporating a material with new technical specifications into an aircraft, all the relevant aeronautical studies must be carried out to verify that the material complies with all the regulations (aeroelasticity or fatigue tests), and these studies are usually expensive and take time to be approved. So from the discovery of new composite material to the incorporation into some critical system of an aircraft, it can take more than a decade [8].

The great technological leap that composite materials represented, in particular fiberglass, which was the first super lightweight composite material used in aircraft, was a determining factor in making significant improvements over aircraft. The first applications of fiberglass in commercial aircraft were in the late 1940s when they were installed in the fairing, nose, and cabin. In addition to conventional aircraft, fiberglass also began to be used in other aircraft, such as helicopters, in which one of the first parts that were manufactured in fiberglass was rotor blades.

The development of new synthetic fibers at that time (carbon, polyester, etc.), with the appearance of sandwiches or cores, allowed to increase the specific resistance of these new materials remarkably, obtaining higher yields with lower consumption. Today, in commercial aviation, most non-structural panels that were previously made of polymers or fiberglass are being made using carbon fiber-reinforced polymers (CFRP) and honeycomb panels such as Dupont's aramid-based Nomex [8].

Today, composite materials are a fundamental part of the aeronautical industry and are used in UAVs of micro or small categories to large commercial aircraft with a higher content of composite materials of 50%. The main manufacturers of carbon fiber are in Japan, with Toray, Mitsubishi, and Toho as maximum representatives; the next major global manufacturer is in the United States: Hexcel; in Europe the big manufacturers are SGL Carbon and Solvay, especially to support the demand of the automotive industry, BMW, or Mercedes.

The difference between a layer and a laminate in a composite material is that a laminate is fundamentally a set of layers stacked on top of each other achieving a compact assembly. The orientations of the laminate will determine the final properties of the part, and symmetrical laminates are always pursued (concerning the central sheet there is an equal sheet on each side of the plane in terms of distance, thickness, and material) and balanced (for each layer in orientation x there is a sheet in orientation -x, in this case, it is not necessary to be concerning the medium plane) [11].

The isolated carbon fibers manufactured by large manufacturers have a thickness of 5–10 μm, so they are not worked with individually, but they are worked with in groups. The filaments are grouped in rovings, depending on the number of fibers grouped in a roving, and the numbering of the so-called tow can be established. A fabric with 8 K refers to the fact that the roving is formed by 8,000 filaments of carbon fiber. In general, the greater the tow numbering, the greater the weight of the fabric.

Once the fibers are obtained in the form of rovings, they can be in the form of fabric or non-intertwined (unidirectional). In an interlaced fabric, there exist the weft and the underside of the fabric, which, based on their interlacing pattern, result in fabrics exhibiting varying mechanical properties such as resistance, rigidity, and flexibility, among others. Many patterns can be obtained with the interlacing of the weft and the underside (plain fabric, twill, satin, or basket type). These patterns are closely related to the amount of epoxy resin that a fabric can absorb during the curing phase. The rigidity of a laminate is different in each of the layers of the fabric because, in each layer of the laminate, there are fabrics with different orientations. This effect causes different polar stiffness diagrams to be obtained in each section of the laminate, so that for a unidirectional laminate there can be obtained the highest rigidity at 0º, while for a plain weave, type laminate both at 0° and 90° there is a high rigidity.

In CFRP composite materials, carbon fibers are embedded in a polymer matrix and their main function is to keep the fibers in a fixed position acting as an adhesive for the fibers. Because the matrix allows the fibers to be fixed in a fixed position, it allows the fibers to work better together, because the polymer transfers loads from one fiber to another through shear stresses. Consequently, external charges are better distributed in a composite material than in isolated fibers: if one filament fractures, the charge is transmitted to the rest of the filaments [11].

When a part is manufactured in composite materials, both the material and the final structure are produced in a single process, due to using a polymer that needs a curing process during the forming of the part. To ensure that a part is produced with the proper shape, composite production methods are based on the use of models (mold positive) and molds. The direct mold or model can be made from the original material of the part to be manufactured (CFRP molds); however, they can also be made simply from EPS expanded polystyrene or DM plywood.

The quality and dimensional accuracy of a final piece made of composite materials depends largely on the mold with which the piece is obtained, so its importance in the manufacturing process is very high. For the final piece to have the desired precision, the mold must be rigid and maintain its geometry in a high-temperature range, since for the curing of the pieces, temperatures of up to 200 °C are reached (although the most common ranges are usually made over 110–130 °C). The entire curing process must be controlled in pressure and temperature, and for this purpose sensors are used to monitor the process. Pressure sensors are inserted into both the oven and the autoclave to monitor the pressure, and a temperature sensor is placed near the mold to monitor the temperature and ensure that the curing cycle is correct [8].

Most epoxy resins have adhesive properties and once the curing process is complete, the final CFRP parts can be trapped against the face of the female mold. To prevent the parts from being trapped against the mold, release agents and sealants are used against the mold, mainly based on polyvinyl alcohol (PVA) and Teflon tape.

Regarding the type of material used in composite materials, as already mentioned above, the components include epoxy resins and interlaced carbon fiber fabrics. These can be purchased individually and during the manufacturing process impregnate the dry fibers (or, on the other hand, it is possible to acquire materials in a pre-impregnated or prepreg state in which the fibers and resin are already mixed, but the resin is uncured). The negative point of prepreg fabrics is that they must be stored in refrigerators at −18 °C; otherwise, the resin begins to cure and ceases to be in a viscous state and goes to a more rigid state.

4 Hardware: Aerodynamic Analysis and Geometry

For flight conditions under a standard atmosphere, at an operating height of about 200 m and 21 °C temperature, the Reynolds number can be obtained as

$$Re(root) = \frac{pvl}{\mu} = \frac{1,19 * 21 * 0,20}{1,512 * 10^{-5}} = 330.556 \quad (2)$$

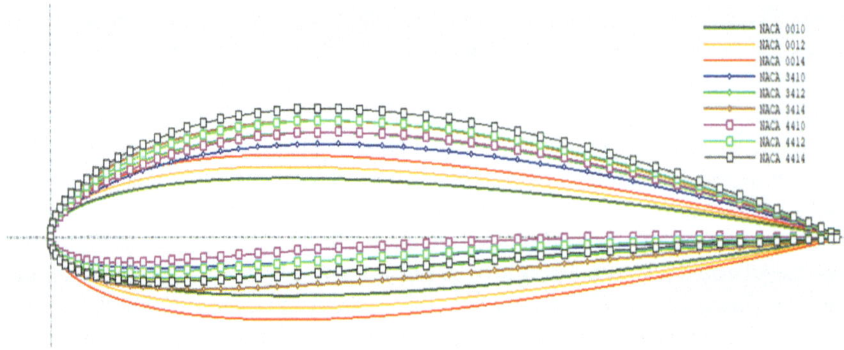

Fig. 2 Airfoils studied

$$Re(wingtip) = \frac{pvl}{\mu} = \frac{1,19 * 21 * 0,15}{1,511 * 10^{-5}} = 247,916 \quad (3)$$

The selection of the airfoil is made by analyzing the behavior of different *airfoils* under the Re conditions indicated above, using a range of angles of attack between $-1°$ and $15°$, because above $15°$ the airfoil can stall. The study of the behavior of the aerodynamic airfoils under flight conditions (Reynolds number and Mach number) is carried out from the XFLR5 program, a suitable tool for studies at low Reynolds numbers.

The selected airfoils belong to the NACA four-digit airfoil family, indicating first, the maximum curvature; second, the position of the maximum curvature; and, last two, the thickness of the airfoil. The nine airfoils studied are shown in Fig. 2.

Airfoils that operate in lower Reynolds numbers for which they were designed have worse efficiency which manifests itself in lower maximum lift coefficient (c_{lmax}), greater drag, and, therefore, worse lift-drag efficiency (cl/cd). Increasing the camber in an airfoil increases the lift available for a given angle of attack while increasing the thickness of the airfoil allows more space to accommodate the internal structure and allows the wing to operate in a higher range of angles of attack without flow separation, which increases stall speed.

The operating cost of a UAV can be greatly reduced with an appropriate and optimized selection of an airfoil that improves aerodynamic efficiency and manufacturing cost. When selecting the airfoil of the aircraft, several factors are considered; maximize the lift coefficient (c_{lmax}), maximize the lift-drag ratio (c_l/c_d), such that the airfoil can operate in low values of Reynolds number, low pitch moment coefficient to minimize the load that the horizontal stabilizer of the aircraft must perform, or consider the stall of the aircraft. Another important parameter when selecting the airfoil is the ease of manufacture of this aerodynamic airfoil: the less complex the airfoil, the easier it is to manufacture and the lower the costs incurred in the process.

The supporting surfaces of unmanned aerial vehicles (UAVs) typically feature several aerodynamic airfoils distributed along the wingspan. Near the root, there's an airfoil designed to stall at a higher speed than the airfoil located near the tip. This design choice is made for controllability reasons. If a stall is detected near the root, the aircraft can still be controlled, thanks to the fact that the airfoils near the tips are designed to stall at a slower rate.

Therefore, the flow conditions in XFLR5 and the criteria when selecting an airfoil for the UAV are as follows:

- Minimize resistance efficiency C_d versus Alpha.
- Maximize lift coefficient C_l versus Alpha.
- Maximize point of maximum fineness C_l/C_d.

In Figs. 4 and 3, the analysis of the airfoils with a Reynolds number of 330,000 for angles of attack between $-1°$ and $15°$ is performed, while in Figs. 5 and 6 the analysis is performed for a Reynolds number of 248,000.

Finally, in Fig. 7 the lift coefficient for a zero angle of attack is displayed.

The NACA 44XX and 34XX airfoil families are studied to carry out the aerodynamic study and select the airfoil that has the best behavior both at the root of the support surface and at the tip. Tables 1 and 2 show a summary of the results obtained: the airfoil selected at the root (Re-330.000) is 4414 and the airfoil selected for the tip (Re-248.000) is 3414 because both airfoils show in $\alpha = 0$ better aerodynamic behavior than for the other thicknesses. Airfoil 4414 is located close to the fuselage because it stalls earlier, at $11°$, while airfoil 3414 stalls at $15°$ (Tables 3 and 4).

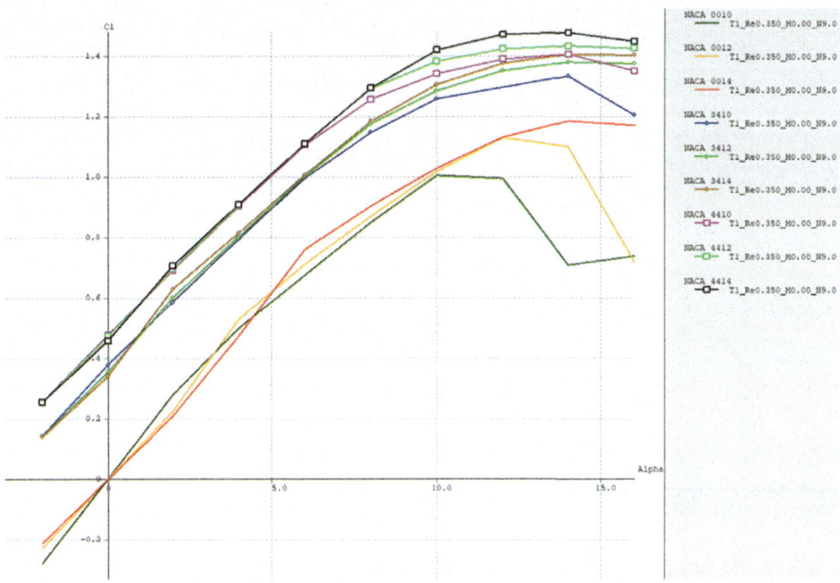

Fig. 3 Efficiency versus angle of attack with a Reynolds number of 330,000

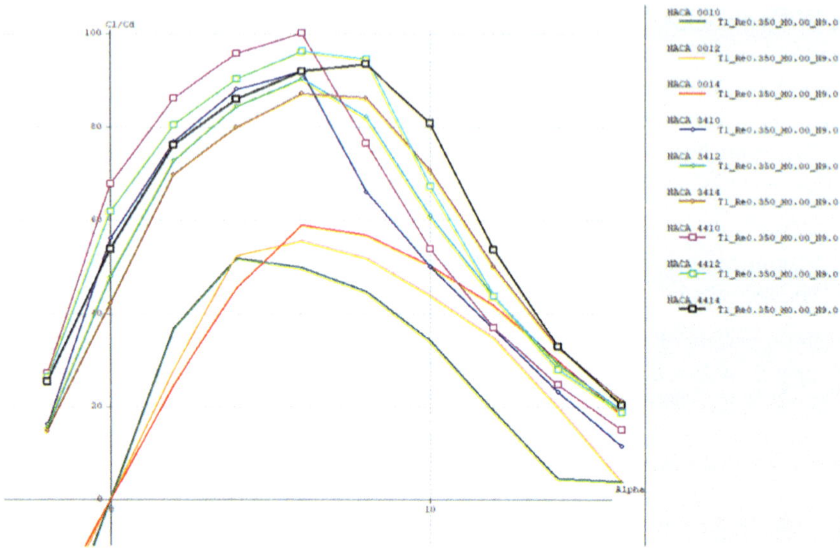

Fig. 4 Lift coefficient versus angle of attack with a Reynolds number of 330,000

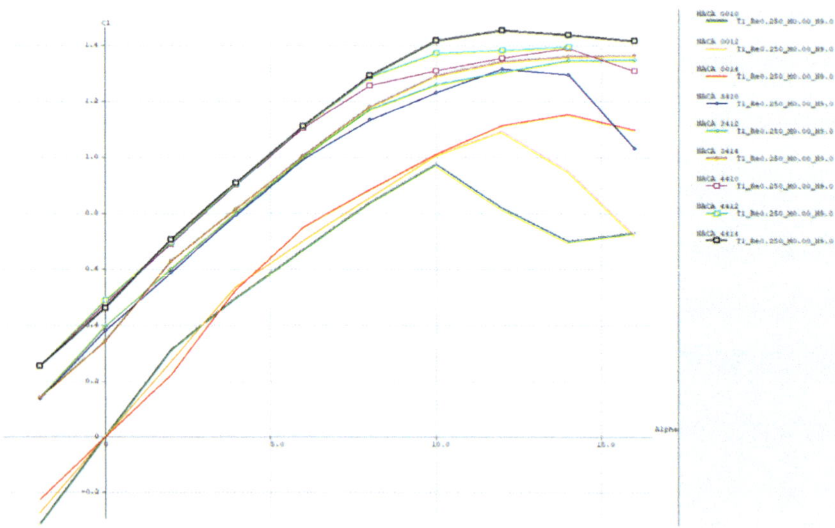

Fig. 5 Lift coefficient versus angle of attack for a 248,000 Reynolds

The aerodynamic airfoil for the empennage is the 0012, a symmetrical airfoil that serves both positive and negative yaw-pitch movements, which is why it is selected for the horizontal and vertical stabilizer.

Through XFLR5 the polars of the three airfoils are obtained in a range of Reynolds numbers between 90,000 and 1,100,000, to analyze the behavior of the airfoils under

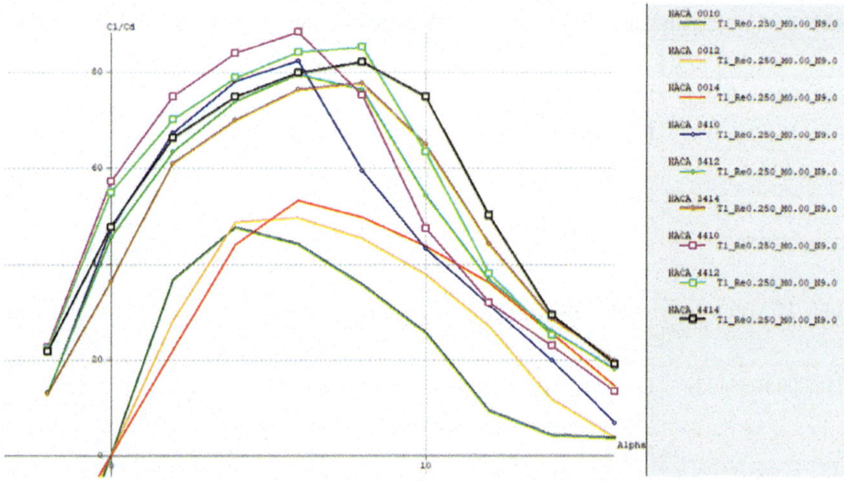

Fig. 6 Efficiency versus angle of attack for a 248.000 Reynolds

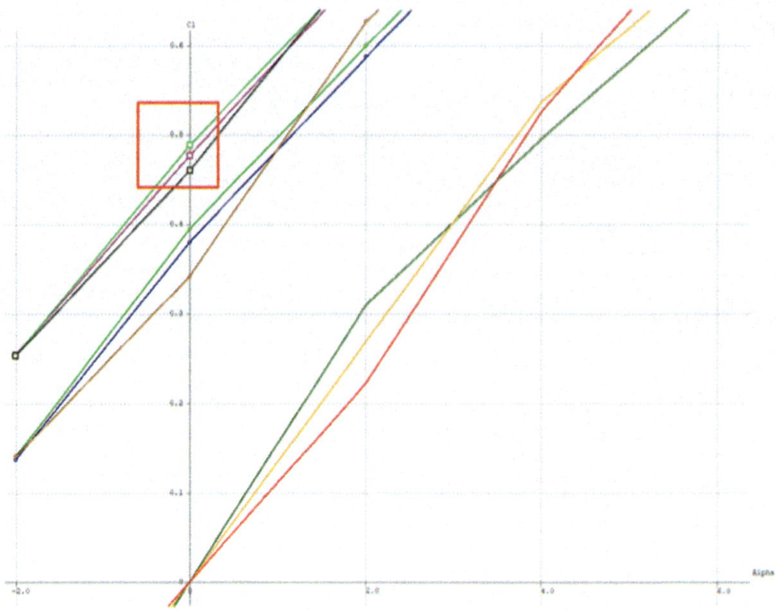

Fig. 7 Lift coefficient versus angle of attack, and zero angle intersection

different flow conditions and angles of attack (−2° to 25°) as shown in Fig. 8. In the design conditions, a cruising speed of 21 m/s has been established; although the aircraft can perform maneuvers at other speeds (landing or takeoff maneuver) higher

Table 1 Airfoil analysis for a 248,000 Reynolds value (44 family)

		Airfoil (NACA)	Value	Angle (°)
248,000 Reynolds	Efficiency versus angle of attack	4410	87	5
	Lift coefficient for zero angle attack	4412	0.51	0
	Lift coefficient versus angle of attack	4414	1.43	11

Table 2 Airfoil analysis for a 330.000 Reynolds value (44 family)

		Airfoil (NACA)	Value	Angle (°)
248,000 Reynolds	Efficiency versus angle of attack	4410	98	5
	Lift coefficient for zero angle attack	4412	0.49	0
	Lift coefficient versus angle of attack	4414	1.49	14

Table 3 Airfoil analysis for a 248.000 Reynolds value (34 family)

		Airfoil (NACA)	Value	Angle (°)
248,000 Reynolds	Efficiency versus angle of attack	3410	83	5
	Lift coefficient for zero angle attack	3412	0.38	0
	Lift coefficient versus angle of attack	3414	1.35	15

Table 4 Airfoil analysis for a 330.000 Reynolds value (34 family)

		Airfoil (NACA)	Value	Angle (°)
248,000 Reynolds	Efficiency versus angle of attack	3410	93	5
	Lift coefficient for zero angle attack	3412	0.39	0
	Lift coefficient versus angle of attack	3414	1.42	14

or lower, for cruising conditions these are speeds compatible with the observation of fire.

The weight range of the aircraft is 2–3 kg, although it is convenient to set the weight at the highest value of the range (in this case three kilograms), as a safety measure. For the study of the geometry of the unmanned aerial vehicle, the values calculated above are introduced, and then the behavior of the aircraft is analyzed. In this case, and for a total area of 0.25 m^2 of the bearing surface:

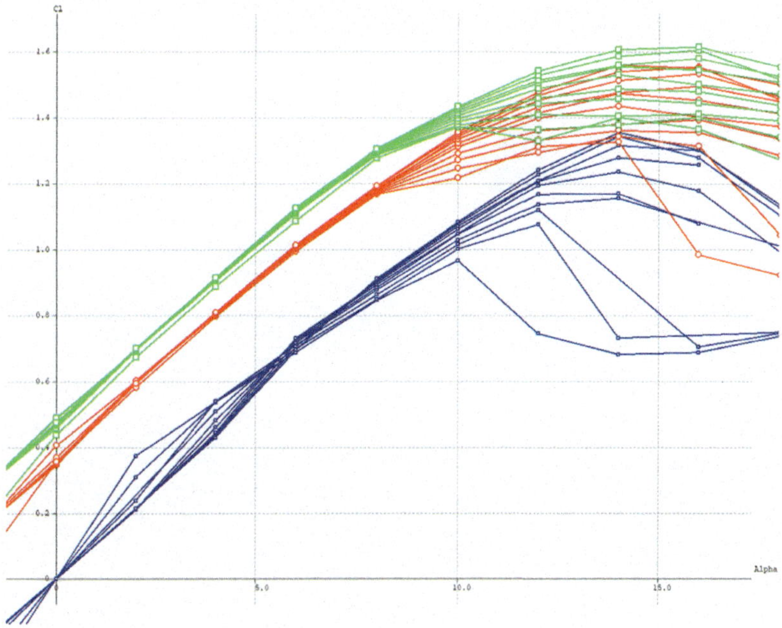

Fig. 8 Lift coefficient concerning the angle of attack

$$L = \frac{1}{2} \rho v^2 c_l S \tag{4}$$

Having $\rho = 1,18\frac{Kg}{m3}$; $v = 22\frac{m}{s}$; $c_l = 0,8$; and, consequently, $S \geq 0,19\,m^2$, which satisfies the previous calculation, 0.25 m^2.

Figures 9 and 10 show the aircraft in plan with the three aerodynamic airfoils.

An important condition when obtaining a stable aircraft is the correct positioning of the longitudinal center of gravity (CoG) of the unmanned aerial vehicle. The longitudinal CoG must be located between the leading edge of the main wing and the aerodynamic center of the airfoil at the root or between the aerodynamic center and the second quarter of the rope for the aircraft to be statically stable. Figure 11 shows the location of the longitudinal CoG located at a quarter of the chord of the root airfoil because it is convenient that both the weight force and the lift force act on the same axis. It is advisable to place the longitudinal CoG in front of the aerodynamic center because it increases longitudinal stability and increases the stall speed of the aircraft. On the other hand, the effects of placing the CoG behind the aerodynamic center of the airfoil are to improve cruising speed and decrease the control of the aircraft if it stalls [12].

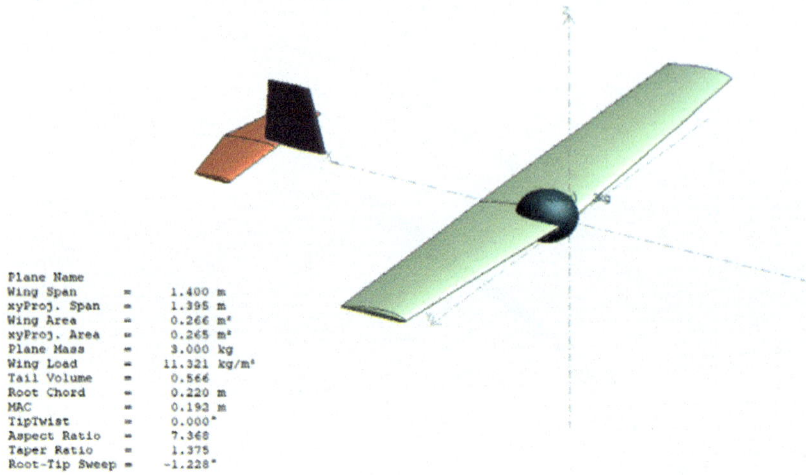

Fig. 9 Lift coefficient versus angle of attack

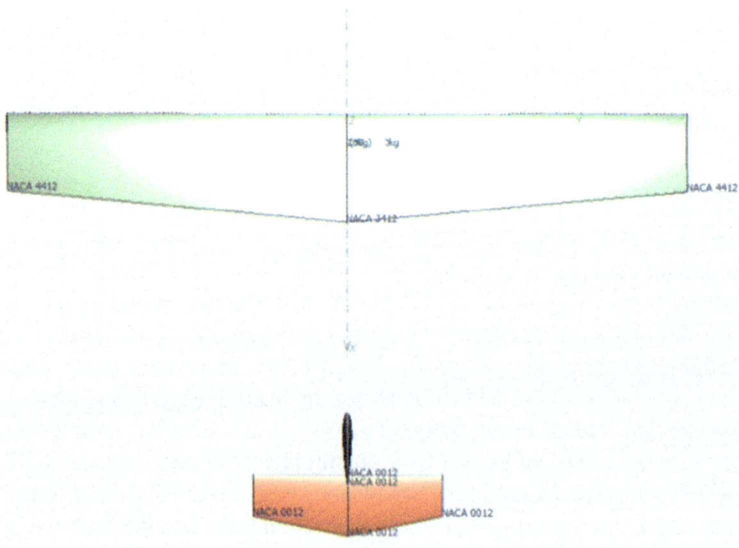

Fig. 10 Platform of the UAV

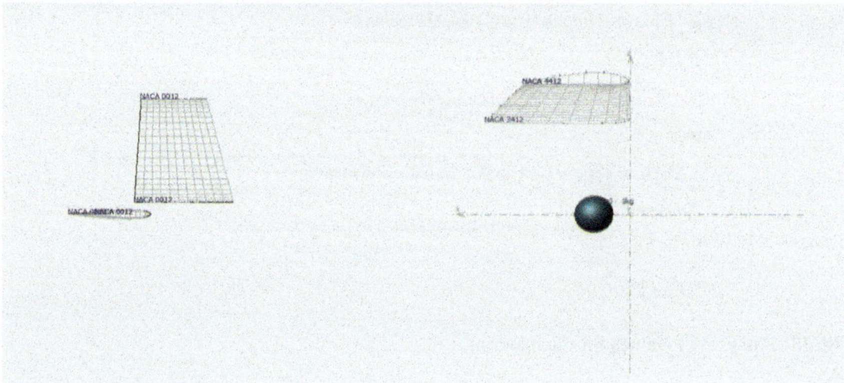

Fig. 11 Longitudinal center of gravity

5 Hardware: Material

As mentioned above, the CFRP composite material used is formed from carbon fiber and epoxy resin as a matrix. The carbon fibers selected are the fibers of type AS4, which are continuous fibers, with high strength and high tension, and are manufactured under aeronautical specifications. Secondly, the selected matrix is 8552 epoxy resin with a curing temperature of 185 °C. The properties of the CFRP material from the manufacturer Hexcel are shown in Fig. 12, highlighting the cured thickness in mm and the grams/square meter.

As can be seen in Fig. 13, the thickness of the material is obtained from the grammage per square meter (rule of composite materials: 100 g/m^2 = 0.1 mm). In this case, the material AS4/8552 has a density of more than 190 g/m^2 and is cured to a thickness of less than 0.2 mm. This approximate thickness per layer is established in Ansys ACP, as illustrated in Fig. 12.

Prepreg Properties - HexPly® 8552 Woven Carbon Prepregs (AS4 Fibre)

Physical Properties

	Units	AGP193-PW	AGP 280-5H
Fibre Type	-	AS4 3K	AS4 3K
Fibre density	g/cm^3 (lb/in^3)	1.77 (0.065)	1.77 (0.065)
Weave	-	Plain	5HS
Mass	g/m^2 (oz/yd^2)	193 (5.69)	286 (8.44)
Weight Ratio, Warp : Fill		50 :50	50 :50
Nominal cured ply thickness @ 37% resin content	mm (inch)	0.195 (0.0076)	0.289 (0.0114)
Nominal Fibre Volume	%	55.29	55.29
Nominal Laminate Density	g/cm^3 (lb/in^3)	1.57 (0.057)	1.57 (0.057)

Fig. 12 Selected material for the UAV

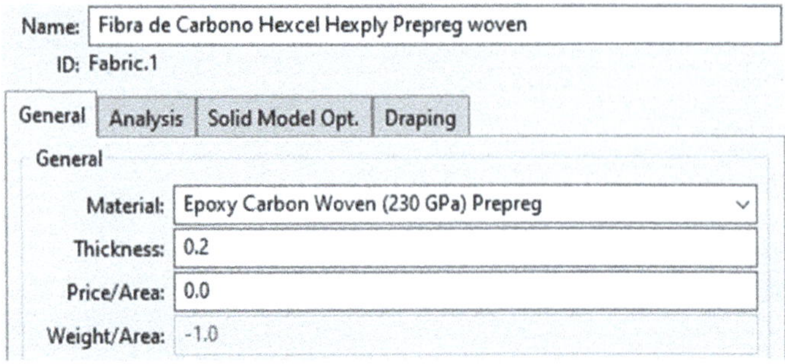

Fig. 13 Ansys ACP values for the material

The selected epoxy resin 8552 is an amine-type resin that in its cured state is very hard and resistant, has good properties against impacts, and resists extreme environmental conditions (corrosion, humidity, and temperature), optimal conditions for aeronautical applications. These characteristics are ideal for application to unmanned aerial vehicle structures.

In Figs. 14 and 15, the final design of the UAV can be seen.

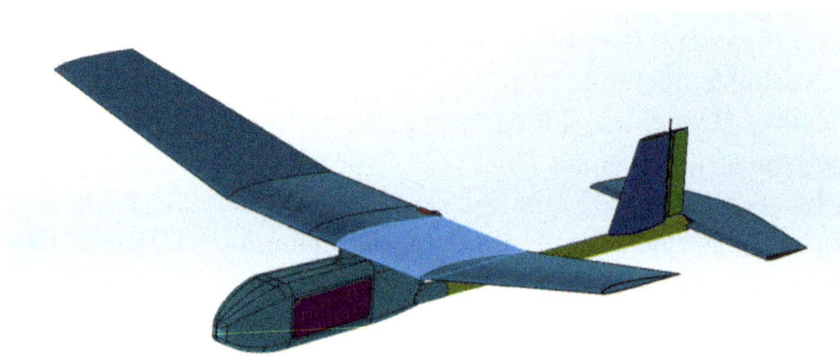

Fig. 14 UAV final design

Fig. 15 Isometric view of the UAV

6 Software: Introduction

The use of UAVs for fire detection and monitoring has been a field of study since the early 2000s. Research teams, such as the Spanish project COMETS [13], have published numerous articles between 2001 and 2005 on the use of UAVs in this field (see [14], and more specifically [15]). The FireRS project [16] further demonstrated the usefulness and necessity of employing UAVs in this domain, leading to the development of the aid system in the current project.

One of the major difficulties faced by the emergency personnel working in firefront surveillance is the monitoring of the firefront in conditions of low or zero visibility. When inspecting the area affected by fire, especially at night, the objective is to provide information about the evolution of the fire (in terms of direction, amplitude, and speed of spreading). In these cases, it is common to use UAVs, which carry different types of cameras that allow an operator to fly them and observe the fire to determine the characteristics. In the case of night fires, the pilot usually encounters difficulties when operating the aircraft: there are high contrasts between the dark environment (due to the night), and the intense light emanating from the fire itself. This subproject tries to overcome these problems by providing an aid layer that is overlapped on top of the camera image, providing cartographic and topographic details of the terrain covered by the UAV, based on satellite data and information provided by the UAV itself.

This software section presents an extensible solution to overcome these problems. The solution involves UAV aid software that includes an augmented reality layer that can be overlaid on the camera image, providing cartographic and topographic details of the terrain covered by the aerial vehicle. This information is based on satellite or terrain sensor data, along with the UAV's data. The solution is designed to, in the short term, amplify the human pilot's skills with an intelligent system.

7 Software: Objectives

Unmanned aerial vehicles have sensor systems that make it possible to obtain information regarding the vehicle, such as its current position and attitude, or more advanced information, such as that obtained by a camera (which makes it possible to visualize a scene from the point of view of the UAV). In the specific domain of the usage of UAVs for fire monitoring, more recent papers, such as [15], highlight the potential drawbacks of utilizing UAVs for visual fire monitoring: "[...] The smoke generated from fires can disrupt the visibility beyond 1 m, which calls for better cameras, such as thermal instead of RGB [...]".

Although the chapter focuses on the use of UAVs in enclosed spaces, the smoke plumes caused by large fires, as well as the light emanating from the flames (especially during the night), cause major visibility problems for the users of UAV-embedded cameras.

The solutions proposed in this chapter involve the use of camera systems adapted to the conditions (in particular, thermal cameras based on infrared light). If such cameras were not available, or to avoid the total loss of the visualization of the visible light spectrum, it is necessary to make use of techniques that allow the reconstruction of the terrain image from the available data (i.e., simulation techniques and the use of a digital twin of the camera via Augmented Reality techniques, which is the approach adopted by this project).

The main objective of this project is the development of a tool to assist in the piloting of drones and UAVs for fire surveillance, through the superposition of computer-generated images with the information received from the cameras of these vehicles.

From a general point of view, the assistance tool resulting from this project allows to improve the efficiency of fire surveillance operations, facilitating the piloting of UAVs by displaying relevant information to the pilot intuitively and clearly. To assist the piloting process in this way, it is necessary to generate a virtual interface (SUB-OBJECTIVE 1) that is properly synchronized with the real situation of the vehicle (SUB-OBJECTIVE 2) and that is displayed in a clear and accessible way for the user (SUB-OBJECTIVE 3).

The subsequent sections elaborate on the fulfillment of these sub-objectives. Sub-Objective 1: Virtual interface, the development and integration of a comprehensive virtual interface are detailed. Sub-Objective 2: Synchronization delves into the seamless synchronization of this interface with real-world vehicle conditions, ensuring accurate and up-to-date data representation. Lastly, Sub-Objective 3: Pilot assistance discusses the user-centric design approach adopted to ensure the interface's clarity, accessibility, and effectiveness in aiding UAV piloting for firefighting purposes.

7.1 *Software: Sub-objective 1 Virtual Interface*

This subsection explains the functionalities integrated into each version of the system related to the generation of the virtual interface.

In the context of software development aimed at enhancing UAV capabilities, this subsection delves into the functionalities integrated into the system's different versions. The focus lies on generating a digital cartographic layer to enhance viewer comprehension and overcome visualization problems. This involves implementing contour lines to depict areas of equal height, aiding terrain understanding. These lines are generated by intersecting horizontal planes, enabling visualization at specified intervals. Additionally, a complementary map display offers a separate top-down view of the terrain to precisely locate the UAV's position. This is achieved by creating a distinct panel that presents a flattened top view with a marker indicating the UAV's orientation. The section also explores motion interpolation to ensure smoother virtual camera movement, minimizing abrupt shifts. Furthermore, contour lines are color-coded according to altitude variations, enhancing their differentiation, and aiding in interpreting terrain changes. This comprehensive software segment advances UAV monitoring capabilities for improved wildfire management.

7.1.1 Contour Lines and Altitude Data

To enhance the observation and understanding of the terrain and its characteristics was implemented the possibility (through an additional layer) of visualizing contour lines on the terrain, as well as the current height of the drone and the height of a point marked with the cursor of the computer. This process entails several distinct steps:

- Description of the contour line: A contour line is a line on a map or on a geographical surface that joins a set of points that are located at the same height. They are used as a reference to know the approximate height of an area, as well as the variations in level and slope of the area.
- Contour line generation process: To generate contour lines in WorldWind, a horizontal plane is created that intersects the surface. The intersection points are stored, and a closed polyline is generated that joins them all. This process is repeated at different heights to generate the set of contour lines.
- Visual demonstration of the contour lines: Fig. 16 shows several examples of contour lines overlaid on the generated visualization.
- Specification and justification of the distance between contour lines: Due to the characteristics of the terrain chosen for the tests, it was decided to establish contour lines every 25 m, since a smaller interval would generate too many contour lines, which would make it difficult to see the terrain. On the other hand, the use of a larger interval would not allow the visualization of an adequate number of curves at the same time, which would prevent obtaining sufficient information about the altitude variations of the terrain.

Fig. 16 Examples of contour lines in the viewer

- However, it is possible to modify the ratio to adjust it to the characteristics of other areas by changing the value of the "*incremento*" variable of the "*dibujarLineasDeNivel()*" method of the AppFrame class.
- Specification and justification of the color choice of contour lines: To select the color of the contour lines, several tests were carried out with different colors, and white was chosen because it is the color that generates the greatest contrast with the most common colors of the terrain. In addition, since the visualization layer can reduce its transparency, the use of contour lines with a high brightness makes it possible to distinguish them better in these situations.

7.1.2 Complementary Display with Map

To facilitate the geographic localization of the UAV, an additional view was implemented through which the position of the vehicle can be observed on a zenithal plane of the terrain, separated from the main system display. The map view is integrated within the "MapFrame" class and encompasses various distinct characteristics:

- Specification and justification for the map inclusion: The inclusion of a zenithal geographic view map allows the position of the vehicle to be more clearly located in longitude and latitude terms.
- Specification and justification for the use of a decoupled panel: A separate display panel has been used to provide greater flexibility to the system. Consequently, the user can arrange the windows in the way they consider most convenient according to the tasks they are carrying out at any given moment, as well as their personal preferences.
- Complementary visualization creation process: To create a supplementary map display, a new panel is generated, and decoupled from the main display panel. In this panel, a flat top view of the terrain is displayed, centered on the currently selected vehicle. A marker is drawn at the position of the UAV to facilitate its localization and the determination of the direction in which it is oriented at any given moment.
- Visual demonstration of complementary view: Fig. 17 shows the panel generated for the complementary map view of this project.

7.1.3 Motion Interpolation in Visualizer

To increase the smoothness of the movement of the virtual camera of the viewer, an interpolation process has been used on the movement of the virtual camera. Instead of immediately changing the camera position after receiving new data from the simulator, animation to the new point is performed for a duration equal to the time between two data requests to the simulator.

For this purpose, the Animator classes of the NASA WorldWind project have been used [17]. In particular, the FlyToFlyViewAnimator class has been used as a starting point [18], but its animators have been modified (this is a composite animator,

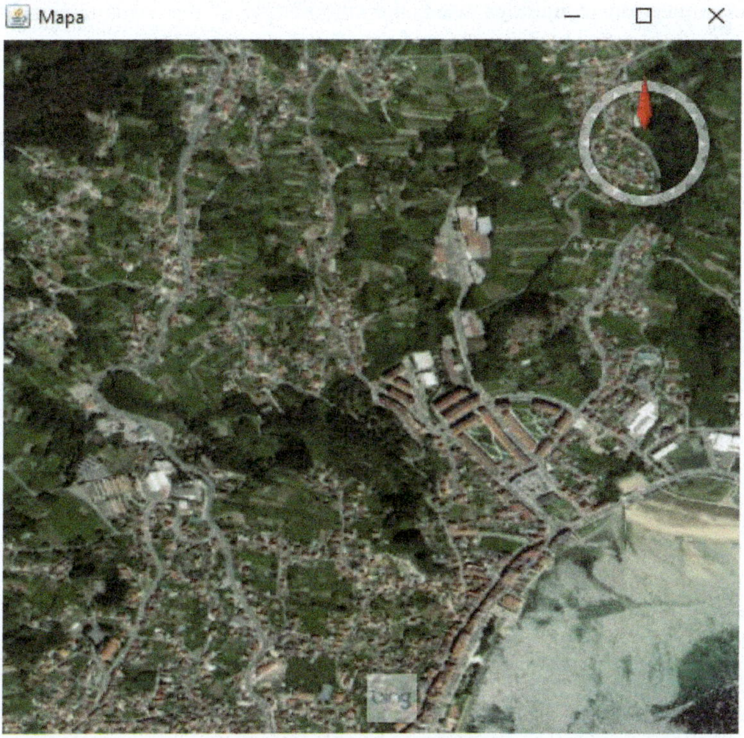

Fig. 17 Complementary map display panel

containing elements for position interpolation, elevation, and attitude of the drone) to avoid the default animations, which include zoom-in and zoom-out effects that were not suitable for the current implementation. Instead, MoveToPositionAnimator animators have been used.

Thus, a smoother visualization is achieved, with the cost of increasing the latency (increase proportional to the time between data requests to the simulator).

Another option explored is to perform interpolations with predictions of the drone's future positions, taking into account the movement made so far and its speed. This technique would avoid additional latency but would generate sudden jumps when the prediction does not match reality, especially in changes in the orientation of the drone.

This process encompasses various distinct characteristics:

- Description of interpolation processes: Interpolation processes use measurements made on a specific event to predict new measurements at positions where no measurements were originally made.
- Specification and justification for the inclusion of interpolation processes: Since the update frequency of position and attitude data may vary depending on the

communication technology used, the smoothness of the visualization may be significantly reduced depending on this factor. Interpolation processes are used to generate intermediate positions and attitudes between the data obtained from the vehicle.

- To generate this intermediate data, it is necessary to have at least one set of margin data to move toward. This generates a minor delay in the visualization, relative to the received data.

7.1.4 Color-Coded Contour Lines with Altitude-Based Color-Coding

As the project progressed and more tests were carried out, it became clear that the monochromatic contour lines were not sufficient to differentiate the changes in altitude, so it was decided that the contour lines should be color-coded according to their relative height. This process encompasses various distinct characteristics:

- Description of the altitude color-coding concept: Topographic maps often use different colors to represent different altitude levels, such that areas of the same color will be at the same altitude, while different colors will be higher or lower, depending on whether they go up or down on the color scale.
- Process of applying altitude color-coding to contour lines: The color code should follow a gradual scale so that slight differences in altitude maintain similar color shades. The total variation of altitudes must also be considered, i.e., if the distance between the minimum and maximum altitude on the map is considerable, a greater number of colors will be necessary to be able to correctly differentiate the changes in altitude.
- Specification and justification of chosen altitude color code: The chosen color code marks blue (#0000ff) as the minimum height (0 m by default) and white (#ffffffff) as the maximum height (2127 m by default). In between these two colors, green (#00ff00) and red (#ff0000) will be found.
- Due to the different types of elevations that the drone could encounter, it was necessary to use a wide color scale, so that the differences in altitude could be recognized regardless of whether the drone was passing through a flat area or a mountainous area. For the maximum altitude, the *Pena Trevinca* peak was chosen since it is the highest point in Galicia and mainland Portugal.
- Contour line coloring implementation process: The color scale is automatically generated in the "makeColorScale" function of the "AppFrame" class in the "AppFrame.java" file, based on the maximum height and the distance between level lines. By modifying these variables, a different color scale could be generated, if necessary. With the default data, 88 level lines will be generated, each of them with a different color.
- Graphical comparison between monochromatic and color-coded elevation contour line displays: Fig. 18 shows how using a color code makes it easier to differentiate which zones are higher than others. In this case, the white-colored zones are the highest, and they gradually go down toward red, yellow, and green.

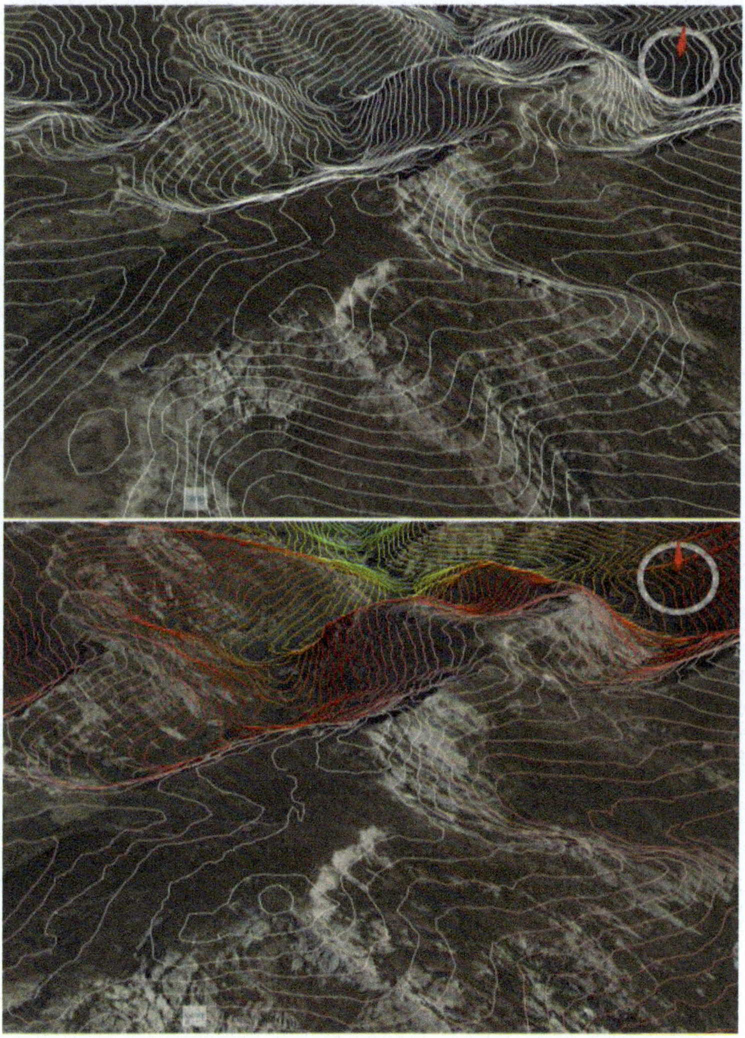

Fig. 18 Comparison between monochromatic and color contour lines

7.2 *Software: Sub-objective 2 Synchronization*

This subsection explains the functionalities integrated into each version of the system related to the synchronized and appropriate positioning and orientation of the UAV camera within the developed virtual environment.

The next subsections outline the initial steps taken in integrating the visualization system using the NASA WorldWind API and the next sections will delve into the specifics of this integration process, detailing key aspects like camera angle configuration, connecting with multiple drones, image synchronization, and synchronization tests.

7.2.1 Initial Integration with the Visualization System

An initial version of the visualization system was developed using the NASA WorldWind API. This API receives the position and attitude data from the simulator implemented in version 0 of the project and uses them to represent the vehicle camera view on a 3D terrain model.

To keep the position and attitude information updated, the visualization system makes periodic requests to the TCP server implemented in the simulator developed in version 0 of the project. Next, the connection process established between the simulator and the visualizer is detailed. This process encompasses various distinct characteristics:

- Connection of simulator with visualizer: The connection between the viewer—developed in Java—and the drone simulator—developed in Python—is made through a TCP socket connection (server implemented in Python on the simulator side, and client implemented inside the viewer, in Java).
- Through this TCP socket connection, text strings representing a JSON object containing the following data are sent:


```
{
"Actitud":
"Attitude:pitch=<num>,yaw=<num>,roll=<num>",
"Elevacion": "<num>",
"GPS": "GPSInfo:fix=6,num_sat=10",
"Localizacion Sistema Global":
"LocationGlobal:lat=<num>,lon=<num>,alt=<num>",
"Localizacion Sistema Global Relativo a Home":
"LocationGlobalRelative:lat=<num>,
lon=<num>,alt=<num>",
"Localizacion Sistema Local":
"LocationLocal:north=<num>,east=<num>,down=<num>",
"Tiempo": "<datetime>"
}
```


Table 5 specifies the meaning of each of the components of this JSON object:

This information requested by the client (viewer) at regular time intervals (parameterized, by default 30 times per second) is necessary to update the position of the drone in its representation in the viewer. An object of class "TimerTaskGetData" has been implemented that implements the requests, while an object of class "Vista" is in charge of updating the position of the vehicle in the augmented view of the terrain.

- Visualizer view update process: The viewer makes periodic requests to the simulator's TCP server, which responds with the information previously indicated.

Table 5 JSON components

COMPONENT	MEANING
{	Delimits the starting point of the JSON object, from which the set of keys and values it contains are included
"Actitud"	The string corresponds to the key of the first element of the JSON object. It refers to the attitude of the UAV, i.e. the angle of the vehicle relative to the three axes of space (pitch, roll, and yaw)
:	Indicates the separation between the key or name of a JSON object element and its value.
"Attitude:pitch=<num>,yaw=<num>, roll=<num>"	The value of the first element of the JSON object is headed with the word "Attitude", which refers to the attitude of the UAV, followed by a delimiter ":" and three elements: "pitch", "yaw" and "roll", corresponding to the three angles that define the attitude of a UAV. Each of them is followed by the symbol "=", which separates it from its value ("<num>"), which is a decimal number. Each of these elements is separated by a delimiter ",".
,	Separator between elements of a JSON object.
"Elevacion"	This string corresponds to the key of the second element of the JSON object. It refers to the altitude of the UAV, i.e. the number of meters above sea level.
"<num>"	It represents the value of the second key of the JSON object and is replaced by a decimal number representing the height of the UAV.
"GPS"	This string corresponds to the key of the third element of the JSON object. It refers to the position information of the UAV, based on the global positioning system or GPS.
"GPSInfo:fix=6,num sat=10"	The value of the third element of the JSON object is headed with the term "GPSInfo", which refers to information about the GPS, followed by a delimiter ":" and two elements: "fix" and "num sat", corresponding respectively to the method of determining the position of the UAV through GPS and the number of available satellites. Each of them is followed by the symbol "=", which separates it from its value, which is an integer. Each of these elements is separated by a delimiter ",".
"Localizacion Sistema Global"	The string corresponds to the key of the fourth element of the JSON object. It refers to the position information about the global system of the UAV, based on the GPS.

(continued)

Table 5 (continued)

"LocationGlobal:lat=<num>, lon=<num>, alt=<num>"	The value of the fourth element of the JSON object is headed with the term "LocationGlobal", which refers to the global position of the UAV, followed by a ":" delimiter and three elements: "lat", "lon" and "alt", corresponding to the three values that define the spatial position of the UAV, its latitude (angular distance between the equatorial line and the UAV's position), its longitude (angular distance between the UAV's position and the 0° meridian), and its altitude (the number of meters in height at which it is located concerning sea level). This last data, the altitude, is obtained repeatedly in two fields. Each of these properties is followed by the symbol "=", which separates it from its value ("<num>"), which is a decimal number. Each of these elements is separated by a delimiter ",".
"Localizacion Sistema Global Relativo a Home"	This string corresponds to the key of the fifth element of the JSON object. It refers to the position information of the UAV with its home position, based on the GPS.
"LocationGlobalRelative: lat=<num>, lon=<num>, alt=<num>"	The value of the fifth element of the JSON object is headed with the term "LocationGlobalRelative", which refers to the global position of the UAV relative to its home position, followed by a ":" delimiter and three elements: "lat", "lon" and "alt", corresponding to the three values that define the relative spatial position of the UAV relative to its point of origin, its latitude (difference between the UAV's current latitude and the latitude of its point of origin), its longitude (difference between the UAV's current longitude and the longitude of its point of origin), and its altitude (the difference between the UAV's current altitude and the altitude of its point of origin). Each of these properties is followed by the symbol "=", which separates it from its value ("<num>"), which is a decimal number. Each of these elements is separated by a delimiter ","
"Localizacion Sistema Local"	This string corresponds to the key of the sixth element of the JSON object. Refers to the relative position information of the UAV to its home position, based on the distance in meters in North, East, and Down orientation.
"LocationLocal:north=<num>, east=<num>, down=<num>"	The value of the sixth element of the JSON object is headed with the term "LocationLocal", which refers to the relative position of the UAV to its position of origin, followed by a delimiter ":" and three elements: "north", "east" and "down", corresponding to the three values that define the relative spatial position of the UAV to its point of origin, the distance from its origin in meters in the north, east and down (vertical) direction. Each of these properties is followed by the symbol "=", which separates it from its value ("<num>"), which is a decimal number. Each of these elements is separated by a delimiter ",".
"Tiempo"	The string corresponds to the key of the seventh and last element of the JSON object. It refers to the temporal information of the received data.

(continued)

Table 5 (continued)

"<datetime>"	Value of the seventh and last element of the JSON object, it contains a timestamp indicating the time at which the information contained in the object was obtained.
}	Delimits the endpoint of the JSON object, after which information about that object is no longer found.

Once this information is received, a new view object is constructed to modify the position and attitude of the UAV in the 3D representation of the viewer.

The specification of the received views is done in the "Vista" class.

- Specification and justification for the periodicity of the updates: Periodic requests are made to the simulator TCP server to obtain UAV position and attitude data at a frequency of 30 times per second, which can be configurable. This time interval between requests is set to try to obtain at least 30 updates per second, allowing a smooth display.

The periodic requests are implemented in the file "TimerTaskGetData.java'.

7.2.2 Camera Angle Configuration

Since the camera angle of the drone is fixed before the flight but can be different in each operation (varying from a view completely toward the ground to a view completely forward), it was chosen to incorporate a configuration option within the visualizer that would allow to adjust the camera angle of the drone. The configuration options are specified in the "OpcionesFrame" class and encompass various distinct characteristics:

- Visual demonstration of the operation of the configuration options:
- Figure 19 shows the options available in the upper left menu of the visualization system:

Figure 20 includes the configuration window that is displayed when selecting the UAV camera angle setting option, which can range from 0.0° to 90.0°.

Figure 21 shows the difference between a 0° angle and a 90° angle in the camera angle setting.

Fig. 19 Camera angle configuration option

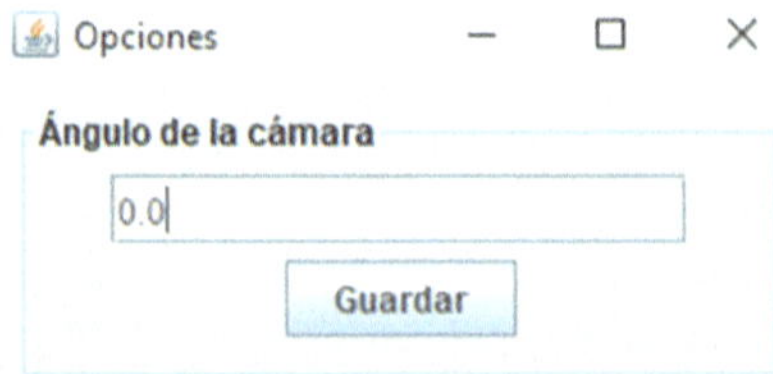

Fig. 20 Camera angle configuration window

Fig. 21 Camera angle configuration results

- Specification and justification for the inclusion of the configuration option: The reason for the inclusion of this option is due to the reality of the UAVs used for the project tests: In these vehicles, the camera angle can be modified before the start of the flight.

The angle can be set between 0° (full downward view) and 90° (forward view). However, during UAV acceleration, the UAV tilts significantly forward, which reduces the target viewing angle to an amount below 90° even if this is the selected setting.

- Specification and justification of the positioning of the configuration option: The option is in the upper left options bar, following the basic configuration of the Java Swing interface menus. This option is included in the "Options" category of this menu, with the possibility of adding new options in the future.

7.2.3 Connection with Multiple Drone Configuration

There is the possibility of monitoring multiple UAVs from the same equipment and display. For this reason, it is possible to configure the reception of data from multiple sources, modifying the display depending on the position and attitude of the drone selected at any time. This process encompasses various distinct characteristics:

- Specification and justification of the possibility of connecting multiple drones: When using the system in real situations, it is sometimes desirable to monitor the work of multiple drones, so it is of interest to have a system to change the connection between multiple vehicles. When the selected UAV is changed, the display is modified to adapt to the position of the new vehicle.
- Description and justification of the connection process: A new option is created in the top menu of the application, which allows choosing among the multiple drones to be configured through the Java code file "AppFrame.java", where the ports of the sockets of the simulator servers created for each of the UAVs can be specified.
- Visual demonstration of the connection selection menu: As shown in Fig. 22, there is a "Change drone" option in the top menu, which allows one to choose among all the configured drones.

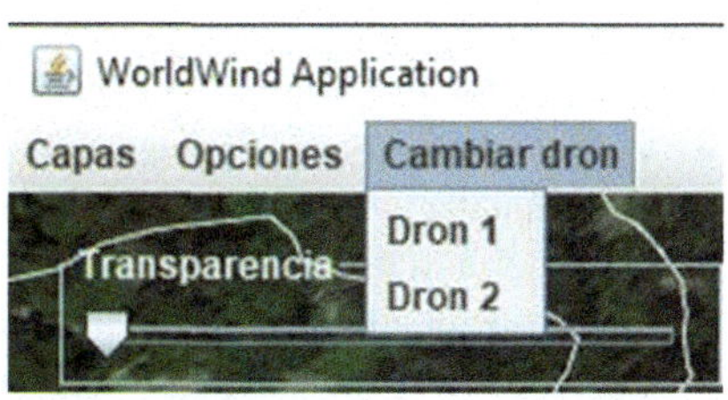

Fig. 22 Active drone change option

7.2.4 Image Synchronization Configuration

A slider has been added to allow the configuration of an additional delay (generated through a data buffer) in the representation of the positions received from the drone in the 3D visualization. The purpose of this delay is to allow coordination of the images received from the UAV camera (which has a higher latency) with the position information received from the drone. This process encompasses various distinct characteristics:

- Description of the latency concept: The concept of latency, in the field of computing and data communication, refers to the delay that exists from the moment specific information is generated until it is available at the destination.
- Description of the image synchronization concept: The images received from the UAV camera will have a higher latency than the position and attitude data given the volume of data to be transmitted, so it is necessary to establish a synchronization process to coordinate the rate of display updates with the camera image updates.
- Description of the buffer concept: A data buffer is a space in the memory of a computer reserved for the temporary storage of data while the information is waiting to be processed.
- Description of the view buffer concept: To delay and coordinate view updates, a data buffer that accumulates data from views that have not yet been processed to modify the current state of the system visualization can be created.
- Graphical demonstration of the view buffer-based image synchronization process: To synchronize view updates with the images received from the UAV camera, a view buffer of variable size is created. The visualization will not be updated until the buffer is full, at which point the oldest view stored in that buffer will be fetched, reducing its size, and updating the visualization with its data.

Figure 23 shows a sequence of images corresponding to the process of configuring the size of the view buffer through the corresponding slider. Once the process is finished, the camera image and the 3D visualization are synchronized, which can be verified through the reference points shown in the figure.

- Specification and justification of view buffer size configuration mode: It is possible to modify the size of the view buffer, and thus the number of updates that the display is delayed, from the second slider component in the top left menu of the main display panel. This process will have to be performed on each connection, as the system latency is unknown each time, and may even vary during the same connection.
- Viewer-Camera Synchronization Tests: Consideration has been given to the possibility that the UAV position data may not be obtained synchronously with the data from the in-vehicle camera. Especially, it is expected that, given the volume of data corresponding to the image transmission, the vehicle images will have a higher latency than the position data. For this reason, a configuration system has been implemented to synchronize the position of the 3D display concerning the vehicle images (delaying the display position update).

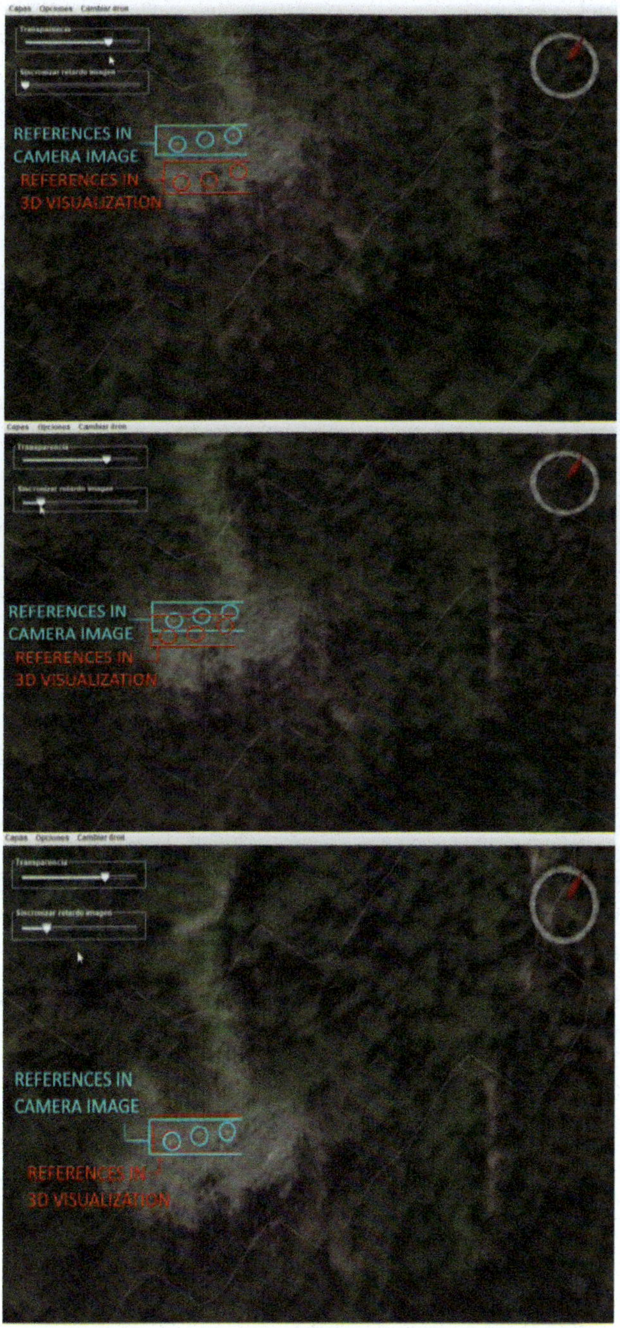

Fig. 23 Sequence of images corresponding to the visual synchronization process

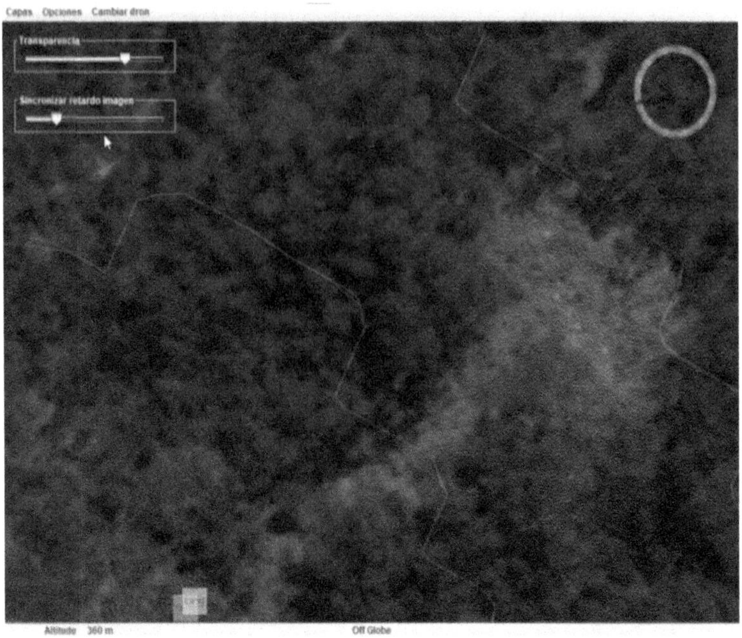

Fig. 24 Synchronization tests

To check the correct functioning of this synchronizer, tests have been carried out with a virtual camera, which simulates a delayed view of what could be observed through the drone's camera. The results obtained from these tests can be seen in the video referenced in Fig. 24.

As can be seen in the videographic document, the system can successfully coordinate the position changes in the 3D visualization with the image updates received from the virtual camera embedded in the UAV.

- Sudden Changes of Position and Attitude Tests: The movement of the UAV is not perfectly uniform in a real environment, as it can be in an idealized simulation environment. Moreover, in unfavorable circumstances, sudden gusts of air can change the position and attitude of the drone abruptly.

To test the effect of these events on the implemented visualizer, an additional layer of code has been added to the simulator, varying the position and attitude of the drone suddenly and randomly. The results obtained from these tests can be seen in Fig. 25.

As can be seen in the referenced video, sudden changes in position affect the coordination between the images received from the drone camera and the position of the display. These effects can be significantly reduced with a higher refresh rate of the information received from the drone.

Fig. 25 Abrupt changes

- Sudden Changes in Altitude Tests: Similar to the previous case, under adverse circumstances, the altitude of the drone can be abruptly modified. More specifically, air pockets with sudden differences in air pressure and air density can cause the drone to lose several meters of altitude in a few seconds.

To test the performance of the display under these circumstances, simulations of a sudden altitude variation of the drone have been performed, and the image obtained has been visualized in the system. The results obtained from these tests can be seen in Fig. 26.

As in the previous case, the desynchronization effects that can be caused by abrupt position changes can be smoothed out noticeably if the drone's data update rate is high.

- Signal Loss Tests: Given the distance between the UAV and the data-receiving equipment, as well as the possible existence of interference, it is possible that under real circumstances the drone's position and camera signals may not be received stably. This can lead to interruptions in data reception.

To test the effects of these transmission cuts, in the simulation phase of the project, random disconnections have been forced in the data transmission process (both position and camera), obtaining the results that can be visualized in the video referenced in Fig. 27.

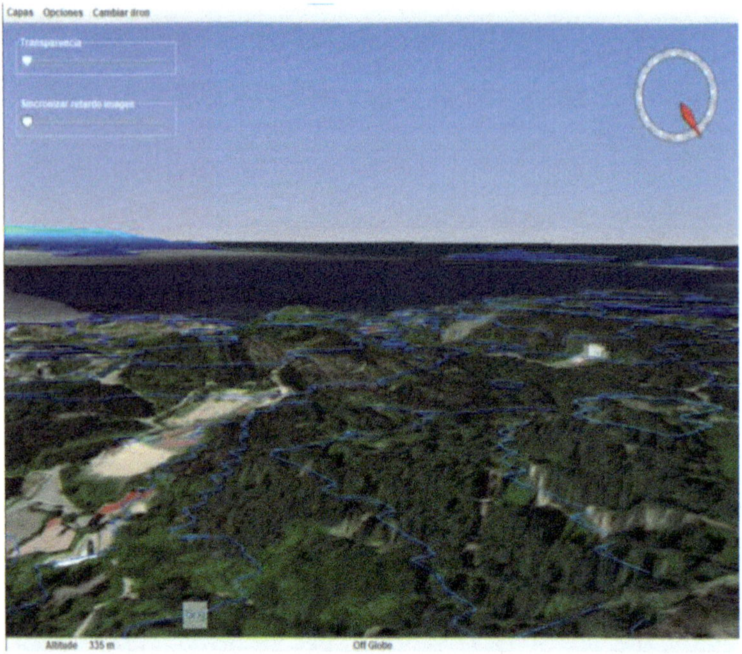

Fig. 26 Altitude changes

Fig. 27 Signal cutoff

As can be seen in the video document, in the event of connection cuts, the display stops, and the position and attitude of the UAV are properly resumed once the UAV signal is recovered.

7.2.5 Integration with Real UAV Equipment

Before the integration tests with the real equipment, the possibility of synchronizing the position and attitude of the display with the gyroscope and GPS data from a smartphone was considered.

However, the tests could not be fully carried out, as the reduced accuracy of the phone's GPS signal prevented the generation of a sufficiently good positioning, especially at low altitudes.

It was possible, however, to check the synchronization between the angular movements of the device and the attitude of the simulated UAV in the visualization (Fig. 28).

In the video referenced in Fig. 20 it can be seen how, by modifying the orientation and inclination of the smartphone in the three axes of space, the visualization adapts the camera attitude so that it is equivalent to that of the phone, thus proving that it is possible to synchronize the visualization with an external gyroscope.

To connect the smartphone device with the developed visualization system, a small intermediary REST API between the two components was created and deployed on a local server. The defined REST API consists of the following resources and methods:

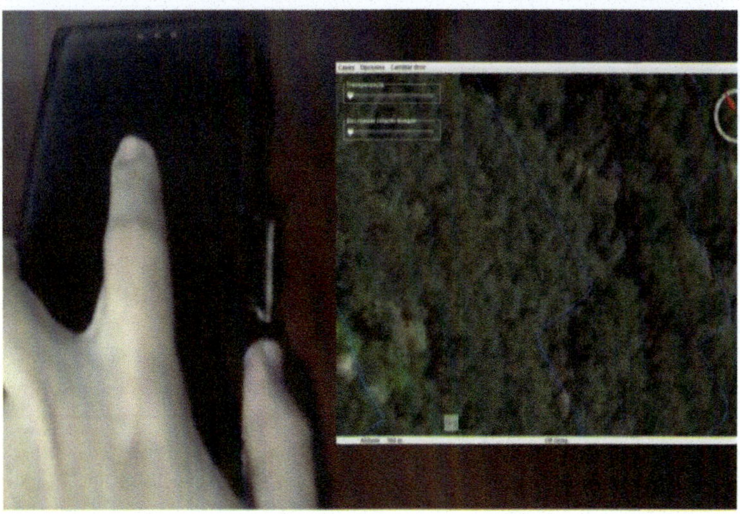

Fig. 28 Smartphone synchronization tests

- PUT api/v1/persistence?key = <key> &value = <value>: This method allows publishing and storing in the API an arbitrary value for any key, replacing the previous value; in case it exists, with the new indicated value.
- GET api/v1/persistence?key = <key>: This method allows obtaining the current value of a specified key through the query parameter "key".

The cell phone, through a web page with JavaScript code, continuously publishes its gyroscope and GPS data to this API. When the UAV position simulator receives a request from the viewer, it queries the persistence API to know the current tilt value of the mobile device in the three axes of space. This information is sent by the simulator to the viewer as the UAV's attitude so that the viewer can then make the necessary camera tilt adjustments to match the actual tilt.

7.3 Software: Sub-objective 3 Pilot Assistance

This subsection explains the functionalities integrated into each version of the system related to the superposition of the virtually generated cartographic layer with the real image captured by the vehicle camera.

7.3.1 Webcam Integration and Layered Visualization

Since the connection to the drone camera is made via a USB port using standard protocols and formats, the drone camera signal can be considered as that of a webcam (if not, it is possible to receive the image and send it to a virtual webcam. Technologies such as the Python module, together with a virtual webcam generator such as OBS, can be used for this purpose).

The image from a webcam can be obtained in Java by using APIs such as Webcam Capture [19]. By capturing the image as a byte buffer, it is available to the viewer to be overlaid with the 3D layer. This process encompasses various distinct characteristics:

- Description of the webcam concept: A webcam is a digital camera that connects to a computer, to which it sends the images it captures so that the computer can retransmit them to other applications on the system.
- Description of the layered visualization procedure: The visualization system generated by the WorldWind API allows for multiple layers. The layers are placed in the order desired by the user, which affects the drawing order of the layers.
- Specification and justification of webcam connection: The visualization system is connected to a webcam, assuming that the UAV cameras will be interpreted by the computer as webcams. If this is not the case, it is possible to transform the information received into a virtual webcam.
- Webcam connection procedure: For the visualization system to access the webcam, use is made of the Java library "Webcam Capture" from Sarxos. As

already explained, this library can connect to a webcam and obtain byte buffers corresponding to the images captured by the webcam.
- Specification and justification of the default order of the layers and the possibility of configuring it: By default, the compass and contour line layers have been placed above the 3D visualization layer, as these layers provide additional information about the 3D visualization. On the other hand, the layer corresponding to the webcam images is placed above all others. This is because, being a 2D layer, it must necessarily be placed on top of the 3D layers within the WorldWind system.

The process of adapting the SDK to display the images below the visualization is possible, but unnecessary to obtain the desired effect. The order of the other layers can be freely modified by the user from the "Layers" menu.

- Specification and justification of the webcam layer transparency settings: It is possible to set the transparency of the webcam image layer from the slider element at the top left of the display. This makes it possible to configure in real time the transparency of the image way easily and simply, but with precision over the level of transparency.

7.3.2 Integration with Real UAV Camera

Once access to the actual UAV equipment was obtained, it was possible to check the operation of the camera incorporated in the UAV. This camera has a wireless receiver that connects to a computer through a USB port.

- Specification and justification of the camera model used: The camera used during the tests is a Foxeer T-Rex Mini. This model was chosen since it met the needs of the project: to allow a medium-distance wireless connection with the equipment running the visualization software, and to be able to dock properly on the unmanned vehicle during the flight.
- Description of the camera connection mode: The camera sends the images via an antenna to a SKYDROID 5.8G receiver, which is connected to the computer via a Micro USB connector.
- Description of the camera identification process from the visualizer: The signal from the video camera is interpreted by the computer as the signal of a webcam. This has made it possible to maintain and reuse the implementation originally developed in version 1.6 when it was assumed that the received signal corresponds to that of a webcam (or that, alternatively, it could be retransmitted in this way a posteriori, after being received in the computer).

7.3.3 Integration with Real UAV Equipment

In the twelfth iteration of the project, focused on the integration of the developed software with the real vehicle, progress was made concerning Sub-Objective 3, related to the superposition of the cartographic layer with the real image captured by the drone camera and testing it.

Additionally, the diagram in Fig. 29 shows the elements and components that are part of the system developed in this project, both for simulated connections (through virtualized equipment), as well as in the case of connections with real UAVs.

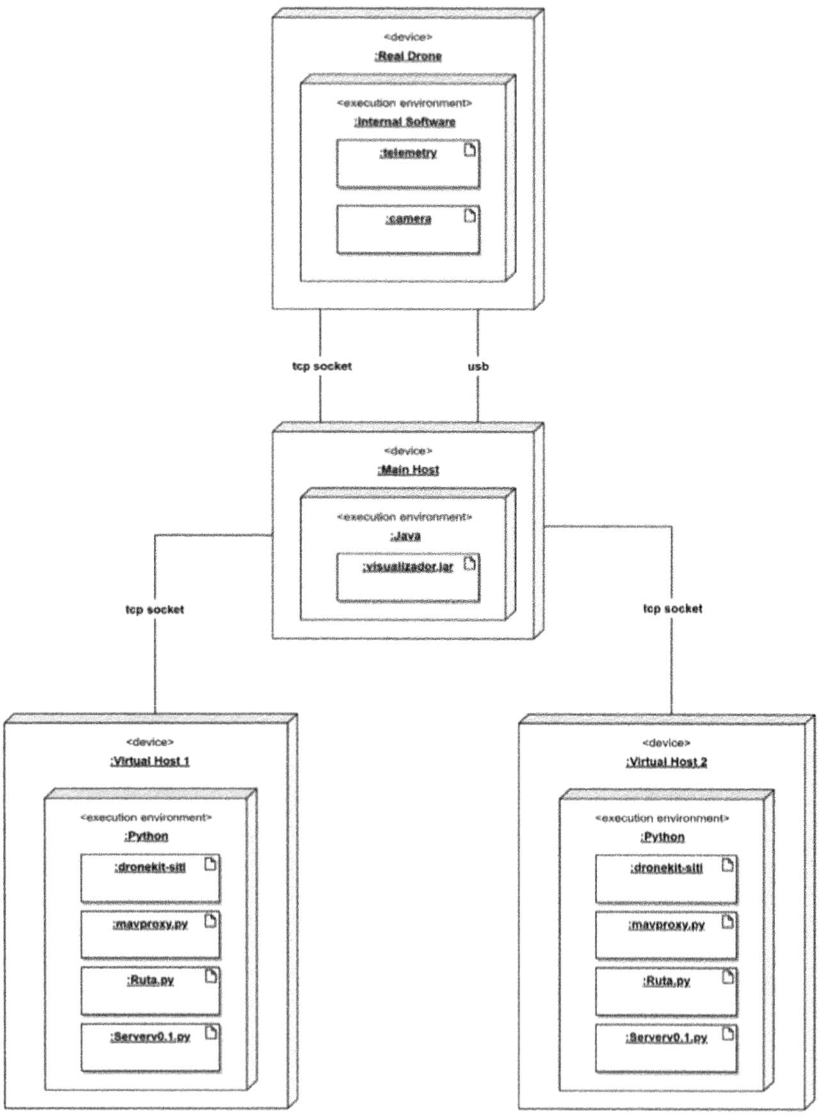

Fig. 29 Component diagram of the system

8 Conclusion

In this chapter, a comprehensive approach to using Unmanned Aerial Vehicles (UAVs) for monitoring and analyzing wildfire evolution was presented. The integration of cutting-edge technologies has enabled the design of new UAV hardware and the development of a sophisticated visualization system that offers invaluable insights into fire surveillance operations.

The software aspect of the project, focusing on generating a digital cartographic layer and enhancing the viewer's functionality, has yielded substantial benefits. The implementation of contour lines, altitude data visualization, and real-time mapping features has enhanced the pilot's situational awareness. The integration of multiple drones within the same system offers a versatile solution for managing and visualizing multiple UAV operations.

Through a rigorous testing process, the system's performance under various conditions, including sudden changes in position, altitude, and signal loss was examined. These tests have provided valuable insights into the system's robustness and its ability to adapt to unpredictable scenarios, crucial for effective decision-making during firefighting missions.

The exploration into integrating the visualization system with real UAV equipment has revealed the potential for enhancing the system's accuracy and relevance by incorporating external sensors, like smartphones. This avenue opens up new possibilities for real-time synchronization and a more immersive user experience.

From a broader perspective, both the new UAV hardware design and the assistance tool developed in this project have the potential to significantly enhance the efficiency of wildfire surveillance operations. Through the provision of an intuitive and clear interface that offers real-time insights into UAV operations, pilots can make informed decisions and swiftly adapt strategies. With the UAV hardware specially tailored for wildfire applications, this project establishes a robust foundation for future advancements in UAV-assisted wildfire monitoring. This contribution aligns with the ongoing endeavors to mitigate the catastrophic impacts of wildfires on the environment and communities.

Acknowledgements This work has been done with the funds of the Interreg V A – E.U. - POCTEP Project [0756_FIREPOCTEP_6_E]: Fortalecimiento de los sistemas transfronterizos de prevención y extinción de incendios forestales y mejora de los recursos para la generación de empleo rural posCovid-19 (FIREPOCTEP).

References

1. Nonami, K. (2007). Prospect and recent research & development for civil use autonomous unmanned aircraft as UAV and MAV. *Journal of system Design and Dynamics, 1*(2), 120–128.
2. Picos, J., Bastos, G., Míguez, D., Alonso, L., & Armesto, J. (2020). Individual tree detection in a eucalyptus plantation using unmanned aerial vehicle (UAV)-LiDAR. *Remote Sensing, 12*(5), 885.
3. Glade, D. B. (2000). *Unmanned aerial vehicles: Implications for military operations* (No. 16). Center for Strategy and Technology, Air War College, Air University.
4. Ghamari, M., Rangel, P., Mehrubeoglu, M., Tewolde, G., & Sherratt, R. S. (2022). Unmanned aerial vehicle communications for civil applications: A review. *IEEE Access, 10,* 102492.
5. Mangewa, L. J., Ndakidemi, P. A., & Munishi, L. K. (2019). Integrating UAV technology in an ecological monitoring system for community wildlife management areas in Tanzania. *Sustainability, 11*(21), 6116.
6. Firepoctep. (n.d.) https://firepoctep.eu/
7. Soutis, C. (2005). Fibre reinforced composites in aircraft construction. *Progress in Aerospace Sciences, 41*(2), 143–151.
8. Kesarwani, S. (2017). Polymer composites in aviation sector. *International Journal of Engineering Research, 6*(6), 518–525.
9. Fahlstrom, P. G., Gleason, T. J., & Sadraey, M. H. (2022). *Introduction to UAV systems.* John Wiley & Sons.
10. Borchardt, J. K. (2004). Unmanned aerial vehicles spur composites use. *Reinforced Plastics, 48*(4), 28–31.
11. Toray Advanced Composites. (2023). Processing guides. Accessed 23 March 2024 from https://www.toraytac.com/resources/processing-guides
12. Learn to Fly Blog. CFI brief: Forward vs aft CG. Accessed 23 March 2024 from https://learntoflyblog.com/cfi-brief-forward-vs-aft-cg/
13. Gonzalez Crespo, R., Piraquive, F., Granados, J. J., Sanjuan, O., & Verdú, E. (2016). Knowledge management model for project management: KM+PMTIC, 55–93.
14. Zander, S., Swertz, C., Verdú, E., Perez, M. J. V., & Henning, P. (2016). A semantic mediawiki-based approach for the collaborative development of pedagogically meaningful learning content annotations. In P. Molli, J. G. Breslin & M.-E. Vidal (Eds.), *Semantic web collaborative spaces* (pp. 73–111).
15. Moreno-Ger, A. M. N. P. (2018). Comparison of clustering algorithms for learning analytics with educational datasets. *International Journal of Interactive Multimedia and Artificial Intelligence, 5*(2), 9–16.
16. Crespo, R. G., Martínez, O. S., Lovelle, J. M. C., García-Bustelo, B. C. P., Gayo, J. E. L., & de Pablos, P. O. (2011). Recommendation system based on user interaction data applied to intelligent electronic books. *Computers in Human Behavior, 27*(4), 1445–1449.
17. Animator (NASA WorldWind). Accessed 23 March 2024 from https://worldwind.arc.nasa.gov/java/v2.1.0/javadoc/gov/nasa/worldwind/animation/Animator.html
18. Overview (NASA WorldWind). Accessed 23 March 2024 from https://worldwind.arc.nasa.gov/autodocs/WorldWindJava/
19. Sarxos. GitHub - sarxos/webcam-capture:. GitHub. Accessed 23 March 2024 from https://github.com/sarxos/webcam-capture

Key Technologies for UAVs in Search and Rescue (SAR) and Surveillance Operations

Silvia Abalo Bóveda, Andrés Ferreiro Gónzález, Lucas Val Terrón, José Antonio Rodríguez Artolazábal, and José Antonio Rodríguez Negro

Abstract In the last years, the use of unmanned aerial vehicles (UAVs) has become particularly relevant in multiple sectors, including search and rescue (SAR) missions, due to their flexibility, precision, speed, and operation range. These systems can dramatically enhance the survival chances of disaster victims while safeguarding and protecting the integrity of rescue teams. UAVs can address these challenges by accessing remotely hazardous areas, performing complex operations, and providing real-time monitoring and environment modeling capabilities. This chapter aims to provide readers with the current state-of-the-art technologies that can be integrated within UAVs to enhance the effectiveness and security of SAR operations.

Keywords UAV · Mission planning · SAR · Surveillance · AI · Payload · Review · UAV communications

1 Introduction

The integration of unmanned aerial vehicles (UAVs) within search and rescue (SAR) operations has brought a significant advantage in terms of efficiency and safety. Their use in SAR missions has shown many positive advantages, including deployment, low response time, high mobility, and versatile capability to mount payloads adapted to specific needs and disasters. This power and autonomy allow these vehicles to work under circumstances and environments where first responder teams may face extremely hazardous conditions.

Traditional procedures and data analysis methods used in SAR operations required manual supervision by rescue teams. In working conditions where no effort is spared to locate victims and the stress load is high, the performance of personnel gradually diminishes. At this point, with the continuous advancement of UAV technology, including automated flights, high-quality images, artificial intelligence capabilities,

S. A. Bóveda · A. F. Gónzález · L. V. Terrón · J. A. R. Artolazábal · J. A. R. Negro (✉)
GRADIANT, Carretera Vilar, 56-58, 36214 Vigo, Pontevedra, Spain
e-mail: jrodriguez@gradiant.org

D. Carou et al. (eds.), *Applying Drones to Current Societal and Industrial Challenges*, Management and Industrial Engineering, https://doi.org/10.1007/978-3-031-55571-8_4

and reliable communications, it does not make sense to overload rescue teams with tasks that a UAV with a specific payload can simplify and address more efficiently.

The final objective of this chapter is to address an overview of cutting-edge UAV technologies that can be leveraged to increase the efficiency, autonomy, and safety of UAV SAR operations. Video analysis techniques (run both on-board and on the edge) empowered by computer vision techniques can greatly automate the process of identification of victims, being capable of adapting to environments and conditions where the human eye would face serious challenges. The mission planning and simulation allow for the creation of a customized route plan, ensuring efficiency in terms of distance and coverage area. Additionally, it ensures that communication conditions and GPS signal levels are satisfactory for safe operations. Finally, UAVs integrate various complementary communication links, enabling not only the provision of real-time data streams but also serving as communication range extenders and providers of emergency communication networks. Without a doubt, this chapter will provide readers with a clear insight into why UAVs and integrated technology are indispensable in SAR missions today.

2 Video Analytics

Nowadays, the enhancement of drone capabilities has ushered in a new era in their utilization across various fields. Their diverse applications now encompass a range of vastly different operations, with some of particular significance being surveillance tasks and search and rescue missions. One of the driving forces behind this notable evolution is the fusion of drones with high-resolution camera technologies, enabling extensive area coverage in significantly reduced periods. However, amidst this technological prowess lies a challenge: the meticulous and labor-intensive task of visually analyzing vast sets of captured images.

This challenge finds its solution in the realm of automatic video analysis, an innovative development that has emerged as an indispensable ally in these environments. These tools, powered by cutting-edge artificial vision and artificial intelligence technologies, not only automate complex tasks but also provide crucial information to coordinate and carry out rescue operations more effectively and safely, enhancing the efficiency of teams on the ground.

The spectrum of scenarios where intelligent video analysis, based on footage captured by UAVs, finds practical applications is as diverse as it is impactful, from precisely locating survivors in disaster-stricken areas to real-time monitoring of natural calamities like earthquakes, floods, or avalanches. Furthermore, its role extends to overseeing rescue teams, aiding in operational coordination, and evaluating potential escape routes, ensuring comprehensive preparedness in crises.

Beyond the mere use of automatic analysis, in these situations, where a swift response is imperative, arises not only the need for continuous real-time monitoring but also the crucial importance of automatically generating instant alarms or precise notifications according to what is detected. This proactive measure, a common functionality in video analysis systems, not only strengthens the effectiveness of response strategies but also ensures uninterrupted operational continuity, capable of functioning 24/7. This eliminates the need for constant visual surveillance, optimizing resource utilization and bolstering overall operational efficiency.

2.1 Algorithms and Methods

The integration of Convolutional Neural Networks (CNNs) into computer vision has significantly affected SAR and surveillance operations. The increase in computational power on-board UAVs has enabled real-time processing of complex algorithms for image analysis. This progress has been pivotal in enhancing capabilities in key areas: object detection, image classification, semantic segmentation, and object tracking, each critical for the effectiveness of SAR and surveillance missions.

2.1.1 Object Detection: A Cornerstone in SAR and Surveillance

Object detection is fundamental in SAR and surveillance, as it involves locating people, vehicles, or other items of interest within an environment. The accuracy and speed of detection directly affect the effectiveness of these operations. Initially reliant on methods like optical flow and basic motion detection, the adoption of CNNs has revolutionized this field. Architectures such as Faster Region-Based Convolutional Neural Network (R-CNN), You Only Look Once (YOLO), and Single-Shot Detector (SSD) have enhanced detection accuracy and speed. Combining these with traditional methods like optical flow creates a hybrid approach, leveraging the strengths of both for improved object tracking and identification in varied and dynamic environments (Fig. 1).

2.1.2 Image Classification: Enhancing Understanding in SAR and Surveillance

Image classification plays a vital role in SAR and surveillance by providing detailed insights into the nature of the objects or areas identified. It aids in distinguishing between different types of objects or environmental features, which is crucial for strategic decision-making. The development of deep CNN architectures like ResNet, Inception, and VGG has revolutionized image classification, offering enhanced accuracy and the ability to process complex scenes. Integrating these models with object detection systems enables more precise identification and analysis of targets, enhancing situational awareness.

Fig. 1 Beach occupancy monitoring using people detection algorithms based on neural networks

2.1.3 Semantic Segmentation: Critical for Environmental Analysis

Semantic segmentation is crucial in SAR and surveillance for detailed analysis of the terrain and environment. It enables the pixel-level classification of images, which is essential for mapping routes in SAR operations and monitoring specific zones in surveillance tasks. Advanced CNN models such as U-Net, Fully Convolutional Network (FCN), and DeepLab excel at segmenting complex scenes, providing essential information for operational planning. Additionally, anomaly detection within segmentation frameworks helps identify unusual patterns or objects, vital for recognizing potential threats or areas of interest.

2.1.4 Object Tracking: Essential for Dynamic SAR and Surveillance

Object tracking is essential in SAR and surveillance, where maintaining continuous observation of moving targets is often necessary. Traditional methods like Kalman filtering and Mean-shift have their limitations in dynamic environments. The advent of CNN-based methods like DeepSORT and GOTURN has provided superior tracking capabilities. These advanced techniques leverage deep learning for robust feature extraction, offering greater accuracy and adaptability in tracking objects or individuals under variable conditions.

2.1.5 Overcoming Training Data Challenges for CNNs

One of the main challenges in deploying CNNs for SAR and surveillance is obtaining sufficient, diverse training data. Synthetic data generation and GANs offer viable solutions. Synthetic data can simulate a range of scenarios, providing a diverse dataset for CNN training. GANs are known for generating realistic images, augmenting training datasets, and improving the models' performance and generalization in real-world scenarios. These methods address the data scarcity challenge, enabling the development of robust CNN models for practical applications.

2.2 Applications

As highlighted in the previous sections, video analysis algorithms leveraging advanced technologies offer numerous advantages in both surveillance and search and rescue operations, delivering automated, precise, and real-time information. These technologies find crucial applications in various scenarios, notably as follows.

Search for missing or stranded people in challenging-to-reach areas: Drones equipped with specialized cameras and powered by AI algorithms efficiently cover vast terrains, automatically identifying movement patterns, temperature disparities, or human forms amid rugged landscapes.

Continuous monitoring of isolated individuals: When the individual is found in remote or inaccessible areas, drones fitted with thermal cameras or other sensors can track changes in body temperature and potentially detect vital signs, ensuring ongoing supervision until rescue is feasible.

Facilitating the strategic planning and execution of rescue missions: Detailed data collected by drones and processed through algorithms aids in comprehensive area analysis, assisting in the formulation of secure rescue routes and evacuation plans.

Surveying areas affected by natural disasters: In instances of natural disasters such as earthquakes, floods, and avalanches, prompt identification of affected regions and assessments of accessibility for rescue teams are imperative. Video analysis algorithms play a pivotal role by identifying environmental changes such as landslides, alterations in water bodies, or any movement posing risks to both rescue operations and the safety of trapped individuals.

3 Mission Planning and Simulation

3.1 *The Importance of Pre-flight Checks in SAR Operations*

Given the importance of SAR operations and considering that the safety of people is at stake, carrying out pre-flight checks is an essential requirement. In general, these operations include verifying the correct condition of the equipment and vehicles that will be part of the operation and carrying out a briefing on it with the personnel involved to plan the mission to be conducted. Although mission planning is carried out in both manned and unmanned operations, in the case of using UAVs it acquires significant importance for the following reasons.

3.1.1 Precision in Search Area Coverage

On many occasions, SAR operations depend on comprehensive search area coverage. On the one hand, mission planning brings precision to this by enabling teams to select specific search zones with precision, thanks to access to a wide range of data: cartography, digital terrain models, regulatory information, etc. On the other hand, the nature of UAVs and the fact that they can fly autonomously greatly increases the degree of precision that can be achieved while following the defined mission path. These factors ensure that no portion of the target area of operations is inadvertently omitted from the search or redundantly scanned, providing a methodical and thorough approach that also affects greatly the efficiency of the operation.

3.1.2 Efficiency and Resource Utilization

During mission planning, UAV flight paths can be optimized, and UAVs are configured to autonomously traverse the search area along the best path possible. This serves to decrease mission execution time and conserve resources like battery life. Such resource optimization is not only cost-efficient but also directly impacts the potential for successful mission outcomes.

3.1.3 Adaptation to Environmental Realities

Mission planning allows SAR teams to adapt the designed flight paths or schedules in response to adverse weather conditions and environmental hazards. Even the UAV model selection or payload specification can be improved using this information, thereby mitigating risks and enhancing operational safety and success probability.

3.1.4 Communication Protocols and Coordination

Effective communication is paramount in SAR operations, more so if UAVs are involved. Mission planning encompasses the establishment of robust communication protocols, addressing issues such as signal strength and interference. SAR teams are enabled to define, model, and obtain information on communication performance over the area of operations for the required radio links. Using this information, the risk of potentially dangerous loss of communication situations can be drastically reduced either by selecting alternative radio links for each scenario or by anticipating such situations and preparing contingency plans with which to react. Furthermore, this also serves to improve coordination among various SAR assets, be they manned aircraft, ground teams, or other actors involved (thereby averting possible conflicts and optimizing operational efficiency).

3.1.5 Data Collection Strategy

The efficiency of data collection during SAR operations is tightly tied to mission planning. Factors such as optimal altitude, sensor selection, and camera angles can be meticulously defined within the mission plan. This, in line with the expected precision in mission execution for UAVs, ensures that collected data is of the highest quality, providing SAR teams with the best information possible to aid situational awareness and decision-making during the operation.

3.1.6 Safety, Risk Assessment, and Mitigation

As previously stated, SAR operations are conducted always under a huge commitment to safety. Mission planning serves as a safeguard of this commitment, allowing SAR teams to consider some considerations that must rule over UAV operations in those scenarios: the identification of no-fly zones, the orchestration of obstacle avoidance strategies, and the establishment of emergency procedures to be executed under unforeseen circumstances. Mission planning is a must-step to identify risks and subsequently manage them. It entails a comprehensive risk assessment that encompasses hazards to UAVs themselves or other crafts, SAR teams, and the public. Such assessments form the foundations upon which robust mitigation strategies are selected and established.

3.1.7 Regulatory Compliance

SAR and UAV operations are both subject to a strict set of regulations. Mission planning serves as a blueprint for ensuring adherence to these rules (such as airspace restrictions and privacy rights) and provides very valuable audit information, thus guaranteeing the lawful execution of SAR operations.

3.1.8 Training and Preparedness

Mission planning extends beyond the operational phase, serving as a valuable resource for training and preparedness. It ensures that all SAR team members are dependable in their roles, responsibilities, and the overarching strategy. This preparedness and training serve to further enhance their readiness to take over the inherent challenges in SAR missions.

In conclusion, mission planning is the foundation where SAR operations rest, even more so considering the involvement of UAVs. The combination of all these factors contributes to the overall success of a mission by enhancing precision, efficiency, quality, operability, and safety.

3.2 Simulation of a Mission as a Part of the Process

As mentioned in the previous section, simulation is one of the key aspects in planning, both as training and to model the behavior and performance of the communication systems to be used in the target area of operations.

This section delves into the second case: the importance of the model and parameterization of communication systems as a core step of the mission planning process for a SAR mission in combination with UAVs.

Firstly, SAR operations are conducted in environments where the safety of people is at stake (both members of the SAR team who participate in the operation and unrelated personnel). Operations must always be conducted safely, so that UAVs contribute to the success of the mission and their presence does not increase the risk to people in the theatre of operations. Considering this basis, it is convenient to add two additional factors that may affect, or compromise, safety in the operation and that must therefore be considered.

On the one hand, it is also common for certain hazardous events to occur in this type of mission, such as unfavorable weather conditions (storms, hurricanes ...) or other adverse situations (floods, fires, earthquakes ...) that not only affect pre-existing communication networks but which can also cause problems when deploying new communication systems in the field. On the other hand, it is essential to consider that not only one UAV and its communication link (or links) are present in operations. Depending on each case, multiple emergency services from different agencies may be deployed: urban and forest firefighters, local, regional, and national police, various medical services, civil protection, army ... Each of them can use or deploy one or more communication systems on the ground to maintain order, communication, and security in the operation through their respective chains of command and for the coordination of the emergency. Of course, each of these actors could also deploy their own UAV systems.

Simulating a communication link or links for a mission before its execution allows all these elements to be combined and considered to establish clear and effective mitigation strategies. To carry out these simulations, the following steps are necessary.

Adequately define the area of operations and the general characteristics of the mission. For example, in certain missions, it may be acceptable to partially lose communication with the UAV if it is performing tasks automatically (something never desirable, but which can be accepted as a compromise for the execution of the mission if its impact has been previously studied and additional cautionary measures have been established). However, in others, it is an essential requirement that communication always remains active, even at the cost of having to alter the course of the mission.

Correctly identify the link under study. Of course, each communication link will require independent analysis to obtain the expected power levels in each zone of the area of operations. For a single UAV, there will be cases in which a Telemetry and Control (TT&C) link will be sufficient to fulfill the mission, while in others additional links will have to be considered (such as links that allow visualization and control of a video payload).

- Adequately characterize each of the ends of the link under study. This characterization includes the identification of the element or elements that are part of the transmission and reception chains in each case: transmission powers, antenna gains and polarizations, miscellaneous equipment losses, and receivers' sensitivities.
- Have access to a tool that implements one or more propagation models that allow modeling the attenuation introduced on the signal when traveling through the communication channel. There are organizations such as the International Telecommunication Union (ITU) or the 3G Partnership Project (3GPP) that define and standardize numerous propagation models suitable for different types of scenarios and frequency bands, as well as commercial software solutions that offer this type of functionality.
- Select a propagation model suitable for the concrete scenario and mission being evaluated from the available ones. If the selected propagation model requires additional information (such as a Digital Terrain Elevation Model or a meteorological dataset), it is necessary to provide such sources.
- Adequately incorporate, or at least consider, information regarding other communication links present in operations of which the SAR team is aware. This applies both to evaluate their impact as sources of interference and to avoid interfering with them to the best extent possible.
- Finally, the consideration of Global Positioning System (GPS) availability is also convenient as a source of information to further improve the decision-making capabilities of the SAR teams. Simulating GPS coverage in the area of operations can have an impact on decisions regarding the Beyond Visual Line of Sight (BVLOS) operation of a UAV or even the allowance for autonomous operation.

By applying these steps, the SAR team can or cannot determine the feasibility of the use of a specific communication system, its adequate integration with the rest of the elements of the system, and even the best location or locations in which to place

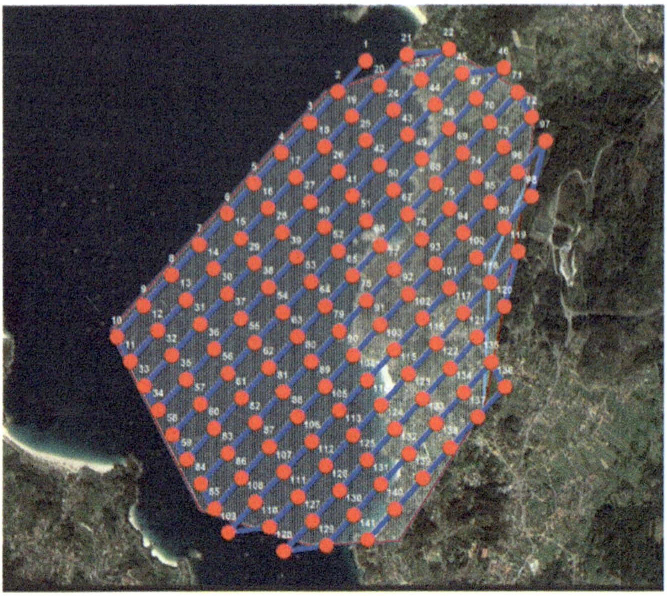

Fig. 2 Example of a photogrammetric sweep mission plan

the UAV base station or communication repeaters if available. Any potential communication impairments are identified and can therefore be considered and evaluated, leading to more effective mitigation strategies. Indeed, this is closely related to other aspects mentioned above such as efficiency, precision in execution, and adaptation to environmental realities (in addition to those already mentioned throughout the section) (Fig. 2).

3.3 SoA Planning

This section will address the state of the art on topics related to UAV mission planning, which in many cases includes.

- The generation of the optimal or sub-optimal route for the execution of the mission.
- Simulation of one or several communication links.
- Consultation with competent authorities on restricted flight zones, no-fly zones, etc.
- Consultation of meteorological information providers.
- Consultation with GIS providers to obtain information on maps, terrain elevations, etc.

3.3.1 Route Generation Algorithms (Motion Planning)

Regarding the generation of optimal routes, numerous alternatives appear depending on the type of route to be generated (ferry, orbit, sweep over an area...) and the method used to generate it, either traditional (graph search algorithms, sampling-based algorithms, interpolating curve algorithms) or based on ML algorithms (supervised learning, optimal value reinforced learning or policy gradient reinforced learning).

Graph search algorithms are useful for designing routes in scenarios where there are obstacles to be avoided or places that must be visited. In this family of algorithms, the literature often refers to the Dijkstra algorithm [1] (used by Google in its Maps tool [2]) or A* (A-star) [3] both of which have multiple variants and optimizations for certain scenarios.

There are two main groups when considering the sampling-based algorithms:

- Random search algorithms such as Rapidly Exploring Random Tree (RRT) [4] or Probabilistic Roadmap Method (PRM) [5] and their variants, which perform random sampling over a given space to generate sub-optimal routes. The RRT algorithm (and its variants) is widely used among other things because it can consider variables such as maximum turning radius and vehicle momentum for the generation and evaluation of the route's candidate points.
- Algorithms for the calculation of distances and angles over the Earth's surface, such as the Haversine [6] (low precision and computational cost), Vincenty [7], and Karney [8] methods. These algorithms are combined with knowledge of the sweep to be performed over an area (e.g. a photogrammetric sweep) which allows precise specification of the distances and angles to be traversed in an orderly manner.

For the previously mentioned case of a photogrammetric scan where the planning-relevant unknowns are the spatial resolution of the images on the ground, the distance between consecutive images, and the separation between consecutive sweeps, there are well-known [9] and widely used algorithms. That allows variables such as flight height, sensor data used, and even flight speed to be considered to solve them.

The last of the traditional family of algorithms for path generation is known as interpolating curve algorithms. These algorithms are used to generate routes based on mathematical rules, generally for path smoothing and curve generation. Some of the most typical algorithms of this type are line and circle [10], clothoid curves [11], polynomial curves [12], Bezier curves [13], and spline curves [14].

Entering the field of ML-assisted algorithms, the ones known as supervised learning algorithms can be addressed first. These include SVM [15] (a well-known algorithm that has been discussed in the supervised learning literature) in its variant Multiclass SVM (MSVM), LSTM [16] (a variant of recursive neural network or RNN), MCTS (a combination of Monte Carlo and search tree methods widely used in games such as chess and Go [17] for motion prediction), and CNN [18].

The second family of ML-assisted algorithms to consider would be the optimal value RL algorithms, among which algorithms such as Q learning [19], DQN learning [20], double DQN learning [21], and dueling DQN learning [22] stand out. All of them have in common that attaching destinations and safe paths with high rewards (positive reward) performs movement planning, while obstacles are attached with penalties (negative reward).

The last family of ML-assisted algorithms mentioned above are known as policy gradient RL algorithms, among which some of their optimized variants can be highlighted: A2C and A3C [23], DPG and DDPG [24], and TROP and PPO [25].

Regarding commercial applications and tools, there are many options and flavors in the market. Matlab provides implementations of many of the algorithms listed before that are implemented in its Navigation Toolbox [26], while other tools, such as UgCS [27] or Mission Planner [28], provide standalone tools for generating missions mostly oriented to photogrammetry and area sweeps. Both include elevation data, photogrammetry algorithm support, some form of no-flight zone indications, and even an embedded database of UAVs.

3.3.2 Radio Propagation Models

As mentioned before, there are plenty of radio propagation models that can be used to estimate the propagation losses in a radio link. The selection of the model should be conducted regarding the operational limits of the model in the first place (in terms of antenna heights, link distances, and frequency ranges in which they are defined) but also considering the technology they model (general, cellular [29]...) and the environment for which are defined (outdoor [30] urban, rural, maritime, mountain...).

There are several authorities and models on the Internet, the most prominent one being the International Telecommunication Union (ITU), which has published several general-purpose models like the ITU-R p.525 [31] (general interpretation of the Friis Free Space Losses) or the more complex ITU-R p.2001–5 [32] (a general-purpose model for terrestrial communications from 30 MHz to 50 GHz). The Institute for Telecommunication Sciences (ITS), part of the United States National Telecommunications and Information Administration, provides comprehensive information and implementation data [33] on many well-known models that may be suitable for a wide range of applications.

Regarding commercial applications, maybe the most prominent ones are Xirio Online [34] (a well-known platform for radio network planning) and CloudRF [35]. There are some free-to-use tools available, like the Signal Propagation, Loss, And Terrain analysis tool (SPLAT!) [36], which implements the Longley-Rice ITM propagation model.

3.3.3 GPS Simulation

To evaluate GPS (or any other GNSS system of the like), precise satellite ephemeris data is required. Luckily, most of the required information (if not all) is for public access. The International Global Navigation Satellite System (GNSS) Service [37] provides updated information on the predicted orbits of GPS and GLONASS satellites in a well-known format, the Two-line Element Set. There are many other providers and technologies available, for example, Celestrak [38] (a non-profit organization), which aggregates data from many satellites including GNSS ones (GPS, GLONASS, Galileo…).

The information contained in the Two-line Element Set (TLE) can be used to predict the orbits of the celestial body to whom they refer for times up to several days ahead. It must be noted however that the TLEs are updated regularly since their accuracy degrades with time, and so caution must be exercised to use the more recent ones when simulating satellites' positions.

With this orbital information and considering some other constraints (like the Field of View to open sky), it is possible to know whether an adequate number of GNSS satellites will be available for use at the given time and location for a specific mission or not.

There are some commercial solutions available online for GPS satellite visibility determination, such as the online tool provided by Trimble [39] (a well-known GPS device manufacturer) or the simulator built by Safran [40], which is available for purchase on their website.

3.3.4 Flight Restrictions

To plan for a mission in terms of restricted flight zones and other navigation constraints, a source of information on this topic is required. Usually, each country has its own set of rules regarding airspace, and, in many cases; there are APIs available to request information. For example, in Spain, the airspace navigation manager known as Enaire provides an API [41] for integrating airspace regulation and information in other services.

3.3.5 Meteorology

Like the previous case, many providers offer meteorological information and predictions in API format to integrate them into other services. The AEMET API [42] in Spain and the National Weather Service one [43], in the USA, are examples of this type of provider. Usually, the offered data is free of charge, requiring only that the sources be cited properly.

3.3.6 GIS Information

More and again, the retrieval of Geographical Information System (GIS) data such as maps and Digital Terrain Elevation (DTM) models can be obtained in most cases through calls to an external API. In Spain, the *Instituto Geográfico Nacional* (IGN) provides an API [44] that grants free access to several products with different levels of accuracy, given that they are cited properly as information providers. Other organizations like the United States Geological Survey [45] provide elevation data with worldwide coverage.

4 Real-Time Decision-Making

4.1 Definition of Real-Time Decision-Making and Requirements

Speaking from a general perspective, real-time decision-making [46] can be defined as the process of performing choices based on the most updated and relevant data, and then taking specific actions over them. To that end, the use of technology based on artificial intelligence and big data strategies is crucial to both analyze a large amount of information and to draw timely and reliable decisions that can have immediate consequences or impact on a context. Nowadays, real-time decision-making is used in a wide range of industries, for example, financial services, retail, manufacturing, and health. The common thing about these sectors is that they manage tremendous amounts of disconnected information, and all of them need to make decisions to guarantee the success of their operations and consequently of their organizations.

Coming back to the scope of the book and, therefore, to what UAV SAR operations are concerned, it makes a load of sense to apply real-time decision-making strategies to tackle this problem. Think about a real catastrophe scenario. An environment with continuous variability and high risks. An event in which many resources of several natures will be provided independently from each other, a large communication channel composed of multiple actors and entities in which information losses or message misunderstandings could occur: A network of conditionings from which a simple action can trigger a domino effect resulting from a significant help for the operation to a complete step backward. Imagine now being inside of the rescue team leadership considering all this information, dealing with a solution to a severe catastrophe, and with time ticking because one wrong step can mean the loss of human lives. Without any doubt, rescue operation teams will benefit from using real-time decision-making as a tool to channel all the information coming from the UAV, efficiently and reliably and to have external support and second opinion to find the most valuable solution to a critical event.

With all this background, now, real-time decision-making related to UAV SAR operations can be defined as the process of making quick and critical decisions during the evolution of SAR operations, with the clear objective of locating, assisting, or helping in any possible way individuals or groups in a potentially dangerous situation, using UAVs to this end. To reach a high-end solution certified to be used in environments where human beings' life is threatened, no information flow, information processing technique, or even decision algorithm is valid. A set of minimum requirements need to be satisfied to guarantee the success and reliability of a real-time decision-making system. Below, the main requirements applicable to the SAR operation decision-making system are defined:

- **Updated and real-time data**: The access to instant data sources is crucial to have inputs to feed up the system. Depending on the context and the means involved in the operation, this data can be gathered from sensors, cameras, databases, IoT devices, satellites, public services, private organizations, etc.
- **Low latency**: Each branch of the workflow needs to be optimized to allow minimum delays in terms of data gathering, content transmission, processing, and decision-making. To achieve this, a low-latency infrastructure must be implemented. And referring to low latency, times of less than a second are expected.
- **Data processing**: The system must have the ability to quickly analyze and process huge information flows, isolating valuable information from the one that only brings noise and confusion. The ability to process and analyze data quickly is essential. This often involves advanced data analytics, artificial intelligence, and big data techniques.
- **Reliable decision models**: This is one of the fundamental cornerstones of a decision-making system. Having instant data, ensuring low latency in every step of the process, along with effective data processing, is of little value if the system has a lack of reliability in the decisions taken. The models used should be effectively trained and tested with an extensive dataset to allow high reliability in changing environmental conditions to generate recommendations or decisions rapidly.
- **Automation**: Any element of the operational flow that can be automated will represent a competitive advantage for the system performance because it will avoid the necessity of relying on manual individual action.
- **Scalability**: The system must be able to handle large volumes of data and scale to accommodate increased data flow or processing demands as needed.
- **Modularity**: The system must be also based on independent and interchangeable modules that can allow for its combination, reconfiguration, or integration of new functionalities with a minimum effort.
- **Redundancy**: Systems should have redundancy and failover mechanisms to maintain uninterrupted operation, 24/7, in case of hardware or software failures.
- **Reliable communications**: Robust communication systems are essential for transmitting information, alarms, notifications, decisions, or any kind of valuable content to the relevant stakeholders of the rescue operation.
- **Security**: Ensuring the security of real-time data and decision processes, as well as compliance with privacy regulations, is mandatory. That is the reason that any stage of the communication or access process must be protected with authentication, authorization, and encryption mechanisms.

4.2 Real-Time Decision-Making System Architecture

System architectures are always designed to solve specific problems in the most efficient way possible. If the priority is on ensuring a fast design, efforts will be made to minimize the bottlenecks of the system, potentially compromising reliability. If the focus is on ensuring solid and reliable analytics, the prize will be related to time and resources. Finally, if a low ratio of errors in decision-making needs to be ensured, a comprehensive analysis of the selected input data needs to be performed.

From the state of the art of decision-making systems [47-50], a wide range of architectures can be seen to approach different kinds of concepts of operations. What is certain is that from these diverse architectures, minimum common system blocks can be extracted to form a general-purpose real-time decision-making system, perfectly applicable to SAR operation missions. Finally, and as previously mentioned, the details and specifications of the blocks detailed below are closely related to proposing a solution to SAR operations decision-making problems. This is only a proposal for an applicable architecture that by no means intends to be qualified as a standard (Fig. 3).

- **Data sources**: The first step in real-time decision-making systems is gathering data from as many sources as possible. Related to SAR operations, one can think about data like UAV on-board sensors, video streaming flows, IoT sensors of the surrounding areas, information of other rescue teams, access to public databases,

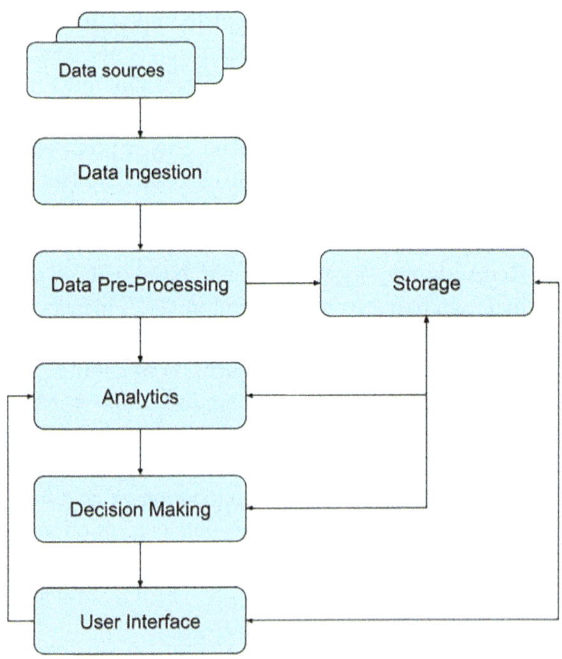

Fig. 3 Real-time decision-making architecture

weather forecasts, terrain maps, etc., in conclusion, any piece of current or past information that can be relevant to the operation. Therefore, these sources produce essential data that must be immediately consumed and analyzed.

- **Data ingestion**: As every input information source belongs to its context, a specific layer to manage the access and supply of information is needed. At this point, think about video streams provided by remote servers, different sensors generating data with different protocols, access to third-party services with different procedures, etc. The role of data ingestion is to unify all these contents to a common entry point to the system.
- **Data pre-processing**: The more data sources the system has, the more volumes of data will need to be managed. And probably, more irrelevant information will also be sneaked into the system. After integrating the proper data streams, it is crucial to build a specific data processing pipeline. To ensure accuracy and reliability in the analysis of the gathered data, this system component needs to clean, filter, and transform the data to be included in the analysis stage.
- **Data storage**: Another important part of a real-time decision-making architecture is the data storage module. Data needs to be stored in fast and scalable storage solutions that can handle large amounts of data generated from the different input sources and uniformized by the pre-processing stage. Furthermore, to avoid bottlenecks in this pipeline, it is essential to optimize the retrieving and insertion of data from its storage media.
- **Analytics**: From the pre-processing and data storage stages, the analytics engine can use various methods, like machine learning, statistical analysis, and data mining, to analyze operational data and generate specific insights. It is mandatory that to be effective within a real SAR operation, an analytics engine needs to extract valuable information to be applied in the resolution of ongoing actions.
- **Decision-Making**: The decision engine is fed by insights or valuable information pieces from the analytics engine. It uses statistics, policies, or even machine learning models to make recommendations or suggestions to the chain of command of the SAR operation. In this way, by being able to leverage artificial intelligence, the decision-making engine can learn about past recommendations and their efficiency, to be able to adjust its internal scheme of rules.
- **User interface**: It is of little use to have a perfectly tuned decision-making system capable of processing large batches of input data, extracting valuable information from other meaningless ones, and providing reliable decisions if the interface does not offer a user-friendly and intuitive way to operate the system. Command chains must be presented with the information and conclusions in an approach that is simple to understand and act using dashboards with utilities like GIS positioning maps, video displayers, notification boxes, actuators commands, easy representation of data based on plots or percentage bars, etc.

Once the basics of any real-time decision-making architecture have been explained, another glowing topic to debate is the location in which these architectures can be deployed. This fact brings one to a very well-known scenario between what are called on-board computing architectures and edge or cloud computing architectures. The reality about this is that there is not only one valid approach applicable to any use case or problem. The final decision of what architecture fits better to solve a specific problem will have to unavoidably ask questions like: What are the requirements to solve the problem? What is the concept of operation? How many resources are available to use? Is there a good connection quality between the elements of my architecture? Answering these kinds of questions would get one closer to making a good choice in terms of selecting the right architecture.

The first thing is to distinguish between on-board, edge, and cloud computing concepts. From a high-level perspective, on-board computing can be defined as the processing or execution of programs on a device itself, for example, a computer or in this case a drone, whereas in the other two concepts, although they both refer to providing IT services through the network, the main difference is the location of the resources. Edge computing relies on the processing performed very near the edge of the network where the data or the communication is generated, rather than counting on a centralized processing warehouse, as cloud computing does. That is the key fact why, in environments where a quick reaction is needed (like SAR operations), on-board and edge computing architectures give the best performance.

To be more descriptive, the examples below will be used to explain the pros and cons of these architectures, always keeping in mind the context of SAR operations. The architecture on the right depicts a deployment of an on-board architecture, where the business logic of conflict resolution is resolved inside a UAV, in contrast with the architecture on the left, where the business logic resides at the edge of the network. Practical examples of applications of these architectures can be an autonomous SAR operations UAV monitored by the chain of command and a SAR operation system used to give support in rescue task decisions (Fig. 4).

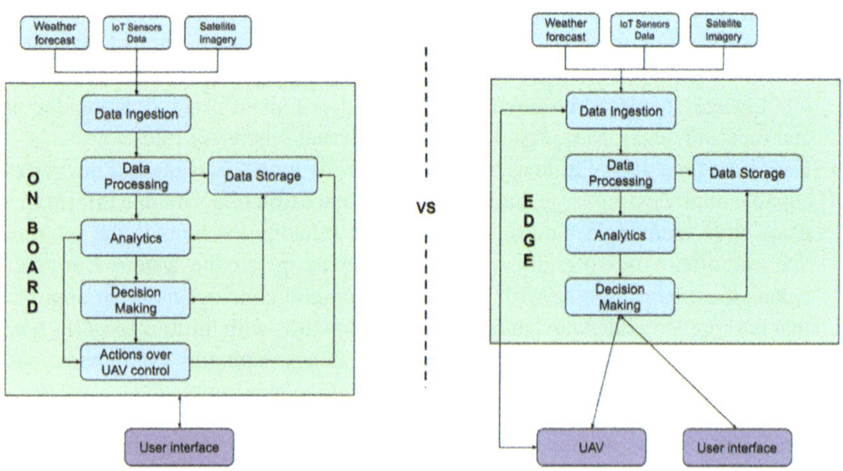

Fig. 4 Comparison between on-board and edge architectures

Thinking about SAR operation requirements, the first thing that comes to mind is time. There is a need to optimize this resource as much as possible to provide quicker and more reliable responses to immediate hazardous events. Nothing is faster than processing the data within the same physical device (UAV) one is using in the operation. The latency will be severely reduced, but, depending on the case, there is no such big difference compared to edge computing. The problem appears when the number of resources to process the data becomes larger and larger. Nowadays, the computational capacity that can be mounted over a drone is impressive, but it is still limited for huge and simultaneous video streams, heavy databases, high-rate sensors, etc. In this case, an on-board computing strategy will produce an increasing bottleneck, whereas edge computing is more scalable and can perfectly deal with massive workloads.

In terms of connectivity dependence, edge computing is very vulnerable to loss of connections in comparison to on-board computing. In hazardous environments, where typical communication technologies can easily stop working or downgrade their performance (a fire and earthquake are examples of disasters that can break down communications), on-board computing is life insurance. It can operate with limited reliance on external sources, making it suitable for emergency scenarios. However, edge computing is not over yet, because, in this type of catastrophe, rescue operatives often deploy their communication networks.

Scalability is another key factor in SAR operations. Depending on the magnitude of a catastrophe, the resources required to respond can increase or decrease easily. In these terms, on-board computing can face severe limitations due to the hardware constraints of the hardware mounted in the system. On the contrary, edge computing takes advantage of leveraging a combination of scalable computing resources and access to broader network capabilities.

Finally, considering all the concepts discussed in this section, it is essential to highlight operational flexibility. Considering that both models are based on software capabilities, it is evident that they can be reprogrammed to respond to different triggers beyond their original commitments. However, on-board computing will be more resilient to change because of its static hardware capabilities, in comparison with the endless array of possibilities that edge-computing capabilities can bring in.

4.3 Data Sources and Real-Time Data Acquisition

Regarding the SAR mission and the technologies available that can be integrated within a UAV, many data sources and sensors can be applied to tackle different critical situations. The key factor to ensure real-time data acquisition at this point is having a direct and reliable interface to access this information. In some cases, sensors or data sources will be decoupled from the command-and-control center, requiring an upper integration and communication layer to deliver data streams. In other scenarios, sensors or data sources will provide themselves with the technology to reach a specific mount point. What is intended to be emphasized in these topologies is that to achieve real-time data acquisition, the delay in each communication step needs to be minimized and optimized.

As a simple illustration, the following bullets outline several sensor options that, depending on the SAR mission, could provide valuable insight into the chain of command.

- **Video cameras.** The most well-known devices within this list capture and record motion pictures one frame at a time. In the SAR context, they are extremely used for surveillance and identification activities.
- **Thermal or infrared camera**. It is a subtype of the video camera. This specific sensor turns infrared radiation into visible images that depict the spatial distribution of temperature differences in a scene. This device is essential in nighttime searches for victims or fire situations.
- **Multispectral and hyperspectral sensors**. Multispectral sensors are devices that can capture and measure electromagnetic radiation in multiple discrete bands across the spectrum. Unlike hyperspectral ones, which capture data in numerous narrow and contiguous bands, multispectral works in a smaller number of border bands. A classic use case for this sensor inside SAR missions is environmental monitoring.
- **Radar**. It is another well-known sensor that usually consists of a synchronized radio transmitter and receiver that emits radio waves and processes their reflections which is typically used for detecting objects. In low visibility conditions, it has a huge value in detecting victims in movements or cavity openings.
- **Gas and fume detector**. It is another advanced sensor that can detect and identify different types of gases. In high-risk operations, where fires or leaks of dangerous substances can be suspended in the air, these devices provide significant safety by being able to distinguish if the air is breathable or not for the rescue teams.

4.4 Real-Time Processing and Analysis

The concept of real-time processing and analysis has evolved remarkably, hand in hand with the advancements in both the state of the art of processing techniques and hardware computational capabilities. Real-time analysis refers to the ability to collect, process, and analyze data to provide insights for making better decisions. In this way, the key aspect is using AI patterns, statistical models, or big data techniques to extract valuable information from a seemingly incoherent and valueless dataset. However, the concept of real time is relative. For some use cases, real-time processing with delays of a few seconds can be appropriate, while, for others, the same approach could be useless.

Going back to the scope of SAR operations performed with UAVs, the relativity of real-time processing becomes even more apparent. In navigation, where instant response to manage the behavior of a UAV is required, latencies of less than a second are indispensable. Any latency beyond this can be a potential danger to the success of a rescue mission, not only in terms of reaching the exact destination but also in terms of safety. Consequently, the integration of high-speed data processing systems, capable of swiftly interpreting and relaying real-time navigation data, becomes imperative.

A different perspective can be drawn involving image processing in SAR operations. For example, for identifying potential survivors in a disaster zone, a slightly larger tolerance in the range of 1 to 10 s can be acceptable. Flying over wide areas at a constant speed gives drones a brief time window to capture high-resolution videos. These streams need to be processed and analyzed with the capabilities currently available on board or the edge, which can take a few seconds. Considering that the importance of this use is to be extremely highly reliable, whether a certain video frame contains survivals, their frame length up to 10 s is not relevant. However, unfortunately, it takes time to be both effective and reliable, and there is no room for negotiation when people's lives are at risk.

In contrast, there are other kinds of extremely complicated scenarios, like structural assessments, in which it is not possible to perform a very responsive analysis due to the enormous number of possibilities that need to be reviewed or the complex consequences that would appear from making a small error in the interpretation. These missions often require the integration of comprehensive analytical frameworks capable of processing and synthesizing diverse datasets to derive actionable insights. Leveraging robust computing systems and optimized data processing pipelines can facilitate timely decision-making, enabling SAR teams to formulate effective strategies and deploy necessary resources promptly.

4.5 Decision-Making Algorithms

In the realm of operations involving Unmanned Aircraft Systems (UAS or drones), conventional route planning algorithms are widely used, allowing UAVs to move between a series of points following a pre-planned trajectory. This planning relies on prior knowledge of the environment; however, in surveillance and search and rescue operations, a rapid response based on the perceived changes in the environment is crucial. This is why decision-making algorithms play a crucial role, distinguishing between those affecting solely internal operations and those directly impacting navigation and, consequently, aviation safety. Internal decisions may pertain to aspects such as energy management, route planning, and coordination among multiple drones in a single mission. On the other hand, decisions affecting navigation address critical issues like avoiding collisions with other aircraft in flight and ensuring the safety of individuals and infrastructure on the ground.

The acceptance of algorithms for decision-making in drones poses a dilemma for aviation authorities, as they face the choice between classical algorithms, which are more deterministic and formula-based, and artificial intelligence (AI) algorithms, presenting unique challenges. In AI algorithms, being black boxes in terms of their internal decision-making, critical aspects of reliability, safety, and explainability need to be addressed, alongside the creation of regulatory frameworks. This approach aims to balance technological innovation with the need to maintain rigorous safety standards, thus ensuring the sustainable and safe development of drone operations in European airspace.

4.6 Applications

Lastly, after all the theory content related to real-time decision-making has been detailed, it is time to give a more applied approach to all this knowledge. The objective of this section is none other than to demonstrate how from a general perspective real-time decision-making capabilities, it is possible to automate and solve real problems occurring within a SAR mission. Below, 3 or many possibilities are shown.

- **Search and fast location of survivors**. Drones equipped with high-resolution video cameras can be perfect data input sources to supply a real-time decision-making system of data. From this information, deep learning models can be applied to trigger an alarm when a possible survival appears in a video frame. This gives SAR teams a powerful tool because instead of taking care of every single frame of a video, they only must focus on a few alerts. This autonomous operation becomes a great advantage, considering that after long periods of work, the human eyes start making an increasing number of errors on both sides (not detecting the obvious and detecting what is not).
- **Surveillance and damage assessment**. Analogous to the previous use case, UAVs equipped with sensors can provide live monitoring of hazardous situations such as natural disasters, collapsed structures, and areas with poor visibility. Real-time data from drones allows SAR teams to make informed decisions about the deployment of resources, the assessment of potential risks, and the adjustment of rescue strategies to ensure the safety of both the victims and the rescuers.
- **Supplies delivery and emergency assistance**. In other missions, where campaign resources or emergency channels are highly automated, it is possible to use real-time decision-making systems to autonomously generate an alarm to perform a specific task. For example, if a shortage of material, measurements, or equipment is detected based on the analysis of available resources from database logs, the system could be available to identify this deficiency and send a drone to the conflict zone with all the materials necessary to solve the issue.

5 Communications

From the inception of SAR missions, reliable and effective communications have been crucial for ensuring continuous coordination between rescue teams and gaining access to critical information. In recent years, the integration of UAVs within SAR missions has brought an endless number of possibilities to improve rescue operations. With enough autonomy to survey large areas, easy deployment, and enough capacity to mount a wide variety of sensors and payloads, UAVs have become another indispensable tool for first responders. However, to take advantage of the vast amount of information these devices can provide, a reliable and stable communication link needs to be established with the coordination and command center. Explaining why these communication links are so important, the different approaches that can be used to make it possible, their weaknesses, and how they are going to evolve will be the goals of this section.

5.1 Why Are Communications so Crucial in SAR Missions with UAVs?

Imagine a hazardous situation where there is a forest fire or an earthquake. Normally, these kinds of disasters cover large areas where access becomes extremely complicated and dangerous. First responders and SAR teams are the ones responsible for navigating through those challenges with the coordination and supervision of the coordination and command center. That this communication channel remains operational and works smoothly is mandatory, as it directly holds the lives of many people in its hands.

Fortunately, with the advancement of UAVs, they have been quickly integrated into SAR teams to enhance security and capabilities in rescue missions, such as enlarging the visibility of an area through aerial video streaming, monitoring, and searching for survivors or relaying data in real time. These kinds of activities involve a new communication link to share the data with the SAR teams and the operation and command center. This new channel brings SAR operations to a new level of action and interoperability for a series of compelling reasons. First, it brings a new real-time input information source to extract, share, and relay complementary data useful for all the members of the mission. Second, this connectivity contributes to an enchanted vision of situational awareness, being able to do both in advance providing updates on the evolution of the mission and identifying potential targets. Third, it helps to optimize search efforts, communicating spots where efforts need to be focused or avoided. Reliable connectivity enables the transmission of this data in real time to the command center, allowing for immediate analysis and decision-making.

Nonetheless, to fully harness the benefits of connectivity between UAVs and the rest of the members of a SAR mission, at least certain strict requirements shall be considered regarding this communication link:

- Reliability: The communication channel must be highly reliable to ensure continuous communication between the command center and the UAV.
- Long-range: Areas to cover within the SAR mission are usually vast. Therefore, distances of kilometers must be guaranteed to make the contribution of UAVs valuable.

- Real-time and low latency: In highly variable scenarios where things can change in an eye, real-time transmission is crucial, requiring being as close as possible to 1-s latencies.
- Bandwidth: UAVs are normally equipped with high-resolution cameras; therefore, a high bandwidth will be necessary to enable proper transmission quality.

At last, it is worth dedicating a brief space to highlight the different types of valuable data that can be shared between UAVs and the command center. In terms of centralizing the operation, it is possible to distinguish between commands (to control the UAV) and telemetry (to monitor its flight and operation). Then, looking at the payloads they can support, the types of data from sensors and actuators increase exponentially. However, their output data can be grouped into two major categories: (1) video streams from cameras and (2) plain/formatted text, the result of the operation of a specific sensor or actuator. Finally, another complementary type of data can be alarming, both incoming and outgoing to UAVs.

5.2 Interconnecting UAVs via Cutting-Edge Communication Technologies

Providing reliable wireless communication during disasters presents a significant challenge, considering that reliable and effective communication between the UAV and the control center will increase the possibilities of a successful rescue. Nowadays, there are countless types of protocols and platforms to develop highly reliable and secure communications. Despite this, not all these technologies are appropriate to integrate over UAVs. Other factors like platform dimensions, weights, consumption, and integrability have a huge impact on the performance of the UAV. Because of this, the following communication technologies are depicted as solutions to create robust and effective connectivity to disaster response.

5.2.1 Radiofrequency Data Links

Radiofrequency data links are communication systems that simply use radiofrequency signals to transmit and receive data between different devices. In these platforms, the digital data is modulated on the frequency carrier wave in the transmitter; this information is sent wirelessly; finally, it is demodulated in the receiver. This process allows a basic communication exchange between the connected platforms.

Depending on the frequency bands used by these devices, one can categorize them as either open or licensed bands. The first one can be used by anyone if they follow the regulations set for the band. Otherwise, the latter requires a specific license from a regulatory organization for its use. Typically, in aerial communications and in UAVs, Industrial, Scientific, and Medical (ISM) bands are very well-known and efficient bands to connect the UAV with the ground. Regarding the coverage and bandwidth constraints, one can select a radiofrequency data link operating in the following bands: 433 MHz, 915 MHz, 2.4 GHz, and 5.8 GHz.

Radiofrequency data links can also be classified into proprietary systems commonly used in civil applications and standardized systems prevalent in the military industry. Civilian data links are commonly developed by manufacturers and designed to meet specific communication requirements. Examples of these platforms are modules like the ones from Microhard [51] or [52]. They both provide miniature, lightweight, and robust OEMs able to cover distances over tens of kilometers, being also able to select a specific frequency band. In addition to these examples, other typical data link references for UAVs include DJI with its communication modules based on Lightbridge and Ocusync and with an open integration Pixhawk, which implements the MAVLink communication protocol underneath.

On the other hand, two widely recognized military standard examples are Link 16 and STANAG. Both developed by NATO, they are established as tactical data links, the first close to the sector of aircraft, ships, and ground vehicles, whereas the second goes to interoperability and standardization between systems.

5.2.2 3GPP Cellular Networks

In simple terms, cellular networks work through a system of base stations and mobile devices. The base stations, geographically distributed, emit radiofrequency signals to serve coverage to specific areas. When a mobile device is powered on or enters a coverage area, it registers in the network, authenticates, and is allocated resources for communications. Communications are bidirectional; the base station sends a signal to the mobile device (downlink) for call and data services, whereas the mobile device transmits uplink signals. The network manages mobility, allowing smooth handover between coverage areas, and the central infrastructure directs traffic and connectivity between the different areas. All these procedures and techniques will vary its operation depending on the kind of cellular technology supported, currently 3G, 4G, 5G, and shortly 6G. For a more detailed understanding of the peculiarities of each technology, it is strongly recommended to consult standards from 3GPP.

Due to being a de facto standard for civil communication in the world, it is not surprising to say that a wide range of cellular communication modules flood the market. In terms of UAV operations, the key actors of embedded modems in the current market are Quectel [53], Sierra Wireless [54], Telit [55], and Neoway [56]. Any of these manufacturers can provide an ultra-compact and versatile cellular modem, easy to integrate into a bunch of different UAV mission computers, and compatible with many world frequency bands. In this simple way, it achieved UAV network connectivity.

It is crucial to emphasize that in disaster areas, there is a high probability that common cellular networks are down. However, this is not a problem for cellular networks, because they can support Device-to-Device (D2D) communications. This is a concept within cellular networks, where devices communicate directly with each other without going through the network infrastructure (base station). Standards like LTE and 5G-NR support D2D communications (since 3GPP release 12, the concept of D2D communications was introduced). Bordering on this concept, a reference can also be made to C-V2X (the communication standard that enables cars to communicate with each other and everything around them). This standard encompasses two interfaces: (1) Uu, a common network interface, and (2) PC5, which again uses device-to-device communication at 5.9 GHz, without requiring the presence of a base station.

5.2.3 SATCOM

Satellite communications transmit information from ground stations to satellites in orbit back to the Earth. The principle of operation is quite simple. The data is transmitted via ground-based stations using radiofrequency signals, which are sent to satellites in a geostationary of low earth orbit, LEO (this is known as uplink). Next, the satellite transponders process the signal and retransmit it back to the Earth (this is known as downlink). Finally, ground-based receivers capture the signals from the satellites and demodulate them to obtain the original information.

Satellite communications applied to UAVs have significant advantages like global coverage and therefore long-distance communications. They are extremely flexible and movable practically without geographic constraints, as they are not attached to ground stations. This fact is extremely important to ensure reliable communications when a UAV flies across a disaster area. On the other hand, in comparison with the previous technologies, these communications are commonly not so strong in terms of rates and latencies. Their development and access costs are quite expensive. Regarding integration portability in UAVs, satellite communication modules are bigger, heavier, and consume a significant amount of energy. This is the reason why they are more convenient to use large-size UAVs.

From a market perspective, there is a growing variety of SATCOM modules compatible with UAV integration. Manufacturers are realizing the significant niche emerging in the UAV sectors and are doubling down efforts to provide more compact and low-consumption systems. Examples of these platforms are the ones provided by manufacturers like Iridium [57], Inmarsat [58], and Honeywell [59].

5.2.4 IEEE 802.XX Standards

Last but not least, it is worth mentioning the IEEE 802.XX family of standards. The Institute of Electrical and Electronics Engineers developed this series of family standards for local area networks (LANs) and metropolitan area networks (MANs). Each standard series defines specific aspects of the network, including the physical layer and the data link layer. Any enthusiast of UAV communications will easily distinguish some of the following standards, with their common names:

- 802.11 (Wi-Fi)
- 802.15 (Bluetooth, Zigbee, etc.)
- 802.16 (WiMAX)
- IEEE 802.22 (Wireless Regional Area Network, WRAN)
- Mesh networks.

In the context of SAR operations, the term mesh networks becomes particularly interesting. Mesh networks, implemented in some of the previous standards, operate on a decentralized architecture where each node is capable of relaying data to the others. These kinds of networks are very resilient, because if a node in the network fails the other ones can continue relaying data, dynamically routing it to find the most efficient path for the transmission.

It is curious to think that at some point this kind of communication was used within real SAR missions. However, the truth is that they have very interesting advantages that allow for their easy integration. 802.11 standards are widespread all over the world, so there is a vast amount of modules, equipment, and devices that support this technology. Yet, despite its low cost, they provide remarkable data rates, allowing an efficient transmission of large volumes of data, like video streaming. Building upon commercial transceivers, extended management and control functions based on open Linux-based software can be easily developed to form an operational communication platform. However, they have other big drawbacks that discourage their use in disaster areas. Their range is very limited to cover large areas, being even more critical when non-line-of-sight communications are involved. 802.11 standards are very sensitive to interference of other bands or equipment, having even severe bandwidth limitations in crowded environments. Finally, yet importantly, 802.11 networks are susceptible to security vulnerabilities, not being a robust communication option in vulnerable environments. These reasons, coupled with the critical nature of SAR mission environments, make it less feasible for these devices to be employed in real operations with drones, reserving them more for recreational or hobbyist activities.

5.2.5 Communication Technologies Comparison

Table 1 provided below offers a straightforward and intuitive comparison of how the previous communication technologies behave for relevant requirements in SAR missions. They will be categorized as high, intermediate, and low impact.

5.3 Future Trends and Challenges

With the current state of health enjoyed by the UAV ecosystem, joined by the growing market perspectives and inversions, everything seems to indicate that UAVs will have a bright future and impact on society. The advancements in technology and the

Table 1 Communication technologies comparison

Categories	RF data link	3GPP networks	SATCOM	IEEE 802.xx
Reliability	High	Intermediate	High	Intermediate
Range	High	High	High	Low
Bandwidth	High	Intermediate	Low	Intermediate
Adaptability	Intermediate	High	Intermediate	High
Security	Intermediate	High	High	Intermediate
Cost	Intermediate	Intermediate	High	Low

expanding use cases display a trajectory of sustained growth for UAVs across various sectors, where SAR operations will not be an exception.

In the next few years, the following future trends represent potential evolution directions for the evolution of UAV communications within SAR missions:

- Hybrid communication systems. In environments as changing and unstable as SAR missions, communication redundancy is essential to ensure continuous operation. To this end, SAR communications need to leverage hybrid communication systems integrating a combination of 5G/6G, satellite, datalinks, or mesh network solutions within a UAV.
- Artificial intelligence for communication optimizations. Related to the previous point, AI can help the communication system to decide depending on the situation, which is the most efficient communication technology to use, always guaranteeing performance and continuous operation. It can also be used to predict interference and other signal-damaging effects, and to dynamically adjust transmission and reception parameters, therefore improving efficiency.
- Smart antennas for enhanced communications. Continuing with the mantra of efficiency and continuous operation, smart antennas will be a key factor to contribute to improving communication signal quality. These intelligent antennas can dynamically adjust beamforming and orientation to optimize signal strength and mitigate interference.
- Standardization. It is a fact that within SAR operations, there are no specific standards and protocols to manage the communication between UAVs and the rest of operations. Standardizing these kinds of tools will have a huge impact on interoperability not only within teams but also with other first responders and agents.

However, to reach these goals and to establish UAVs as a completely integrated member of SAR missions, UAV communications will have to face some specific challenges:

- Miniaturization of communication systems. Typically, SAR UAVs will not exceed 10 kg of weight, leaving a maximum of 3 kg for payloads. Considering that other imaging sensors are vital as primary information providers, it emphasizes even more the need to reduce the size of communication systems on UAVs. Satellite modules are a perfect example of this problem, very powerful in terms of connectivity, but very limited in terms of integrations.

- Energy-efficient communications. A trade-off between power consumption and signal coverage must be balanced to maximize mission duration.
- Integration with SAR existing infrastructure. To make this collaboration useful, UAV communications must be flexible enough to allow proper integration into the SAR team's current communication infrastructure and procedure. Apart from hybrid communication systems, backhaul communication systems can be an interesting communication link option to integrate with other less common or updated communication protocols.
- Privacy and compliance: Due to the sensitive information that will be transmitted through these channels, they must adhere to regulations and compliance standards to safeguard this information.

6 Specific Payloads and Other Use Cases

The traditional objective for a UAV on a SAR mission is usually to provide imaging, either in the visible spectrum or one of the infrared bands (such as thermal), turning the UAV into the eyes of an operator. However, the capabilities of this type of aircraft can go much further, as has already been seen in some of the previous sections, with improved performance in terms of flight autonomy and communication range, as well as the ability to carry a greater load.

One of the first uses that UAVs had in SAR missions, beyond taking images, was to deliver goods. For years, there have been teams of lifeguards on the beaches that are equipped with UAVs capable of transporting and launching a life jacket to avoid drowning. It is also common to deliver first aid kits or medicines in isolated areas or accidents in remote areas, where it is faster to move a UAV than a human team.

Another common payload in SAR missions is the inclusion of sensors to collect information from the theater of operations: it is possible to integrate synthetic aperture radars to obtain data in conditions of low visibility or geo-radars that can be used to detect trapped people under the rubble. It is also possible to integrate gas sensors for the analysis of dangerous atmospheres or water sampling equipment as a prior step to the deployment of human resources on the ground, especially useful in situations with risk (chemical, biological, radiological, and nuclear) CBRN (Fig. 5).

As previously mentioned, communications play a determining role in the operation of UAVs, but in SAR missions the availability of a reliable means of communication in the theater of operations that allows communication between the teams involved in the operation is also essential, and/or the people who are in that area; for this reason, it is also possible to integrate communication equipment repeaters and cellular communication cells (4G and 5G), thus creating a connectivity bubble in the area. There is also communication equipment whose objective is to locate people on the ground, for example, Lifeseeker, from Centum Research & Technology [60],

Fig. 5 Lifeseeker integrated into a UAV

is an innovative airborne system capable of locating mobile phones accurately even in regions lacking network coverage and under adverse weather conditions. This system harnesses the significant social impact of mobile communication technologies in recent years, transforming a mobile phone into an emergency beacon that precisely guides rescue teams to its location.

7 Conclusions

Throughout this chapter, a review has been made of the technologies that have the most impact on the use of UAVs in SAR missions. Thus, the review comprises from the pre-flight phases, where mission planning has a determining impact on its success and carrying out the simulation can identify potential risks or critical situations for the aircraft, to the use of AI-based video analytics solutions. Such technologies facilitate the operators' mission and allow information to be provided to decision-making systems in real time to raise the level of automation in the operation. Moreover, it highlights the importance of the communication systems, a fundamental element for the correct operation of the UAV. Moreover, it addresses long-distance missions or the different payloads, beyond the cameras, which can be integrated into a UAV, which allows the use of these vehicles in increasingly diverse situations.

It is undeniable that UAVs represent a revolution in different sectors, and, in the case of SAR missions, they are a tool with great potential. By the improvement of communication systems and systems designed to increase their autonomy in operation (such as video analytics or mission computers with real-time decision-making capacity), UAVs will reduce response times in this type of situation and improve the safety of human teams, reducing the risk they have to face.

References

1. Dijkstra, E. W. (1959). A note on two problems in connexion with graphs. *Numerische Mathematik, 1*, 269–271.
2. Karur, K., Sharma, N., Dharmati, C., & Siegel, J. E. (2021). A survey of path planning algorithms for mobile robots. *Vehicles, 3*(3), 448–468.
3. Hart, P. E., Nilsson, N. J., & Raphael, B. (1968). A formal basis for the heuristic determination of minimum cost paths. *IEEE Transactions on Systems Science and Cybernetics, 4*(2), 100–107.
4. LaValle, S. M. (1998). *Rapidly-exploring random trees: A new tool for path planning. Technical Report*. Iowa State University.
5. Kavraki, L. E., Svestka, P., Latombe, J.-C., & Overmars, M. H. (1996). Probabilistic roadmaps for path planning in high-dimensional configuration spaces. *IEEE Transactions on Robotics and Automation, 12*(4), 566–580.
6. de Mendoza y Ríos, J. (1795). Memoria sobre algunos métodos nuevos de calcular la longitud por las distancias lunares: y aplicación de su teórica á la solución de otros problemas de navegación. Imprenta Real, Madrid.
7. Vincenty, T. (1975). Direct and inverse solutions of geodesics on the ellipsoid with application of nested equations. *Survey Review, 23*(176), 88–93.
8. Karney, C. (2013). Algorithms for geodesics. *Journal of Geodesics, 87*, 43–55.
9. Wynne, J. B., & Campbell, R. H. (2011). *Introduction to remote sensing* (5th ed., p. 261). Guildford Press.
10. Reeds, J. A., & Shepp, L. A. (1990). Optimal paths for a car that goes both forward and backward. *Pacific Journal of Mathematics, 145*(2), 367–373.
11. Funke, J., Theodosis, P., Hindiyeh, R., Stanek, G., Kritayakirana, K., Gerdes, C., Langer, D., Hernandez, M., Müller-Bessler, B., & Huhnke, B. (2012). Up to the limits: Autonomous Audi TTS. In *IEEE Intelligent Vehicles Symposium (IV)*, Alcalá de Henares.
12. Xu, W., Wei, J., Dolan, J. M., Zhao, H., & Zha, H. (2012). A real-time motion planner with trajectory optimization for autonomous vehicles. In *IEEE International Conference on Robotics and Automation*, St Paul.
13. Bautista, D. G., Perez, J., Lattarulo, R. A., Milanes, V., & Nashashibi, F. (2014). Continuous curvature planning with obstacle avoidance capabilities in urban scenarios. In *IEEE International Conference on Intelligent Transportation Systems*.
14. Farouki, R. T., & Sakkalis, T. (1994). Pythagorean-hodograph space curves. *Advances in Computational Mathematics, 2*(1), 41–66.
15. Evgeniou, M. P. T. (2001). Support vector machines: theory and applications. In *Machine learning and its applications*. Springer.
16. Hochreiter, S., & Schmidhuber, J. (1997). Long short-term memory. *Neural Computation, 9*(8), 1735–1780.
17. Silver, D., Huang, A., Maddison, C. J., Guez, A., Sifre, L., van den Driessche, G., Schrittwieser, J., Antonoglou, I., Panneershelvam, V., Lanctot, M., Dieleman, S., Grewe, D., Nham, J., Kalchbrenner, N., Sutskever, I., Lilicrap, T., Leach, M., Kavukcuoglu, K., Graepel, T., & Hassabis, D. (2016). Mastering the game of Go with deep neural networks and tree search. *Nature, 529*(7587), 484–489.
18. Gao, W., Hus, D., Lee, W. S., Shen, S., & Subramanian, K. (2017). Intention-net: integrating planning and deep learning for goal-directed autonomous navigation. In *1st Annual Conference on Robot Learning, Mountain View*.

19. Smart, W. D., & Kaelbling, L. P. (2002). Effective reinforcement learning for mobile robots. In *IEEE International Conference on Robotics and Automation* (Vol. 4). Minneapolis.
20. Mnih, V., Kavukcuoglu, K., Silver, D. G. A., Antonoglou, I., Wierstra, D., & Riedmiller, M. (2013). Playing Atari with deep reinforcement learning. In *NIPS Deep Learning Workshop*, Lake Tahoe.
21. Hasselt, V. H., Guez, A., & Silver, D. (2015). Deep reinforcement learning with double Q-learning. In *Thirtieth AAAI Conference on Artificial Intelligence*, Phoenix.
22. Wang, Z., Freitas, N., Lanctot, M. v. H. H., Hessel, M., & Schaul, T. (2016). Dueling network architectures for deep reinforcement learning. In *33rd International Conference on Machine Learning*, New York.
23. Bai, H. S., Ye, N., Hsu, D., & Lee, W. S. (2015). Intention-aware online POMDP planning for autonomous driving in a crowd. In *IEEE International Conference on Robotics and Automation (ICRA)*, Seattle.
24. Everett, M., Chen, Y., & How, J. P. (2018). Motion planning among dynamic, decision-making robots with deep reinforcement learning. In *IEEE/RSJ International Conference on Intelligent Robots and Systems (IROS)*, Madrid.
25. Paxton, C., Raman, V., Hager, G. D., & Kobilarov, M. (2017). Combining neural networks and tree search for task and motion planning in challenging environments. In *IEEE/RSJ International Conference on Intelligent Robots and Systems (IROS)*, Vancouver.
26. Mathworks. Matlab Navigation Toolbox. Accessed 23 March from https://es.mathworks.com/products/navigation.html
27. SPH Engineering. Universal Ground Control Station. Accessed 23 March 2024 from https://shop.ugcs.com
28. ArduPilot. Ardupilot's Mission Planner. Accessed 23 March 2024 from https://ardupilot.org/planner
29. Sun, S., Rappaport, T. S., Rangan, S., Thomas, T. A., Ghosh, A., Kovács, I. Z., Rodriguez, I., Koymen, O., Partyka, A., Järveläinen, J., & Sun, S. (2015). Propagation path loss models for 5G urban micro-and macro-cellular scenarios. In: *IEEE 83rd Vehicular Technology Conference*, Nanjing.
30. Sati, G., & Singh, S. (2014). A review on outdoor propagation models in radio communication. *International Journal of Computer Engineering & Science, 64*(2), 64–68.
31. ITU-R. (2019). Calculation of free-space attenuation. In *Recommendation ITU-R P.525* (Vol. 4).
32. ITU-R. (2023). A general purpose wide-range terrestrial propagation model in the frequency range 30 MHz to 50 GHz. *Recommendation ITU-R p.2001* (Vol. 5).
33. Institute of Telecommunication Sciences. ITS Open Source Software. Accessed 23 March 2024 from https://its.ntia.gov/software/its-open-source-software
34. Xirio Online. Accessed 23 March 2024 from https://www.xirio-online.com/web/home/welcome.aspx
35. CloudRF. CloudRF online radio propagation simulator. Accessed 23 March 2024 https://cloudrf.com
36. Magliacane, J. Signal propagation, loss, and terrain analysis tool (SPLAT!). Accessed 23 March 2024 from https://www.qsl.net/kd2bd/splat.html
37. International Association of Geodesy. International GNSS Service. Accessed 23 March 2024 from https://igs.org
38. Kelso, T. S. Celestrak. Accessed 23 March 2024 from https://celestrak.org
39. Trimble. GNSS Planning Online. Accessed 23 March 2024 from https://www.gnssplanning.com
40. Safran. Multi-antenna GNSS Simulation for Unmanned Aerial Vehicles (UAVs). Accessed 23 March 2024 from https://safran-navigation-timing.com/multi-antenna-gnss-simulation-for-unmanned-aerial-vehicles-uavs
41. ENAIRE. Servicio de Información Aeronáutica: AIP España. Accessed 23 March 2024 from https://aip.enaire.es/AIP

42. AEMET. Sistema para la difusión y reutilización de la información de AEMET (AEMET OpenData). Accessed 23 March 2024 from https://opendata.aemet.es/centrodedescargas/inicio
43. National Weather Service. NWS API Service. Accessed 23 March 2024 from https://www.weather.gov/documentation/services-web-api
44. Instituto Geográfico Nacional. Infraestructura de Datos Espaciales. Accessed 23 March 2024 https://www.ign.es/web/ide-area-nodo-ide-ign
45. United States Geological Survey. Earth Resources Observation and Science (EROS) Center. Accessed 23 March 2024 https://www.usgs.gov/centers/eros
46. González, C. (2005). Decision support for real-time, dynamic decision-making tasks. *Organizational Behavior and Human Decision Processes, 96*, 142–154.
47. Delic, K. A., Douillet, L., & Dayal, U. (2001). Towards an architecture for real-time decision support systems: Challenges and solutions. In *Proceedings 2001 International Database Engineering and Applications Symposium*, Grenoble.
48. Ortiz, G., Caravaca, J. A., Garcia-De-Prado, A., Chavez de la, O. F., & Boubeta-Puig, J. (2019). Real-time context-aware microservice architecture for predictive analytics and smart decision-making. *IEEE Access, 7*, 183177–183194.
49. Hyun Joong, Y., & Shen, W. (2006). Simulation-based real-time decision making for manufacturing automation systems: A review. *International Journal of Manufacturing Technology and Management, 8.1*(3), 188–202.
50. Waharte, S., & Trigoni, N. (2010). Supporting search and rescue operations with UAVs. In *International Conference on Emerging Security Technologies*, Canterbury.
51. Microhard. Microhard - Wireless Innovation. Accessed 23 March 2024 from https://microhardcorp.com/
52. S. Technologies. Silvus Technologies. Accessed 23 March 2024 from https://silvustechnologies.com/
53. Quectel. Quectel. Accessed 23 March 2024 from https://www.quectel.com/
54. S. Wireless. Sierra Wireless. Accessed 23 March 2024 from https://www.sierrawireless.com/
55. Telit. Telit Cinterion. Accessed 23 March 2024 from https://www.telit.com/
56. N. Technologies. Neoway. Accessed 23 March 2024 from https://www.neoway.com/
57. Iridium. Iridium Satellite Communications. Accessed 23 March 2024 from https://www.iridium.com/
58. Inmarsat. Inmarsat. Accessed 23 March 2024 from https://www.inmarsat.com/en/index.html
59. HoneyWell. HoneyWell. Accessed 23 March 2024 from https://www.honeywell.com/us/en
60. Centum. Lifeseeker Centum Research & Techonolgy. Accessed 23 March 2024 from https://www.centum-rt.com/

C-UAS in the Protection of Critical Infrastructures

Higinio González-Jorge, Luis Miguel González-deSantos, Enrique Aldao, and Gabriel Fontenla-Carrera

Abstract Democratic societies base their socioeconomic activities on a series of so-called critical infrastructures, which in the event of a catastrophic failure can make it unviable to live in cities. These infrastructures are often located in remote places and without any specific protection, turning them vulnerable to attacks. Currently, drones, although showing great applicability in civilian operations, also present a technological option not only for armed forces but also for terrorism and smuggling. Low-cost drones can load several kilograms of explosives and act as a loitering munition against critical infrastructure, making it relatively easy for a terrorist organization to carry out this type of action. This danger makes it necessary to develop defense systems such as the so-called Counter-Unmanned Aircraft Systems (C-UAS), which can detect and identify an aerial threat, and then decide the type of interception (soft or hard kill). This work develops a review of the different C-UAS technologies on the market, their strengths, weaknesses, and main operational capacities. It is also intended to serve as a promoter of the technological opportunity that is opening up for companies developing this type of technology.

Keywords C-UAS · Critical infrastructure · Loitering munition · Drone

1 Introduction

Democratic societies are based on respect for citizens and property, public or private. In these countries, there are many infrastructures, owned by the government or by private companies, which are key for the socioeconomic development and well-being of the population. Due to this importance, they are called critical infrastructures [1], being some examples power plants, water systems, rail transport, banking systems, satellite technology, airports, telecommunications, and government systems.

H. González-Jorge (✉) · L. M. González-deSantos · E. Aldao · G. Fontenla-Carrera
Institute of Physics and Aerospace Science, Campus of Ourense, University of Vigo, 32004 Ourense, Spain
e-mail: higiniog@uvigo.gal

D. Carou et al. (eds.), *Applying Drones to Current Societal and Industrial Challenges*, Management and Industrial Engineering, https://doi.org/10.1007/978-3-031-55571-8_5

The European Directive 2008/114/EC of December 8, 2008, establishes as critical infrastructure "The element, system or part thereof located in the Member States which is essential for the maintenance of vital societal functions, health, physical integrity, safety, security, and the social and economic well-being of the population, the disruption or destruction of which would seriously affect a Member State in failing to maintain those functions" [2].

At the national level, each state transposes the European regulations. In the case of Spain, the National Center for Infrastructure Protection and Cybersecurity (CNPIC) [3] provides the following useful definitions for essential service, critical infrastructures, and strategic infrastructure:

- *Essential Service*: It encompasses the infrastructures necessary for providing basic social functions such as health, security, and citizen economic welfare, as well as ensuring the operation of State Institutions and Public Administrations.
- *Critical Infrastructure*: These are strategic infrastructures, which provide essential services and whose operation is vital. The disruption of their activities would have a great impact on essential services, as there are no alternative solutions to replace or compensate for their functions.
- *Strategic Infrastructure*: It covers all the facilities, technologies, and systems that enable the operation of essential services.

Critical infrastructures are all those physical or digital entities that provide basic functions and services to support essential socioeconomic, environmental, and political systems. Disruptions or interruptions in their operation [4, 5] due to natural causes (e.g., a wildfire or hurricane affecting an airport) or man-made causes (e.g., a terrorist attack on a power plant) can have serious consequences. Threats to these infrastructures would affect any state by preventing the normal continuation and development of the basic activities of society. However, the problem worsens when critical infrastructures are interdependent [6, 7]. The failure of one critical infrastructure would mean the shutdown or impairment of the services of both, which makes the protection of these infrastructures even more important. Even though critical infrastructures are similar in most countries, their management depends on many factors such as the population needs, economic resources, and level of development of each country.

Information on critical infrastructure inventory and location is classified as secret due to the high sensitivity for national security. As an example, Spain has more than 3,500 recognized critical infrastructures. The detailed strategic areas are next described:

- Financial and Tax System (e.g., banking entities, information, securities, and investments).
- Administration (e.g., basic services, facilities, information networks, main assets, and national heritage monuments).
- Water (e.g., reservoirs, storage, treatment, and networks).

- Food (e.g., production, storage, and distribution).
- Power plants and networks (e.g., production and distribution).
- Installations related to Outer Space.
- Nuclear power plants (e.g., production, storage, and transport of dangerous goods, nuclear and radiological materials).
- Chemical Industry (e.g., production, storage, and transport of dangerous goods, chemical materials).
- Research: laboratories that due to their idiosyncrasy have or produce critical or hazardous materials, substances, or elements.
- Health (health sector and infrastructure).
- Information and Communications Technologies (ICT, whether they are critical infrastructures in themselves, such as telecommunications networks, or provide information and communications services to other critical infrastructures).
- Transportation (airports, ports, intermodal facilities, railroads and public transport networks, traffic control systems).

Continuing with the Spanish example, the National Security Strategy [8] describes a map of risks and threats to National Security with an approach that highlights their dynamism and interdependence, in a security environment where hybrid strategies are gaining prominence. Therefore, it identifies as one of the components of this map the threats to critical infrastructures, insofar as they enable the normal development of socioeconomic activity and are the target of threats, both physical and digital [9], which could lead to an interruption or denial of services. The National Critical Infrastructure Protection Plan (PNPIC) [10] is a state programming instrument whose main purpose is to keep Spanish infrastructures that provide essential services to society safe. To this end, it contains a series of measures to prevent and protect critical infrastructures from all threats that may affect them. It is a classified document registered with the National Center for the Protection of Critical Infrastructure (CNPIC) and was approved by the Secretary of State for Security in May 2007. Its main objectives include the following:

- It obliges all critical infrastructures to have a security manager to liaise with the National Center for the Protection of Critical Infrastructures if necessary.
- It promotes the exchange of information among the companies and institutions affected by this classification.
- It defines the measures to be implemented by critical infrastructures in the event of being victims of an attack.
- It includes the opinions and directives followed by the State to mobilize its operational capabilities in the event of deliberate attacks, as well as the coordination mechanisms established with critical operators in these cases.

Threats that could jeopardize Spain's vital and strategic interests have increased in recent years, and this is applicable to any democratic country. Security and government officials are concerned about the threats and vulnerabilities to which critical infrastructures are exposed: terrorism, organized crime, weapons of mass destruction proliferation, espionage, natural causes, and cyberspace, maritime space, terrestrial and outer space vulnerabilities [11–13]. Focusing on the last point, it is possible to observe that airspace can be compromised by state and non-state actors. Examples are actions against commercial aviation, navigation control systems, and illicit traffic. Drones represent a security threat as they can be exploited for malicious purposes such as espionage or attacks on individuals and critical infrastructure. It is therefore essential to have a deep understanding of this threat to develop effective response protocols. This problem motivates the present study, which reviews the different existing drones as a threat to critical infrastructures and the different existing technologies for their detection and identification, called counter unmanned aircraft systems (C-UAS).

This work is structured as follows: Sect. 2 is focused on drones as a threat to critical infrastructures, Sect. 3 depicts the counter-unmanned aircraft systems technology, Sect. 4 exhibits a discussion of the topic, and Sect. 5 the conclusions.

2 Drones as a Threat to Critical Infrastructures

Critical infrastructures are vital for the development of socio-economic activities in a country as shown in the introduction and must be protected in a special way. Figure 1 shows examples of critical infrastructures in Spain. Many critical infrastructures do not have special surveillance systems. This is mainly due to the fact that in peacetime they are hardly ever attacked, and installing these systems presents high costs. However, the lack of surveillance, coupled in many cases with their remote location, and the current escalation of warfare since the conflict in Ukraine, makes it worthwhile to change this paradigm and start worrying more about their surveillance. In addition, there are new threats such as drones [14–18], whose different versions allow attacks on this type of infrastructure. Figure 2 shows a division of the different drone threats and how they can act as a tool for damaging critical infrastructures. In this section, the same scheme will be followed in each subsection, focusing on three existing products on the market. The characteristics that best define the systems will be summarized in a table.

2.1 Extended Range Drones

Extended range drones are defined as those that can operate at distances of several tens of kilometers. In a war situation, this allows one to cross a battle front and, in case of a terrorist act, it would allow attacking an infrastructure from a large distance. They are complex and expensive systems, which are mainly oriented to the military field, but there are also civilian solutions that could adapt a low-cost explosive warhead and attack a critical infrastructure. Due to the working ranges,

Fig. 1 Examples of critical infrastructures in Spain. Repsol refinery in A Coruña[1] (left) and Vandellós nuclear power plant[2] in Tarragona (right)

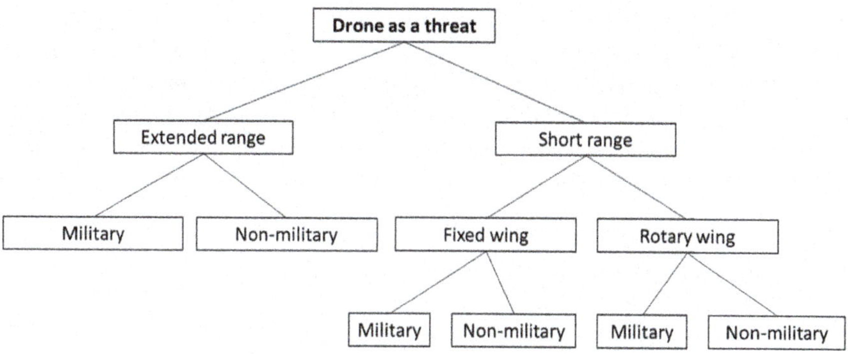

Fig. 2 Drone threat classification

these devices/artifacts are typically fixed wing systems. In addition, this facilitates flying at higher speeds with smaller sizes, which makes the use of countermeasures more difficult. In this work, the focus is given to drones acting as loitering munitions in the military field. Other systems in which the drone can carry a munition such as a guided missile are left aside. Therefore, some examples such as the well-known Reaper, Predator, Global Hawk, or Bayraktar TB2 are left out of this context and are not discussed in this paper [19].

2.1.1 Military Drones

Military drone data focuses on three systems with capabilities for extended range, the AeroVironment Switchblade 600 [20], HESA Shahed 136 [21] (Fig. 3), and IAI Mini Harpy [22]. All of them are quite well known, but especially the Switchblade 600 and the Shahed 136 due to their extensive use in the Ukrainian-Russian conflict.

[1] https://commons.wikimedia.org/wiki/File:Pastoriza,_Arteixo_Galiza_08b.jpg.
[2] https://www.flickr.com/photos/ckr84/4174434343.

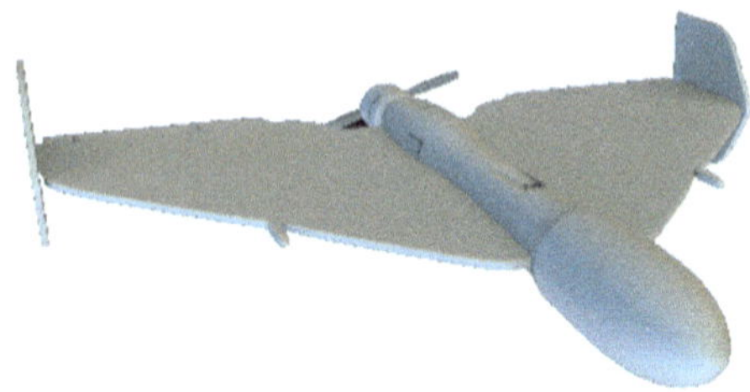

Fig. 3 Example of an extended range military drone: HESA Shahed 136[3]

The former is used mainly as an anti-tank weapon by Ukraine and the latter to attack critical infrastructures, such as the power grid, by Russia [23]. These systems have gained great usability in recent war conflicts because they can avoid traditional anti-aircraft defenses, and they are very low cost, which makes them suitable for mass use.

These drones are quite complex systems, since they use, for example, encrypted and long-range communications, and military-grade inertial navigation systems (Table 1). They could attack critical infrastructures, as is happening in case of conflict, but it does not seem feasible that they could be used by terrorist organizations. Since they are considered weapons, their sale is restricted to a limited number of customers.

2.1.2 Non-military Drones

Non-military extended range drones under study are Dronamics Blackswan [25], Deltair DT26 [26], and SCR Atlantic I [27]. Their technical specifications are shown in Table 2.

These systems are accessible to many users, and any terrorist or smuggler organization could acquire them. It is remarkable that their price is high, and their deployment is not easy as they require the use of catapults or runways. On the other hand, they are systems with action ranges of several tens of kilometers (between 100 km to 2500 km). In addition, some of them, like the Dronamics system, could carry a large amount of explosives or smuggling (350 kg). All these drones present the possibility of autonomous guidance based on GNSS waypoints and/or remote control, accompanied by payloads with observation systems that could help in target selection.

2.2 *Short Range Drones*

Short range drones are defined as those with a maximum operative range in the order of ten kilometers. In war situations, these units are not typically used to cross the front lines and conduct operations behind enemy lines. Instead, they are used to supporting close-range tactical attacks or defense operations, such as sporadic

[3] https://commons.wikimedia.org/wiki/File:Shahed_136_rendering.png.

Table 1 Main characteristics of military drones under study (extended range)

	AeroVironment Switchblade 600	HESA Shahed 136	IAI Mini Harpy
Developer country	USA	Islamic Republic of Iran	Israel
Endurance	40 min	10 h	2 h
Wingspan	No data (foldable)	2.5 m (flying wing type)	2.9 m (foldable)
Length	1.3 m	3.5 m	2.5 m
Flight altitude	150 m	No data	304–1524 m
Operational range	90 km	1,800–2,500 km	100 km
Guidance system	Autonomous (GPS, INS) Manual	Autonomous (GLONASS, INS)	Autonomous (GPS, INS) Manual
Speed (maximum)	113 km/h	185 km/h	370 km/h
Maximum take-off weight	22.7 kg	200 kg	40 kg
Warhead	Tandem-charge High-explosive anti-tank [24]	Explosive (impact-detonated) warhead	WDU-17/B
Warhead weight	8.4 kg	30–50 kg	8 kg
Engine	Electric	MD-550 piston	Electric
Launch platform	Portable tube Multipack	Rocket Assisted Take Off	Canister
Unit cost	€30,000–€40,000	€10,000–€50,000	n/a
Imagery sensors	High-resolution EO/IR gimbaled sensors	n/a	Dual EO/IR and anti-radiation seeker

attacks on the front line or to cover a retreat. Their ammunition-carrying capacity is less than that of high-range systems. They are very dangerous from the point of view of terrorist actions since many civilian systems could be converted to a weapon system with modifications that are within the reach of many organizations. Due to the great variety of existing systems in this field, the equipment has been classified into fixed and rotary wing systems, military, and non-military.

2.2.1 Fixed Wing Military Drones

Fixed wing military drones under study are three loitering munitions: AeroVironment Switchblade 300 [28], Uvision UAV Hero-30 [29], and STM Alpagu [30]. Models are illustrated in Fig. 4 and their technical specifications are shown in Table 3.

Table 2 Main characteristics of non-military drones under study (extended range)

	Dronamics Blackswan	SCR Atlantic I	Delair DT26
Developer country	Bulgaria	Spain	France
Endurance	12 h	8 h	135 min
Wingspan	16 m	3.8 m	3.3 m
Length	8 m	2.8 m	1.6 m
Flight altitude	6000 m	3500 m	n/a
Operational range	2500 km	100 km	135 km
Guidance system	Autonomous (GNSS, INS) Manual	Autonomous (GNSS, INS) Manual	Autonomous (GNSS, INS) Manual
Speed	200 km/h	120 km/h (cruise) 170 km/h (maximum)	60 km/h (cruise)
Maximum take-off weight	n/a	55 kg	17 kg
Payload weight	350 kg	5 kg	3.5 kg
Engine	Piston	Piston	Electric
Launch platform	n/a	n/a	Catapult
Unit cost	n/a	n/a	€300,000
Imagery sensors	EO/IR, RADAR (included as payload)	EO/IR (included as payload)	EO/IR (included as payload)

Fig. 4 Fixed wing military drones (short range). AeroVironment Switchblade 300[4] (left), Uvision UAV Hero-30[5] (center), STM Alpagu[6] (right)

These systems, as well as those of high range, have the property of being specifically designed as a weapon to carry an explosive warhead. They also have other characteristics such as rapid deployment, the use of image sensors, and encrypted communications. Due to their size, materials of construction, flight altitude, and speed, they would be very difficult to detect by a defense system. These drones, intended as

[4] https://commons.wikimedia.org/wiki/File:Switchblade_300_in_flight_(200902-M-EU630-1102).jpg.

[5] https://commons.wikimedia.org/wiki/File:Uvision_Hero-30_ILA-2022.jpg.

[6] https://commons.wikimedia.org/wiki/File:STM_Alpagu_Tactical_Attack_UAV,_Kyiv,_2019_02.jpg.

Table 3 Main characteristics of fixed wing military drones under study (short range range)

	AeroVironment Switchblade 300	Uvision UAV Hero-30	STM Alpagu
Developer country	USA	Israel	Turkey
Endurance	15 min	30 min	15 min
Wingspan	n/a	n/a	1250 m
Length	49.5 cm	80 cm	65 cm
Flight altitude	152 m	3100 m	400 m
Operational range	10 km	40 km	10 km
Guidance system	Autonomous (GPS, INS) Manual	Autonomous (GPS, INS) Manual	Autonomous (GPS, INS) Manual
Speed	101 km/h (cruise) 161 km/h (maximum)	100 km/h (cruise) 160 km/h (maximum)	58 km/h (cruise) 80 km/h (maximum)
Maximum take-off weight	2.5 kg	3 kg	3.7 kg
Warhead	Anti-personnel	Anti-personnel	Anti-personnel
Warhead weight	n/a	0.5 kg	0.5 kg
Engine	Electric	Electric	Electric
Launch platform	Portable tube Multipack	Portable tube Multipack	Portable tube Multipack
Unit cost	€6,000	n/a	n/a
Imagery sensors	EO and IR cameras	EO and IR cameras	EO and IR cameras

loitering munitions, are distinguished by being hardly accessible to terrorist organizations. Moreover, their warheads are fundamentally designed as an anti-personnel munition, not to attack armored vehicles or infrastructures. From the perspective of this work, they do not seem to be systems to be worried about, except in case of warlike conflict.

2.2.2 Rotary Wing Military Drones

Rotary wing military drones under study are three loitering munitions: IAI Rotem L [31], Rafael Firefly [32], and STM Kargu [33]. These models are illustrated in Fig. 5 and their technical specifications are shown in Table 4.

These systems are intended for urban warfare scenarios, providing observation and attack capabilities. They also present encrypted communications, and allow, for example, to attack an enemy entrenched in a building by entering through a window. They are classified as anti-personnel weapons and are not suitable for attacking critical infrastructure. On the other hand, being a military grade system, their sales are limited and their access to criminal organizations is difficult.

Fig. 5 Rotary wing military drones (short range). IAI Rotem[7] (left), Rafael Firefly[8] (center), STM Kargu[9] (right)

Table 4 Main characteristics of rotary wing military drones under study (short range range)

	IAI Rotem L	Rafael Firefly	STM Kargu
Developer country	Israel	Israel	Turkey
Endurance	30 min	15 min	30 min
Flight altitude	305 m	n/a	3000 m
Dimensions	n/a	80 × 80 × 400 mm (LxDxH)	707 × 707 × 409 mm (LxDxH)
Operational range	10 km	1000 m	10 km
Guidance system	Autonomous (GPS, INS) Manual	Autonomous (GPS, INS) Manual	Autonomous (GPS, INS) Manual
Speed	93 km/h	70 km/h	72 km/h
Maximum take-off weight	5.8 kg	3 kg	7.8 kg
Warhead	2 × M-67 hand grenades	Omnidirectional fragmentation warhead	Proximity fuse type; 840 pieces 4 mm cube
Warhead weight	1.2 kg	0.35 kg	1.3 kg
Engine	Electric	Electric	Electric
Launch platform	Vertical Take-off / Landing	Vertical Take-off/ Landing	Vertical Take-off/Landing
Unit cost	n/a	n/a	n/a
Imagery sensors	EO/IR cameras	EO/IR cameras	EO/IR cameras

[7] https://commons.wikimedia.org/wiki/File:IAI_Rotem-L_Loitering_Munition_at_ADAS_2018.jpg.

[8] https://commons.wikimedia.org/wiki/File:SPIKE_IMG_7898.jpg.

[9] https://commons.wikimedia.org/wiki/file:stm_kargu.png.

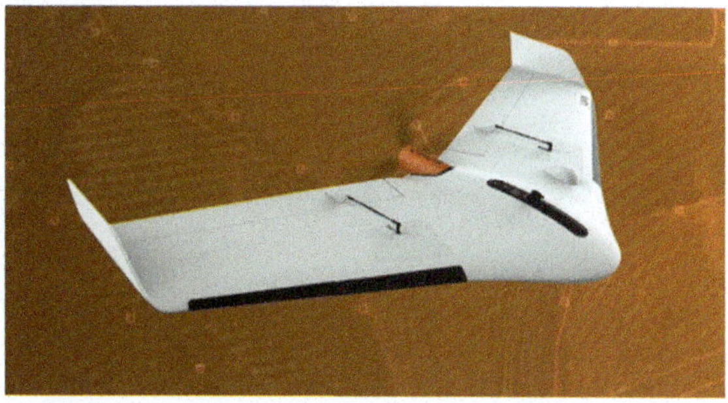

Fig. 6 Example of a fixed wing non-military drone (short range): Delair UX11[10]

Table 5 Main characteristics of fixed wing non-military drones under study (short range)

	UAV Instruments Cies	Delair UX11	SCR Tucan
Developer country	Spain	France	Spain
Endurance	2 h	50 min	90 min
Wingspan	2.2 m	1.1 m	2.74 m
Length	0.73 m	n/a	1.44 m
Flight altitude	n/a	6000 m	n/a
Operational range	15 km	53 km	25 km
Guidance system	Autonomous (GNSS, INS) Manual	Autonomous (GNSS, INS) Manual	Autonomous (GNSS, INS) Manual
Speed	50 km/h (cruise)	54 km/h (cruise)	65 km/h (cruise) 100 km/h (maximum)
Maximum take-off weight	4 kg	1.5 kg	5 kg
Payload weight	500 g	200 g	750 g
Engine	Electric	Electric	Electric
Launch platform	Catapult	Hand-launched	Catapult
Unit cost	€12,000	€15,000	n/a
Imagery sensors	EO/IR cameras (option)	EO camera (option)	EO/IR cameras (option)

2.2.3 Fixed Wing Non-military Drones

Fixed wing non-military drones under study are UAV Instruments Cies [34], Delair UX11 [35] (Fig. 6), and SCR Tucan [36]. Technical specifications are shown in Table 5.

These systems are affordable and open for purchase to any user, which could make them accessible to any terrorist organization. However, the main drawback is that their maximum payload weight is less than 1 kg in all the studied cases. If

[10] https://www.quadricottero.com/2020/06/delair-riceve-la-prima-autorizzazione.html.

Fig. 7 Rotary wing non-military drones (short range). DJI Agras T30[11] (left), Yamaha R-max[12] (right)

comparing them with the warheads of military equipment, they could be considered anti-personnel weapons, but it seems unlikely that they would be powerful enough to cause significant damage to critical infrastructures.

2.2.4 Rotary Wing Non-military Drones

Rotary wing non-military drones under study are DJI Agras T30 [37] (Fig. 7, left), Multirotor Cargo Robot [38], and Yamaha R-max [39] (Fig. 7, right). Their technical specifications are shown in Table 6.

This category shows civil systems accessible to any buyer without any restriction. Depending on the size, these systems can carry several tens of kilograms of payload (e.g., explosive/weapon) and extend their range of operation to several kilometers. These can be especially useful from the point of view of attacks on critical infrastructures by terrorist organizations due to the high payloads they can carry, their relatively low price, and the lack of limitations on the sale of the equipment. In addition, since these are rotary wing systems, no special conditions are required for take-off systems, such as catapults. This means that deployment can be very fast. From the point of view of critical infrastructure protection, the authors believe they are among the drones to be taken into account, especially those that can carry payloads above 10 kg, such as the DJI Agras T30 or the Yamaha R-max.

---

[11] https://www.quadricottero.com/2021/08/i-droni-per-lagricoltura-dji-agras-t30.html.

[12] https://commons.wikimedia.org/wiki/File:JAEA_YAMAHA_RMAX_G1_at_Fukushima_Sky_Park.jpg.

Table 6 Main characteristics of rotary wing non-military drones under study (short range range)

	DJI Agras T30	Multirotor cargo robot	Yamaha R-max
Developer country	China	Germany	Japan
Endurance	8 min	22 min	60 min
Flight altitude	n/a	n/a	n/a
Dimensions	2858 × 2685 × 790 mm (LxDxH)	Ø 85 cm, H 49 cm	3130 mm (main rotor); 535 mm (tail rotor) 2750 × 720 × 1080 mm (LxDxH)
Operational range	3 km	15 km	40 km
Guidance system	Autonomous (GNSS, INS) Manual	Autonomous (GNSS, INS) Manual	Autonomous (GNSS, INS) Manual
Speed	25 km/h	55 km/h	40 km/h
Maximum take-off weight	76.5 kg	7.5 kg	94 kg
Payload weight	30 kg	1.5 kg	16 kg
Engine	Electric	Electric	Piston; 2-stroke, horizontally opposed 2-cylinder
Launch platform	n/a	n/a	n/a
Unit cost	€16,000	n/a	€60,000
Imagery sensors	EO camera	EO camera	n/a

3 Counter Unmanned Aircraft Systems

Critical infrastructures can be attacked by different types of unmanned aircraft. The usual way to protect them against this type of threat is by means of the so-called C-UAS (counter unmanned aircraft systems) [40]. C-UAS actuation is divided into three main phases: detection, identification, and interception (multisensory multilayered concept). In the detection phase, technologies based on EO/IR, acoustic, RADAR, or LiDAR sensors, among others, are used [41]. They could be assembled in static form or in mobile platforms as a van/all-terrain vehicle. In the identification phase, signal processing algorithms are used to identify and classify the type of threat. The third and last phase, related to threat interception, is where direct action is taken against the target. This phase is divided into two main categories, called soft kill, when the countermeasures used are based, for instance, on the use of electromagnetic radiation; or hard kill, when, for example, munitions or high-energy lasers are used [42]. In this chapter, the attention is mainly given to the first two stages of a C-UAS.

3.1 Detection Phase in C-UAS

Several commercial C-UAS type systems are currently available. The following is a review of them, focusing on the types of sensors used in each. The cases under study are Northrop Grummann M-ACE [43], Indra ARMS [44], IAI Drone Guard [41], Kaspersky Antidrone [45], Thales Eagle Shield [46], and ESG Guardion [47]. Table 7 shows the main technical specifications and in Fig. 8 some examples are illustrated.

It can be concluded that since these are systems developed by military companies, the specifications shown are limited. For example, the sensitive range of the sensors, the algorithms used for target identification or classification, the relationship between the size and material of the target, the distance and the signature obtained on the sensor, among other characteristics, are not specified. However, it can be observed that all the solutions are based on multisensor approaches where RADAR systems (2D and 3D), RF analyzers and EO/IR imaging systems are combined in all of them. Some also use acoustic sensors, such as the ESG Guardion system and the Kaspersky Antidrone. The latter also integrates LiDAR sensors.

The following works depict, from a scientific outlook, the more relevant advances related to C-UAS technology:

Lykou et al. [48] show a classification of drones as threat in function of weight: micro drones (<2 kg), mini drones (2–25 kg), small drones (25–150 kg), medium drones (150–600 kg), and large/tactical drones (>600 kg). They also exhibit a summary of the main incidents of drones in airports between 2018 and 2019 (a total of 10 incidents located in the United Kingdom, Ireland, Germany, Singapore, United Arab Emirates, Japan, Canada, and USA (three)). They indicate the main detection sensors and technologies.

RADAR detection works either at 35 GHz or at 9.4 GHz to detect and track nearby drones, exhibiting significant results. Nevertheless, RADAR is an expensive equipment and requires national frequency spectrum licensing and environmental compatibility study.

Table 7 Technical specifications of different C-UAS available on the market

		Northrop Grummann M-ACE	Indra ARMS	IAI Drone Guard	Karpesky Antidrone	Thales Eagle Shield	ESG Guardion
Developer country		USA	Spain	Israel	Russia	France	Germany
Sensors	RADAR	Yes	Yes	Yes	Yes	n/a	Yes
	EO/IR	Yes	Yes	Yes	Yes	n/a	Yes
	RF	Yes	Yes	Yes	Yes	n/a	Yes
	LiDAR	No	No	No	Yes	n/a	No
	Acoustic	No	No	No	Yes	n/a	Yes
Detection range		n/a	n/a	n/a	n/a	7 km	n/a

Fig. 8 Some examples of C-UAS available on the market. IAI Drone Guard[13] (top), Kaspersky Antidrone[14] (bottom left), Northrop Grummann[15] (bottom right)

Radio Frequency (RF) scanners use passive detection technology and provide a cost-effective solution for detecting, tracking, and identifying drones based on their communication signature. However, validation of detection accuracy remains unverified in environments crowded with WIFI signals and other emitting smart devices.

Acoustic detection offers the capability to identify the audio pattern transmitted by the drone propellers, and it can be used for drone positioning and classification. Usually, a microphone captures a sound made by the drone and calculates the location using the time difference applying the arrival technique, while more sets of microphone arrays can be used for rough triangulation of drones.

Visual detection technologies, both in the visual and infrared spectrum, offer means for detecting and categorizing drones. They are not typically used as a primary detection source. Electro-optical sensors rely on visual signatures to detect the drone, while infrared sensors use a heat signature. They are often equipped with a high zoom capability to show small objects at large distances. This technology can classify

[13] https://logdef.blogspot.com/2020/07/drone-guard-la-contromisura-iai-ai.html.

[14] https://www.flickr.com/photos/e_kaspersky/50568875436/.

[15] https://desarrollodefensaytecnologiabelica.blogspot.com/2015/12/northrop-grumman-demuestra-la-utilidad.html.

and identify drones; however, the signature shows similarities with birds, and false positives are common.

LiDAR technology shows a high detection accuracy in the range of a few tens of meters. In addition, they provide accurate coordinates of the object. False positives and confusion with birds are also common.

Seng Tan et al. [49] classify/divide drones into five main groups depending on weight: Group 1 (0–9 kg.), Group 2 (9.5–25 kg), Group 3 (<599 kg), Group 4 (weight >599 kg and FL <180) and Group 5 (weight >599 kg and FL >180). They indicate the sensor systems typically found in C-UAS technology as RADAR, RF, acoustic, EO, and IR. Some C-UAS detect and identify the drone using the RADAR signature. This device can also be used for drone tracking. Unlike RF, RADAR signatures for UAS must be distinguished from those for birds. RF scanning supports the detection and geospatial location of drones based on the communication link frequencies. In some cases, drones can also be identified based on RF behavior. Engines from the drones can contribute to the acoustic signatures. EO cameras detect and identify the drone in the base of the visual signature, and the IR cameras use the heat signatures. The manuscript also indicates the different systems used for soft and hard kill interception: RF jamming, GNSS jamming, spoofing, dazzling, Laser, microwave, projectiles, or collision with another drone.

Castrillo et al. [50] classify the drones based on the weight, following the NATO methodology: Class I (<150 kg), Class II (150–600 kg), and Class III (>600 kg). They explain the sensing technologies with pros and cons:

Acoustic sensing. The detection range is higher than 150 m, with medium classification capability and low global feature characterization. It shows the possibility of moving close to the target and improving the identification task. On the other hand, it needs proper ego-noise cancellation due to the noise of the propellers.

Optical. Detection ranges higher than 150 m for the EO–VIS sensors, with high classification capability and high global feature characterization. In the case of IR sensors, the detection range is lower than 150 m, with low classification capability and low global feature characterization. Optical systems have the possibility to change the perspective and to operate close to the target with a higher resolution and better identification capabilities. However, computational requirements are high and need efficient video stabilization.

RF. Detection range higher than 150 m, with low classification capability and low global feature characterization.

LiDAR. Detection ranges between 50 to 150 m, with low classification capability and low global feature characterization. Possibility to move close to the target and improve the detection phase. Computational requirements are high.

RADAR. Detection ranges higher than 150 m, with medium classification capability and medium feature characterization.

Besada et al. [51] detail the specifications of some commercial C-UAS active RADAR: the ART Midrange 3D high-resolution surveillance RADAR, the ARMS FMCW RADAR, the HENSOLDT Xpeller Counter UAV solution, the Echodyne Counter-drone RADAR and the Ranger R8SS-D RADAR. They show a maximum detection range between 1.2 and 9 km, a minimum detection range around 10 m, range accuracy between 3 and 0.05 m, angular resolution between 0.8° and 6°, minimum detection velocity around 0.1 m/s, azimuth coverage between 120° and 360°. In relation to the passive detection RADARs, they show the specifications of the Doruk RADAR, with a detection range of 6 km, a minimum detection velocity of 0.2 m/s, an azimuth accuracy of 1°, azimuth coverage of 360°, range accuracy of 5 m, scanning rate of 90°/s, target tracks above 200, and clutter suppression >45 dB. Other systems such as the DedroneSensor RF-160 based on RF detection exhibit a detection rate between 1.6 and 6 km, with omnidirectional, passive detection and classification. This system also includes an audio detector with a detection range of 500 m, azimuth angle coverage between 10° and 90°, an audio spectrum between 0 and 96 kHz, and a microphone range between 50 and 80 m. The study is also composed of different detection systems based on video imagery: Axis Q6215-LE PTZ, Triton PT-Series HD, ARMS, and Spinel series. All systems show a horizontal field of view of 360°, vertical field of view of 20°, frame rate up to 2 Hz, spectral band between 3 and 12 m, image resolution between 3 and 30 Mpixel, and detection range around 400 m. In all cases, the final specifications of the products can be improved using sensor integration and data fusion.

3.2 *Identification Phase in C-UAS*

Bernardini et al. [52] research focuses on the parametrization in the time and frequency domain of environmental audio data to develop a machine learning based drone warning system that employs support vector machines to understand and recognize the drone audio fingerprint. The study includes preliminary experimental results.

Zhang et al. [53] emphasize the significance of drone classification. They employ the micro-Doppler signatures depending on the rotation of rotor blades. A dual-band RADAR classification scheme is proposed. It is based on spectrograms obtained by performing a short-time Fourier Transform on data collected by K-band and X-band RADAR sensors. Then, a principal component analysis (PCA) is applied to extract the features from the spectrograms. Finally, the classification results are obtained using a Support Vector Machine.

Unlu et al. [54] adopt a computer vision approach for the detection and classification of drones as threats. They used two-dimensional scale, rotation, and translation of invariant Generic Fourier Descriptor features and classified targets as a drone or bird by a neural network. For training the system, a large dataset composed of birds and drones sourced from open platforms, achieving a classification accuracy of 85.3%.

Diamantidou et al. [55] detect and classify potential drone threats based on a fusion neural network framework, merging features extracted from unimodal modules to enhance accuracy in deductions. The method is validated by thorough application to drone detection and classification tasks.

Al-Sa'd et al. [56] present work focused on constructing a database of different RF signals of various drones under diverse flight models. Then, they use three deep neural networks (DNN) to detect the presence of a drone, type, and flight mode. The DNN is validated using various metrics, with averaged accuracy ranging between 99.7 to 46.8%.

Djurek et al. [57] are focused on the data analytics of the noise emitted by quadcopters, which is connected to the changing revolutions per minute of the electric quadcopter motors. The suggested algorithm uses signal processing with a nonlinear function, which results in new spectral lines that correspond to rotational speed differences between the front and rear motors.

Leonardi et al. [58] compare various methods of utilizing micro-Doppler analysis across different stages of a multifunctional RADAR. These approaches include classic spectrogram-based classification, spectrum-based basification in which the received signal from the target is picked up after the moving target detector, and features-based classification where the received signal from the target undergoes the detection step, after which discriminating features are extracted and used as input to the classifier.

4 Discussion

First, it should be noted that the discussion will be focused on possible cases of attack on critical infrastructures through terrorist-type actions, where military equipment is most likely to be unavailable. On the other hand, the ease of deployment of the attack system must be contemplated. In this case, rotary wing systems have great advantages over fixed wing systems that need a launching system such as a catapult or a runway. This complexity means that in a terrorist act, it will be more complicated to use a fixed wing drone than a rotary wing drone. It seems that the group of systems to take into account is that of civilian rotary wing drones used for tasks such as cargo or spraying (see Sect. 2.2.4). They can be deployed from a van in a few minutes, with vertical takeoff and landing, and with a cargo capacity over than 10 kg of explosives. In addition, this equipment can be easily purchased on the market, which facilitates subsequent modification to integrate an explosive payload.

The aerial systems are shown in Sect. 2.2.4 are divided into two main groups: systems based on an internal combustion engine and systems based on electric motors. The former will produce a larger sound wave and will be easier to detect with C-UAS systems that integrate acoustic sensors, with their corresponding signal processing. In the case of electric systems, which are more silent, other types of sensors should prevail.

All the aircraft discussed in this section are above a certain size (more than 2.5 m in length), especially the DJI Agas T30 and the Yamaha R-max. This makes them detectable by a proper RADAR system, despite the fact that much of the materials used to manufacture them are plastics and carbon fibers. Furthermore, their approach speeds to the target are not too high, between 25 and 40 km/h, which means that there is a certain response time to activate a soft kill or hard kill countermeasure.

According to the bibliographic review carried out, RF scanner equipment seems to be very useful for detecting and identifying this type of system. However, it may happen that the system avoids radio communication with the pilot and navigates to the target completely autonomously using its inertial systems and sensor fusion, even in GNSS denied environment [59, 60]. In this case, in addition to RADAR, the literature shows how image sensors should be used, both active and passive, in the visible, infrared, or/and thermal spectrum. The thermal spectrum can be of special interest for detecting drones with combustion engines that have an intense heat source.

In the case of all systems based on imagery, it is of particular interest the use of computer vision-based techniques for automatic detection, tracking, and classification of the target. However, it must be taken into account that digital cameras may have problems such as sensor saturation due to incident solar radiation. Moreover, it is also easy to generate false positives due to spurious elements such as a bird. In this case, the application of algorithms related to the trajectory analytics must be considered, since a bird has very different flight dynamics than a quadcopter. This fact could allow its differentiation simply based on the coordinates of the flight. It could be of special interest to use cameras located in different spatial zones but at a known distance between them and timely synchronized. They could apply photogrammetry equations to detect and track the motion of an intruder drone in the airspace of critical infrastructure [61–63].

Nowadays, there are low-cost solid-state LiDAR systems on the market that can be utilized to detect, position, track, and classify targets at distances up to 150 or 200 m. However, there have not been many direct applications of these devices in C-UAS equipment yet. Only the Karpesky system, which integrates a LiDAR sensor, has been found on the market, although its characteristics are not specified. This seems to be a development path to be studied for C-UAS systems due to the large amount of work that has been done in recent years oriented to LiDAR data processing [64, 65].

5 Conclusions

This chapter provides an exhaustive review of the existing challenges for the protection of critical infrastructures against threats such as unmanned aircraft acting as loitering munition. The study reviews military and non-military threats from fixed and rotary wing systems. It concludes that the most dangerous systems from a terrorist action point of view are rotary wing with high payload capacity, which are typically used for spraying and cargo. They also allow carrying more than 10 kg of explosives/weapons and navigating autonomously to the target.

The different C-UAS solutions presented on the market are also under study and revised from a scientific point of view. The study does not address the elimination of the threat via hard kill or soft kill countermeasures. The review shows the need to use multi-sensor solutions to take advantage of the optimal working range and detection specifications of each sensor. In this way, most of the systems used a combination of RF sensors with RADAR, acoustic, and image systems, both in the visible and IR spectrum. The information obtained is processed using different algorithms, but this is a field of work in which there is undoubtedly still a high margin for improvement. Likewise, it is also interesting to experiment with the use of other remote sensing systems such as LiDAR, especially nowadays that solid-state sensors show low prices, robustness, easily deployable, and integrable with other sensing systems.

The accessibility of drone technology, due to its relatively low cost, is a dangerous threat to preventing terrorist activities. This makes it necessary to continue advancing in the development of C-UAS technology to curb possible misuse of this type of aircraft.

Acknowledgements Authors thank the Ministries of Universities and Science and Innovation for grants ref. FPU21/01176, PID2021-125060OB-I00, TED2021-129756B-C31, and University of Vigo for grant ref. 585507. University of Vigo financial support is given by the Recovery, Transformation and Resilience Plan of the European Union—NextGenerationEU.

References

1. Rinaldi, S. M., Peerenboom, J. P., & Kelly, T. K. (2001). Identifying, understanding, and analyzing critical infrastructure interdependencies. *IEEE Control Systems Magazine, 21*(6), 11–25.
2. Council Directive 2008/114/EC of 8 December 2008 on the identification and designation of European Critical Infrastructures and the assessment of the need to improve their protection. Accessed 21 March 2024 from https://www.ccn-cert.cni.es/publico/InfraestructurasCriticaspublico/DirectivaEuropea2008-114-CE.pdf
3. National Center for Infrastructure Protection and Cybersecurity. Accessed 23 March 2024 from https://cnpic.interior.gob.es/en/home/
4. Mitsova, D., Escaleras, M., Sapat, A., Esnard, A. M., & Lamadrid, A. J. (2019). The effects of infrastructure service disrruptions and socio-economic vulnerability on hurricane recovery. *Sustainability, 11*(1). Article number 516.
5. Eid, M., & Rosato, V. (2016). Critical infrastructure disruption scenarios analyses via simulation. *Studies in Systems, Decision and Control, 90*, 43–61.
6. Ouyang, M. (2014). Review on modeling and simulation of interdependent critical infrastructure systems. *Reliability Engineering and System Safety, 121*, 43–60.
7. Brown, G., Carlyle, M., Salmerón, J., & Wood, K. (2006). Defending critical infrastructure. *Interfaces, 36*(6), 530–544.
8. National Security Strategy. Accessed 21 March 2024 from https://www.boe.es/boe/dias/2021/12/31/pdfs/BOE-A-2021-21884.pdf
9. Ten, C. -W., Manimaran, G., & Liu, C. -C. Cybersecurity for critical infrastructures: Attack and defense modeling. *IEEE Transactions on Systems, Man, and Cybernetics Part A: Systems and Humans, 40*(4), 853–865.
10. National Critical Infrastructure Protection Plan. Accessed 23 March from https://cnpic.interior.gob.es/en/naic/
11. Stewart, M. G. (2011). Life-safety risks and optimization of protective measures against terrorist threats to infrastructure. *Structure and Infrastructure Engineering, 7*(6), 431–440.

12. Shafieezadeh, A., Cha, E. J., & Ellingwood, B. E. (2015). Adecision framework for managing risk to airports from terrorist attack. *Risk Analysis, 35*(2), 292–306.
13. Kaewunruen, S., Alawad, H., Cotruta, S. (2018) A decision framework for managing the risk of terrorist threats at rail stations interconnected with airports. *Safety, 4*(3). Article number 36.
14. Hassanalian, M., & Abdelkefi, A. (2017). Classifications, applications, and design challenges of drones: A review. *Progress in Aerospace Sciences,* 91, 99–131.
15. Page, J. M., & Williams, J. (2022). Drones, Afghanistan, and beyond: Towards analysis and assessment in context; *European Journal of International Security, 7*(3), 283–303.
16. Zhang, X., & Kusrini, K. (2021). Autonomous long-range drone detection systems for critical infrastructure safety; *Multimedia Tools and Applications, 80*(15), 23723–23743.
17. Lv, H., Liu, F., & Yuan, N. (2021). Drone detection by the drones's RF communication. *Journal of Physics: Conference Series, 1738*(1). Article number 012044.
18. Wolfendale, J. (2018) Technology as terrorism: Police control technologies and drone warfare. *Advances Sciences and Technologies for Security Applications.* 1–22.
19. Grant, R. (2007). The drone war. *Air Force Magazine, 90*(7), 36–40.
20. AeroVironment Switchblade 600. Accessed 12 December 2022 from https://www.avinc.com/tms/switchblade-600
21. HESA Shahed 136. Accessed 12 December 2022 from https://special-ops.org/hesa-shahed-136-kamikaze-drone/
22. IAI Mini Harpy. Accessed 12 December 2022 from https://www.iai.co.il/p/mini-harpy
23. What are kamikaze drones and why is Russia using them in Ukraine? Accessed 12 December 2022 from https://www.theguardian.com/world/2022/oct/17/russia-targets-ukraines-energy-grid-as-winter-approaches
24. The same used in the FGM Javelin. Accessed 12 December 2022 from https://missilethreat.csis.org/missile/fgm-148-javelin/
25. Dronamics Blackswan. Accessed 12 December 2022 from https://www.dronamics.com/
26. Deltair DT26. Accessed 12 December 2022 from https://delair.aero/delair-commercial-drones/dt26-open-payload/
27. SCR Atlantic I. Accessed 12 December 2022 from https://scrdrones.com/producto-item/atlantic-i/
28. AeroVironment Switchblade 300. Accessed 12 December 2022 from https://www.avinc.com/tms/switchblade
29. Uvision UAV Hero-30. Accessed 12 December 2022 from https://uvisionuav.com/portfolio-view/hero-30/
30. STM Alpagu. Accessed 12 December 2022 from https://www.stm.com.tr/en/alpagu
31. IAI Rotem L. Accessed 12 December 2022 from https://www.iai.co.il/p/rotem
32. Rafael Firefly. Accessed 12 December 2022 from https://www.rafael.co.il/press/elementor-6147/
33. STM Kargu. Accessed 12 December 2022 from https://www.stm.com.tr/en/kargu-autonomous-tactical-multi-rotor-attack-uav
34. UAV Instruments Cíes. Accessed 12 December 2022 from https://www.uav-instruments.com/wp-content/uploads/2020/03/Ficha-cies-2.2-SP-ES.pdf
35. Delair UX11. Accessed 12 December 2022 from https://delair.aero/delair-commercial-drones/professional-mapping-drone-delair-ux11/
36. SCR Tucan. Accessed 12 December 2022 from https://scrdrones.com/producto-item/tucan/
37. DJI Agras T30. Accessed 12 December 2022 from https://www.dji.com/es/t30
38. Multirotor Cargo Robot. Accessed 12 December 2022 from https://www.multirotor.net/en/products/aircraft/cargo-robot
39. Yamaha R-max. Accessed 12 December 2022 from https://www.yamahamotorsports.com/motorsports/pages/precision-agriculture-rmax
40. Michel, A. H. (2018). Counter-Drone Systems. Center for the Study of the Drone at Bard College, Feburary 20, 2018. Accessed 12 December 2022 from http://dronecenter.bard.edu/counter-drone-systems/

41. IAI Drone Guard. Accessed 12 December 2022 from https://www.iai.co.il/p/eli-4030-drone-guard
42. High Energy Laser Weapon. Accessed 12 December 2022 from https://www.raytheonintelligenceandspace.com/what-we-do/advanced-tech/lasers
43. Northrop Grummann M-ACE. Accessed 12 December 2022 from https://www.northropgrumman.com/what-we-do/land/counter-unmanned-aerial-systems-c-uas/
44. Indra ARMS. Accessed 12 December 2022 from https://counteruas.indracompany.com/en
45. Kaspersky Antidrone. Accessed 12 December 2022 from https://antidrone.kaspersky.com/en/
46. Thales Eagle Shield. Accessed 12 December 2022 from https://www.thalesgroup.com/en/markets/defence-and-security/air-forces/airspace-protection/counter-unmanned-aircraft-systems
47. ESG Guardion. Accessed 12 December 2022 from https://esg.de/en/solutions-services/air/counter-uas-system
48. Lykou, G., Moustakas, D., & Gritzalis, D. (2020). Defending airports from UAS: A survey on cyber-attacks and conuter-drone sensing technologies. *Sensors, 20*(3537), 1–40.
49. Tan, C. S., Van Bossuyt, D. L., & Hale, B. (2021). System analysis of counter-unmanned aerial systems kill chain in an operational environment. *Systems, 9*(79), 1–27.
50. Castrillo, V. U., Manco, A., Pascarella, D., & Gigante, G. (2022). A review of counter-UAS technologies for cooperative defensive teams of drones. *Drones, 6*(65), 1–36.
51. Besada, J. A., Campaña, I., Carramiñana, D., Bergesio, L., & de Miguel, G. (2022). Review and simulation of Counter-UAS sensors for unmanned traffic management. *Sensors,* 22(189), 1–30.
52. Bernardini, A., Mangiatordi, F., Pallotti, E., & Capodiferro, L. (2017). Drone detection by acousting signatura identification. In *IS and T International Symposium on Electronic Imaging Science and Technology,* 60–64.
53. Zhang, P., Yang, L., Chen, G., & Li, G. (2017). Classification of drones based on micro-doppler signatures with dual-band RADAR sensors. In *Progress in Electromagnetic Research Symposium,* 638–643.
54. Unlu, E., Zenou, E., & Riviere, N. (2018). Using shape descriptors for UAV detection. In *IS and T International Symposium on Electronic Imaging Science and Technology,* 2761–2766.
55. Diamantidou, E., Lalas, A., Votis, K., & Tzovaras, D. (2019). Multimodal deep learning framework for enhaced accuracy of UAV detection. *Lecture Notes in Computer Science,* 768–777.
56. Al-Sa'd, M. F., Al-Ali, A., Mohamed, A., Khattab, T., & Erbad, A. (2019). RF-based drone detection and identification using deep learning approaches: An initiative towards a large open source drone database. *Future Generation Computer Systems, 100*, 86–97.
57. Djurek, I., Petosic, A., Grubesa, S., & Suhanek, M. (2020). Analysis of a quadcopter's acoustic signature in different flight regimes; *IEEE Access, 8*(8954678), 10662–10670.
58. Leonardi, M., Ligresti, G., & Piracci, E. (2022). Drones classification by the use of a multifunctional RADAR and micro-doppler analysis. *Drones, 6*(124), 1–18.
59. Zahran, S., Mostafa, M. M., Masiero, A., Moussa, A. M., Vettore, A., & El-Sheimy, N. (2018). Micro-RADAR and UWB aided UAV navigation in GNSS denied environment. *International Journal of the Photogrammetry, Remote Sensing and Spatial Information Sciences—ISPRS Archives, 42*(1), 469–476.
60. Du, H., Wang, W., Xu, C., Xiao, R., & Sun, C. (2020). Real-time onboard 3D state estimation of an unmanned aerial vehicle in multi-environments using multi-sensor data fusion. *Sensors, 20*(3). Article number 919.
61. Sun, B., Edgar, M. P., Bowman, R., Vittert, L. E., Welsh, S., Bowman, A., & Padgett, M. J. (2013). 3D computational imaging with single-pixel dtectors. *Science, 340*(6134), 844–847.
62. Tola, E., Lepetit, V., & Fua, P. (2010). Daisy: An efficient dense descriptor applied to wide-baseline stereo. *IEEE Transactions on Pattern Analysis and Machine Intelligence, 32*(5), 815–830.
63. Yang, Q., Wang, L., Yang, R., Stewenius, H., & Nister, D. (2009). Stereo matching with color-weighted correlation, hierarchical belief propagation, and occlusion handling. *IEEE Transactions on Pattern Analysis and Machine Intelligence, 31*(3), 492–504.

64. Aldao, E., González-deSantos, L. M., Michinel, H., & González-Jorge, H. (2022). UAV obstacle avoidance algorithm to navigate in dynamic building environments. *Drones, 6*(1). Article number 16.
65. Aldao, E., González-deSantos, L. M., & González-Jorge, H. (2022). LiDAR based detect and avoid system for UAV navigation in UAM corridors. *Drones,* 6(8). Article number 185.

Transport of Light Parts for Logistics Supply in Industrial Manufacturing Plants by Means of UAV

Pedro Orgeira-Crespo, Guillermo Rey, Pablo Rouco Pousada, and Fernando Aguado-Agelet

Abstract Unmanned Aerial Vehicles (UAVs) have revolutionized various industries by offering innovative solutions for tasks ranging from surveillance to logistics. This chapter presents a comprehensive study on the development and deployment of a hovercraft UAV system tailored for efficient light part delivery within a manufacturing plant environment. The system leverages a combination of technologies including computer vision, electromagnetic tags, ultrasonic sensors, accelerometers, and gyroscopes for precise corridor navigation. The indoor localization aspect is addressed through the implementation of ArUco markers, fused with Ultra-wideband (UWB) tags. These visual markers enable the UAV to accurately determine its position and orientation in complex indoor settings. The integration of ultrasonic sensors enhances the system's capability to detect obstacles, enabling real-time adjustments to the flight path to ensure safe and reliable navigation. Additionally, the incorporation of the MPU6050 sensor provides crucial feedback for corridor navigation, allowing the UAV to maintain stability and orientation while maneuvering through tight spaces. Results demonstrate the successful localization accuracy and the ability to operate within confined pathways, mitigating the risk of disruptions in the manufacturing plant environment. The outcomes of this research contribute to advancing UAV-assisted logistics and operations within industrial settings.

Keywords UAV · Logistics · Positioning · Indoor navigation

P. Orgeira-Crespo (✉) · G. Rey · P. R. Pousada
Aerospace Area, Department of Mechanical Engineering, Heat Engines and Machines, and Fluids, Aerospace Engineering School, University of Vigo, Campus Ourense, 32004 Ourense, Spain
e-mail: porgeira@uvigo.es

F. Aguado-Agelet
Telecommunication Engineering School, atlanTTIC, University of Vigo, 36310 Ourense, Spain

D. Carou et al. (eds.), *Applying Drones to Current Societal and Industrial Challenges*, Management and Industrial Engineering, https://doi.org/10.1007/978-3-031-55571-8_6

1 Introduction

Unmanned Aerial Vehicles (UAVs) are present in all areas of people's lives, from their first military uses [1], to recreational uses for the public [2], through industrial uses [3]. Many studies have been developed over the last few years concerning specific problems (route analysis [4], increase in autonomy [5]) with a special emphasis on the literature on civil uses for a specific purpose, such as logistics distribution [6]. The expectation is the reduction of land transport for delivery, both in cities (to avoid traffic jams, pollution, and noise), as well as in remote areas (to deliver, for example, medicines to areas that have suffered natural disasters [7]).

Within the scope of the use of UAVs for logistics, there are different studies from different points of view. On the one hand, there are investigations based on route planning so that the vehicle can cover the deliveries [8], which include the division of the delivery in the initial delivery from a central location to different offices, to conclude the last-mile delivery. On the other hand, some projects deal specifically with last-mile delivery exclusively, focusing on decision-making from an operational point of view [9]. There is also a series of investigations with emphasis on the problems of route optimization, especially in the joint use of UAVs and trucks, considering the difficulty that arises in the interaction between these two means of transport [10].

Manufacturing industries are witnessing a paradigm shift in their operational strategies, driven by technological advancements. UAVs have emerged as a viable solution for addressing challenges related to efficiency, speed, and cost-effectiveness in intralogistics and material movement within manufacturing plants. This chapter explores the use of these vehicles for delivery within manufacturing plants and its implications for enhancing overall operational efficiency. UAVs offer multifaceted applications in manufacturing plants, including just-in-time material delivery, tool and spare part transportation, and inspection of hard-to-reach areas. These applications streamline supply chain management, reduce downtime, and enhance safety by minimizing human involvement in potentially hazardous tasks. UAVs can facilitate quick and accurate material movement, improving production line efficiency and overall productivity. The adoption of aerial vehicles for intra-facility logistics presents several key benefits. These vehicles enable real-time tracking of materials, reducing delays and enhancing production-planning accuracy. Automated delivery systems using UAVs can operate around the clock, minimizing production downtimes and expediting operations. Furthermore, the reduced need for traditional ground-based transportation contributes to a decrease in operational costs and carbon footprint. Despite their potential, the deployment in manufacturing plants comes with challenges. Technical limitations, such as payload capacity, flight range, and battery life, need to be addressed to ensure efficient and sustainable operations. Regulatory compliance and safety measures, especially in environments with humans and sensitive equipment, require rigorous attention. Moreover, integrating UAV operations into existing plant infrastructure and information systems can present compatibility and cybersecurity challenges.

There are (for a long time) well-studied technologies that minimize friction between two surfaces. An example of this can be the magnetic levitation train "Maglev" present in Japan [11]. Another way to minimize friction is to sandwich a low-viscosity fluid (air, in most cases) between the relevant surfaces that will have relative motion. This is the case of amphibious vehicles known as hovercrafts, capable of moving on any sufficiently regular flat, being therefore especially useful for the field of military transport (although also used at the service of civilians in coast guard operations, humanitarian aid, and tourism). Hovercrafts can constitute, precisely, a solution for the transport of light parts in the industry, replacing the small robots for the movement of merchandise that are already being used in the logistics field [12]. The standard of hovercraft is understood as that with a skirt and represents most ships of this type. The operation is simple: an impeller forces a large volume of air to the bottom of the vehicle and before the resistance imposed by the ground to the passage of air, it is retained, and the pressure increases. This increase in pressure in a sufficiently wide sand generates a force capable of lifting the ship [13]. To create a receptacle in which to confine the air is normally used what is known as a skirt that, in addition, must be made of a flexible and resistant material, capable of withstanding the cycles of swelling and deflating. The passage of air from the skirt to the bottom of the vehicle is possible thanks to a series of grooves and holes present on the inside of the skirt. Finally, the air exits to the outside through the space between the skirt and the ground created by the pressure of the same air that is evacuated throughout the periphery of the ship. Although the armed forces of Great Britain were the first to use hovercrafts with a war purpose (in the 60s, with the SR series), the Soviet Union further developed this type of fleet in the military field. The Soviet Union provided designs that cover from small ships used by coast guards to the largest hovercraft in the world, the Zubr LCAC class [14]. The United States for its part would buy the license of the SR-N5 through the Bell Company and sell the hovercraft under the name "Bell SK-5", which would be used as a test in the Vietnam War [14]. Soon, the opening of this type of mobility was transferred to the civilian sphere; only one year after confirming the viability of hovercrafts with the operation of the SRN1 in 1961, the first ship of this type, the VANA-3 (Vickers-Armstrong) would be designed and tested. The test consisted of taking passengers from one bank to another of the Solent (Southampton, England) in what would be the first commercial trip in hovercrafts in history. Years later, in 1965, the SRN6 would be making regular routes crossing that same strait with 38 people [15].

In parallel, France was trying to exploit this technology by creating the organization SEDAM (Société d'Etude et de Développement des Aéroglisseurs Marins) that would be able to design and create the N300, N102, and N500 hovercrafts. On the other side of the globe, the MV-PP10 of Mitsui Engineering and Shipbuilding, of Japanese origin, was the most successful model, making routes between the numerous islands of the archipelago until the year 2000 [16]. Like any other vehicle, hovercrafts also have a motorsport niche with a dedicated hobby. Faced with this interest on the part of individuals, some companies manufacture (with professional and competitive techniques) small hovercrafts, already assembled or to be assembled by the buyer. The competitions, mainly present in Europe, are not sponsored, and are funded solely by competitors and the public interested in seeing drivers reach speeds of up to 150 km/h.

The specific type of hovercraft to be analyzed will be the case of "skirt-less", which lacks inflatable structures (skirt) to achieve the intermediate fluid layer between the ship and outer surface that allows movement. Although there is (to the best of the researcher's knowledge) no active skirt-less hovercraft (except for small experiments carried out and financed by model airplane enthusiasts) curiously, the first hovercraft lacked a skirt [15]. The Swedish philosopher and scientist, Emanuel Swedenborg, had already devised the concept of generating pressure to reduce the drag of a ship sailing centuries ago, but it would only remain as a concept to be tested [17]. The British engineer Christopher Cockerell invented the SRN1, which was considered by many to be the first practical hovercraft, after convincing the relevant authorities of the interest in the amphibious vehicle [18]. After the cancelation of the project to build an interceptor fighter (the SR-177), the Sanders-Roe company showed special interest in the nautical world. In parallel, Cockerell was carrying out different tests of a different fleet concept than the one existing to date, trying to somehow integrate emerging technologies such as centrifugal fans and the use of the air cushion as a friction-minimizing medium: a hovercraft [18].

Hovercraft platforms provide UAVs with the ability to maneuver over diverse terrains, making them suitable for applications requiring versatile movement. However, ensuring stable flight control for UAVs using hovercraft technology demands a deep understanding of the complex dynamics governing their behavior. Controlling the stability of hovercraft-based UAVs poses a significant challenge due to complex aerodynamic interactions with the ground. The ground effect, which causes a change in aerodynamic forces as the vehicle approaches the ground surface, influences lift, drag, and control authority. This effect becomes more pronounced as the vehicle descends closer to the surface, potentially causing unintended roll, pitch, or yaw motions. Hovercraft-based UAVs require tailored control strategies to account for the specific challenges introduced by ground effects and aerodynamic interactions. Conventional control algorithms may need significant modifications or even redesign to accommodate the dynamic behavior during the transition from hovering to flight. Adaptation mechanisms must be implemented to mitigate instability arising from abrupt changes in aerodynamic forces during takeoff and landing. Maintaining stability necessitates accurate sensing and rapid feedback loops, although sensors may encounter inaccuracies due to turbulent airflows near the ground. Integration of sensors that are robust to these conditions is crucial for ensuring reliable stability control and preventing potential crashes caused by erroneous sensor readings.

The proliferation of UAVs has extended their deployment into indoor environments where GPS signals are often unavailable. Ultra-wideband (UWB) technology presents a promising solution to address the challenges of indoor navigation by offering high-precision ranging and positioning capabilities [19]. UWB technology utilizes short-duration, low-power pulses with extremely wide frequency bandwidth to accurately measure time-of-flight (ToF) between transmitted and received signals. This principle enables UWB systems to achieve centimeter-level ranging accuracy, making them well-suited for indoor positioning applications. UWB technology holds the promise of revolutionizing indoor UAV navigation, enabling precise and reliable localization in GPS-denied environments [19]. As technology has advanced and challenges have been addressed, UWB is playing a pivotal role in expanding the capabilities of UAVs for indoor applications [19]. UWB-based positioning relies on precise distance measurements between a UAV and multiple ground-based anchors. By trilateration of these distances, the UAV's position can be accurately determined. UWB's immunity to multipath effects and ability to penetrate obstacles contribute to its accuracy in indoor environments where signal reflections are common. UWB-based indoor navigation offers several benefits, including high accuracy, low latency, and minimal susceptibility to environmental factors. These advantages make UWB an ideal choice for applications requiring precise UAV localization, such as search and rescue operations, warehouse inventory management, and indoor inspections. Real-world implementations of UWB-based indoor navigation for UAVs demonstrate its effectiveness in enhancing positioning accuracy, thus this technology can enable autonomous flights in complex indoor environments. Case studies highlight successful deployments across various industries, displaying the potential impact of UWB technology [20]. Despite its potential, UWB-based indoor navigation presents challenges. Signal propagation characteristics, interference from other wireless devices, and the need for anchor calibration are critical considerations. Furthermore, integrating UWB technology into UAVs requires addressing power consumption, hardware limitations, and algorithm complexities.

The rest of the chapter is organized as follows: Sect. 2 describes the materials and methods for the design of a hoverboard UAV; this will include not only the design of the structure of the UAV, but also the indoor localization system using computer vision fused with UWB tags, and the control of the motion of the hoverboard. Section 3 displays the results, and Sect. 4 depicts the conclusions.

2 Materials and Methods

2.1 Environment

Figure 1 depicts the overall floor plan arrangement employed for testing within this project's scope. The UAV flight corridor is notably positioned at an approximate height of 1.5 m, enclosed by a safeguarding net, thoughtfully installed to preemptively mitigate any potential risks associated with drone-related incidents. Ensuring safety is a priority, the UAV's flight path conforms to the perimeter of the facility, adopting a rectangular trajectory that aligns with the designated flight aisle. Additionally, to fulfill immediate operational demands, the UAV approaches the operator's workstation exclusively when an urgent requirement arises for a specific component.

Navigation is a key capability that needs sensing, determination of positioning, a plan for the motion, and the implementation of the motion itself. To face navigation, the three questions are: "What is my location?", "where should I go", and "How to

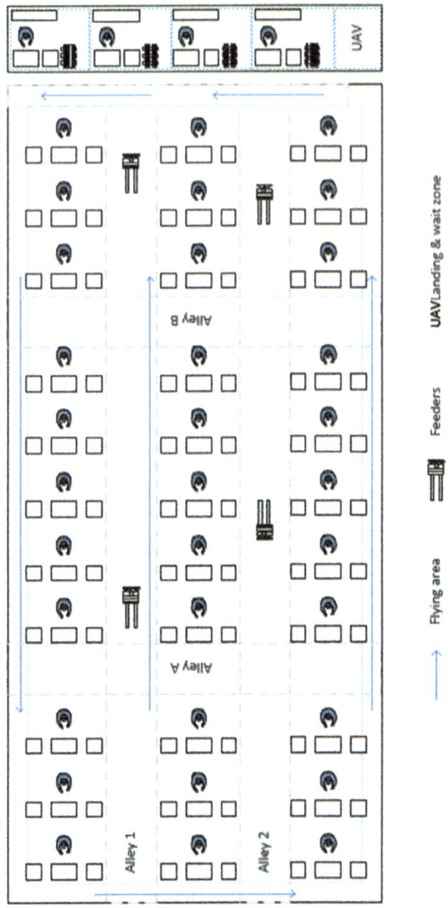

Fig. 1 Manufacturing plant for the indoor logistics scenario

get there?" In many cases, there is also a need to address the environment and check mission requirements, to avoid potential obstacles while succeeding in the whole mission.

UAV navigation has been very intense in the literature since the early 2000s, first from the point of view of outdoor applications [21], thanks to the availability of GNSS services like GPS. Early, the interest increased in indoor and other GNSS-denied scenarios, which needed a different paradigm to implement navigation. As in outdoor scenarios, indoor navigation needs state information (sometimes already present before a flight, like a map of the facilities and an initial state of the UAV, and sometimes obtained during the execution of the flight, like the determination of the distance with regards walls in a corridor). Positioning needs the determination of the state during the whole flight and/or initial state information. For the UAV to plan its trajectory, indoors is more challenging than outdoors, since the number of restrictions is greater, and the stabilization of the aircraft gets complicated because of wall effects.

2.2 Visual Localization

Selecting the right sensors to achieve autonomous navigation is a complex task since it provides the initial data and is a milestone for the choice of the solution algorithm. Sensor selection criteria include dynamic range, error characteristics, bandwidth, vehicle payload constraints, resolution, response time, operating environment, flight settings, computational resources, and power availability. Throughout the literature, vision sensors have been applied to localization, and mapping, translational velocity estimation, obstacle detection, attitude estimation, and landing pad detection and landing [22].

Visual localization utilizes vision sensors and computer vision to extract location information from visual scenes. Some advantages include the low weight and power consumption of vision sensors, and the rich environmental representation, which includes shapes, color, and texture that can be achieved. However, several drawbacks must be considered, for instance, vision algorithms are affected by illumination variations, low feature density, and motion blurring and consume relatively high computational resources [22].

ArUco markers, a type of fiducial marker, have gained prominence as a cost-effective solution for enabling precise indoor navigation of UAVs. Since indoor navigation of UAVs presents unique challenges due to the absence of GPS signals, ArUco markers offer a visual solution by enabling precise localization. ArUco markers are 2D binary patterns that can be easily detected and recognized by computer vision algorithms. Pose estimation involves calculating the position and orientation of a camera relative to a detected ArUco marker. This process requires minimal computational resources and facilitates real-time navigation, using the ArUco library.

ArUco is an OpenSource library for camera pose estimation using squared markers. ArUco is written in C++. Its main features include: it detects markers with a single line of C++ code, and detects various dictionaries (ARUCO, AprilTag, ArToolKit+, ARTAG, and CHILITAGS); it is faster than any other library for detection of markers; it has few dependencies OpenCV (≥2.4.9) and eigen3 (included in the library); it has a trivial integration with OpenGL and OGRE; it is a fast, reliable, and cross-platform solution (Windows, Linux, Mac OS, Android) [23, 24].

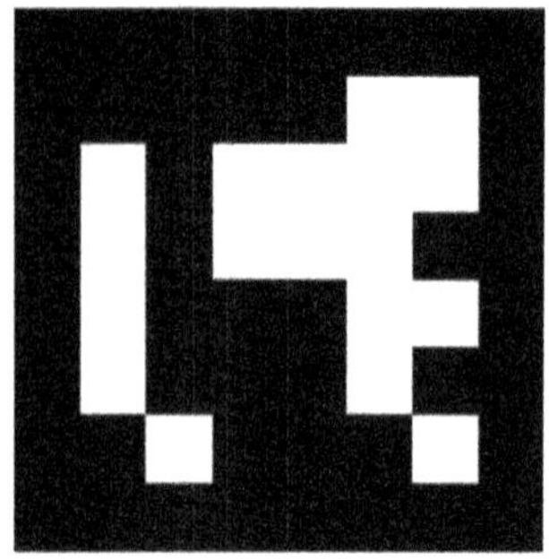

Fig. 2 ArUco marker in a 7 × 7 dictionary, with a 100 mm size

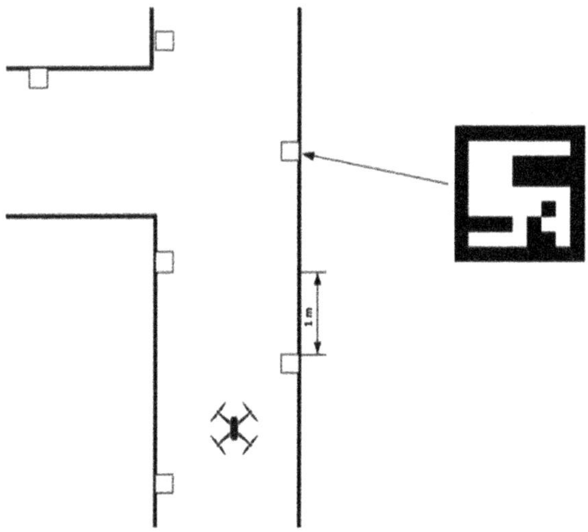

Fig. 3 Position of visual markers along navigation corridors

Rafael Muñoz and Sergio Garrido at the University of Córdoba [23] originally developed the ArUco bookstore. It is based on OpenCV and allows the detection of various types of tags. Its detection process is much more effective and faster than other libraries, which makes it especially suitable for real-time motion analysis applications. ArUco markers are square markers composed of a wide black border with an internal binary matrix, as can be seen in Fig. 2. The black border allows a quick detection of the image, and the binary coding, allows the identification of markers, and the application of error detection and correction techniques [24].

In this case, a 7 × 7 ArUco marker dictionary has been used, displaying markers at one meter, as shown in Fig. 3.

Fig. 4 Raspberry Pi camera module

For the capture of the markers, two Raspberry Pi camera modules have been used. ArUco markers, coupled with onboard cameras, offer a cost-effective solution for precise visual navigation of Unmanned Aerial Vehicles. The Raspberry Pi camera module offers a compact and accessible camera solution for UAVs; the setup is quite simple, and it just involves connecting the camera module to the Raspberry Pi board, configuring camera settings, and enabling the necessary interfaces. By capturing images using the Raspberry Pi camera, markers can be identified and localized within the image frame. By knowing the location of the markers within the navigation corridors of the factory and recognizing the marker on whose side you are navigating, the position of the UAV can be determined in a computationally short time. The detected ArUco markers provide the necessary information to estimate the UAV's position and orientation relative to the markers. The obtained pose estimates are integrated with the UAV's control system for navigation. By updating the UAV's position and orientation based on marker detection, the UAV can follow a desired trajectory and navigate within the environment. Figure 4 displays a Raspberry Pi camera module before integration into the UAV platform.

The implementation requires the OpenCV installed on the embarked Raspberry Pi, and a simple constant reading of the video stream, to analyze when a marker is present, and to identify which one is it. Listing 1 provides a simplified version of the implementation.

```
#include <opencv2/opencv.hpp>
#include <opencv2/aruco.hpp>
#include <iostream>

using namespace cv;
using namespace std;

int main() {
   // First, we initialize the camera
   VideoCapture cap(0); // 0 for the first camera, 1 for the other

   if (!cap.isOpened()) {
      cerr << "Error accessing the camera." << endl;
      return -1;
   }

   // Using the marker
   Ptr<aruco::Dictionary> dictionary = aruco::getPredefinedDictionary(aruco::DICT_7X7_100);

   while (true) {
      Mat frame;
      cap >> frame; // we get a frame from the camera

      // Marker detection
      vector<int> markerIds;
      vector<vector<Point2f>> markerCorners;
      aruco::detectMarkers(frame, dictionary, markerCorners, markerIds);

      if (markerIds.size() > 0) {
         // Obtaining ID
         for (size_t i = 0; i < markerIds.size(); ++i) {
            cout << "Detected marker with ID: " << markerIds[i] << endl;
         }

         // displaying a bounding box
         aruco::drawDetectedMarkers(frame, markerCorners, markerIds);
      }

      // we now show the frame
      imshow("Marker detected ", frame);

      // how to get out
      if (waitKey(1) == 27) {
         break;
      }
   }

   cap.release();
   destroyAllWindows();

   return 0;
}
```

Listing 1 Simplified implementation of the marker identification

2.3 UWB

The Decawave DW1000 UWB transceiver was used because of its unique combination of lightweight characteristics, low power consumption, and precision. Using a data rate over 6 Mbps with a frequency of 64 MHz for pulse repetition, and 1 kb of preamble symbol repetition, a 1 ns sampling period could be achieved. An implementation of the DW1000 is shown in Fig. 5.

By deploying UWB tags in the same positions as the ArUco markers, a combination of localization using computer vision and positioning by received signal values can be obtained.

A derivation of an approximate maximum likelihood estimation for the localization l can be built around the channel impulse response (CIR). If c is the vector corresponding to the CIR value, the pulse can be written as:

$$s(\Delta) = [s(0 \cdot T_s - \Delta), s(1 \cdot T_s - \Delta), \ldots, s((T-1) \cdot T_s - \Delta))] \tag{1}$$

with a noise measure given by:

$$n = [n(0 \cdot T_s), n(1 \cdot T_s), \ldots, n((T-1) \cdot T_s))] \tag{2}$$

Shifting the pulse with a delay of $\Delta(s(t-\Delta))$, the model of the vector of the CIR can be expressed as:

$$c = \sum_{k \in K} \alpha_k s(\Delta_k) + n = S(\Delta)\alpha + n \tag{3}$$

where $\Delta = [\Delta_1, \ldots, \Delta_k]$, $S(\Delta) = \left[s(\Delta_1), \ldots, s(\Delta_k)\right]$, $\alpha = [\alpha_1, \ldots, \alpha_k]$, and k is the number of deterministic multipath components that are to be estimated to reach a precise measurement.

Fig. 5 DW1000 compliant chip implementing UWB localization

At this point, a simplification of the model can be accomplished by estimating the noise in the channel as white Gaussian noise; the likelihood of the UAV's position considering amplitudes and delays can be found as:

$$l(c|\Delta, \alpha)\alpha \exp\{-\|c - S(\Delta)\alpha\|^2\} \tag{4}$$

Since for a specific Δ, it is possible to obtain the amplitudes using the linear least squares solution given by:

$$\hat{\alpha} = \left(S(\Delta)^H S(\Delta)\right)^{-1} S(\Delta)^H c \tag{5}$$

where the H is the transposition due to the Hermitian.

Now, to maximize the likelihood of Eq. (4), it can be switched the deterministic multipath components and delays to formulate the estimation for the UAV as:

$$\hat{l} = \text{argmax}\, l(c|l, \{a_v\}v \in V) \tag{6}$$

where m_v is the location of the virtual anchors that can be seen because of the reflections that happen every time an anchor is detected, for every multipath component V. The solution of the previous equation provides the real distance from the UAV to every anchor (UWB tag).

2.4 Sensor Fusion

Combining Ultra-Wideband (UWB) localization with computer vision-based localization using ArUco tags can provide a more robust and accurate positioning system. Both technologies have their strengths and weaknesses, and by combining them, some of their limitations can be mitigated and it is possible to achieve better overall positioning accuracy. Potential approaches to take advantage of both methods include:

- Data Fusion: the key concept here is to fuse the data from both UWB and ArUco localization systems. The idea is to develop a fusion algorithm that combines the position estimates from both technologies to obtain a more accurate and reliable position.
- Kalman Filtering or Extended Kalman Filtering: Kalman filters are widely used for sensor fusion. A Kalman filter that takes the UWB position estimate and the ArUco position estimate as inputs and outputs a fused position estimate can provide optimal positioning. Kalman filters are particularly good at handling noise and uncertainty in measurements from different sources.
- Weighted Averaging: another simple fusion approach is to calculate a weighted average of the UWB and ArUco position estimates. The weights can be dynamically adjusted based on the reliability of each technology's estimate. For example,

if the UWB signal is strong and stable, a higher weight can be assigned to its estimate.

- Redundancy and Validation: the idea behind this approach is to use the redundancy of having two localization systems to validate each other's measurements. If the UWB and ArUco estimates are in close agreement, better confidence in the accuracy of the fused estimate can be obtained. If there is a significant discrepancy between the two, it might indicate a measurement error or outlier.
- Real-time Correction: since both UWB and ArUco measurements might have latency and delay, a fusion algorithm that considers the timestamps of each measurement and performs corrections in real time can be crucial for accurate tracking, especially in dynamic environments.
- Adaptive Fusion: the core point of this approach involves crafting an adaptive fusion algorithm that dynamically adjusts the fusion strategy based on the quality of measurements and the specific scenario. In certain instances, UWB might perform better, while in others, ArUco might offer greater accuracy.

In this case, the approach was to provide real-time localization developing an Extended Kalman Filter (EKF) to improve positioning. Other common alternatives, such as the particle filter, demand higher computational resources, potentially hindering their application on lightweight UAV platforms. The EKF is an advanced version of the traditional Kalman Filter that can handle non-linear system models by linearizing them around the current estimate. Typically employed for state estimation in systems characterized by non-linear process and measurement models. In the context of positioning estimation of a UAV, the EKF can provide more accurate estimates by incorporating sensor measurements and system dynamics, even in cases where the relationships between variables are not linear. The EKF assumes that the system's behavior can be described using a state vector and non-linear state transition and measurement functions. For UAV positioning estimation, the state vector might include parameters such as position (x, y, z), velocity, and orientation. At each time step (s), the EKF uses the previous state estimate and the system's dynamics to predict the next state. The prediction encompasses two main steps:

- State Prediction: this step uses the non-linear state transition function to estimate the new state based on the previous state estimate and control inputs (if applicable).
- Covariance Prediction: the covariance matrix represents the uncertainty in the state estimate. It is updated using the system's dynamics to account for how uncertainties evolve.

After making a prediction, the EKF incorporates measurements to correct the state estimate. This involves two main steps:

- Measurement Residual: the difference between the actual and the predicted measurement based on the current state estimate is calculated. This difference is called the measurement residual.
- Kalman Gain and Update: the Kalman gain is calculated to determine the weight given to the prediction and the measurement. Thus, a high gain gives more weight to the measurement, while a low gain gives more weight to the prediction. The

state estimate and covariance matrix are then updated using the Kalman gain and measurement residual.

To provide the mathematical model that will build the filter that fuses both information, first a description of the motion of the UAV must be developed. The physical law that explains the velocity of the UAV in a position l can be written as:

$$l(t) = l(t-1) + \Delta l(t) = l(t-1) + \int_{-t1}^{t} \alpha(\tau) d\tau \tag{7}$$

Accordingly, when writing the model for the Kalman Filter, using the discrete-time expressions:

$$x_{s+1} = x_s + \Delta T v_{xs} + \frac{\Delta T^2}{2} a_{xs} \tag{8}$$

$$y_{s+1} = y_s + \Delta T v_{ys} + \frac{\Delta T^2}{2} a_{ys} \tag{9}$$

where s represents the step in discrete time, and v and a are the velocities in a two-axis reference system.

The position of the tag can be obtained using trilateration having $L = (l_x, l_y)$ as the location of the UAV and m as the location of the anchor as follows:

$$P \cdot \begin{pmatrix} t_x \\ t_y \end{pmatrix} = f \tag{10}$$

where

$$P = \begin{pmatrix} 2 \cdot a_{u,1} - 2a_{u,2}2a_{v,1} - 2a_{v,2} \\ \cdots \\ 2 \cdot a_{u,1} - 2a_{u,n}2a_{v,1} - 2a_{v,n} \end{pmatrix} \tag{11}$$

and

$$f = \begin{pmatrix} d_2^2 - d_1^2 + a_{u,1}^2 - a_{u,2}^2 + a_{v,1}^2 - a_{v,2}^2 \\ \cdots \\ d_n^2 - d_1^2 + a_{u,1}^2 - a_{u,n}^2 + a_{v,1}^2 - a_{v,n}^2 \end{pmatrix} \tag{12}$$

If x_s is the a priori state and y_s is the observation in each step s, to estimate the state variable x in that given step, it is a must to add the a priori estate estimation with the correction. x_s is obtained according to the data already known before entering the next step s; the a posteriori estimate of the previous step $\widehat{x_s}$ is calculated (estimated) using the a priori state estimation at that step, and the measurement of the variable in that step. The state vector that holds the prediction of the position and speed in two dimensions is:

$$x_s = \left(x_s \; y_s \; v_{xs} \; v_{ys} \right) \tag{13}$$

It is now possible to define the measurements at a stage s of the localization as:

$$f_s = Px_s + h_s \tag{14}$$

where h_s represents the measurement noise.

2.5 *Hovercraft Flow Design*

For an aircraft of these characteristics to maintain flight, at a short distance from the ground simply to avoid friction, a lift force is required (as in any flying system). In this case, the pressure difference between the base of the UAV and the outside gives rise to this lifting force following the simple law:

$$F = \Delta P \cdot S \tag{15}$$

where F is the lift force, P is the pressure, and S is the area. It is required (in a situation of stability) that the force is constant, and since the surface will not vary during the flight, it is precisely the difference in pressures that must be determined. The pressure (generated by the airflow), the height concerning the ground and the weight to be lifted are the design parameters of the system.

As for the physical construction and distribution, the first idea that arises is the pumping of air directly against the ground surrounded by a surface that supports pressure, the configuration of an "open plenum", as shown in Fig. 6.

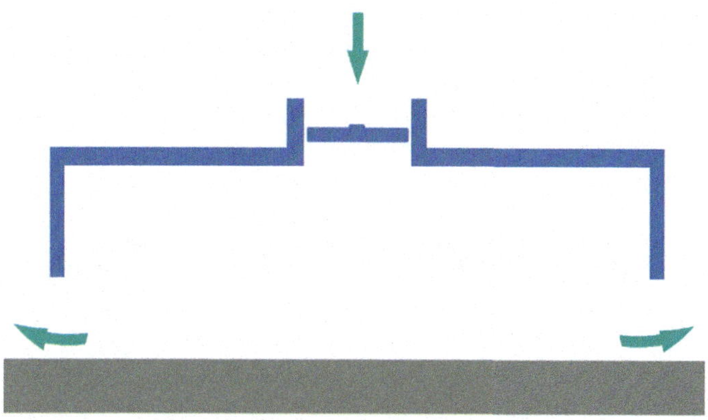

Fig. 6 Visualization of airflow in an open plenum hovercraft. The top arrow represents the exhaled airflow, and the side arrows indicate the outgoing airflow

Although it is a valid solution, this design requires a much higher airflow than other options that will be shown below. Since the air can easily escape, therefore a heavier engine would be necessary; capable of supplying the required power.

A first improvement concerning this design is to pass the air through a compartment that guides the air to the periphery of the UAV (with an inclined angle toward the interior of it), thus causing the air to be trapped by the said flow of incoming air. This phenomenon is known as the "impulse curtain effect", and because of it, the necessary airflow is greatly reduced, leading to greater load capacity for the same engine power, compared to the previous case of the "open plenum". The reason is that the air fed through the periphery circulates at high speed, which infers a great moment, and this is a barrier for the air that is inside and tends to escape from the area of greater pressure (below the aircraft) to the zone of lower pressure (the environment). Consequently, the leakage that exists in front of the "Open Plenum" construction is effectively reduced, and therefore, the necessary flow and the required power of the engine. The flow is depicted in Fig. 7.

Concerning the propulsion of the system, given the configuration of the UAV in the form of a hovercraft, it is required that a mass of air generates the cushion that gives rise to levitation. To also have an orientation capacity of the aircraft, deflectors have been used between the interior of the UAV and the ground to have the ability to move in two dimensions (forward, reverse, left, and right), as well as control of these maneuvers. Figure 8 shows the operating scheme of the valved propulsion system.

In Fig. 8, it is possible to see the longitudinal section of the design, and how the movement created by the opening of certain baffles causes movement. In Fig. 9, a plan representation of the same concept is displayed.

All movements seen in the plan can be modeled and implemented with this simple configuration, including forward–backward, left–right, and turning movements, activating the upper left and lower right valves, or vice versa.

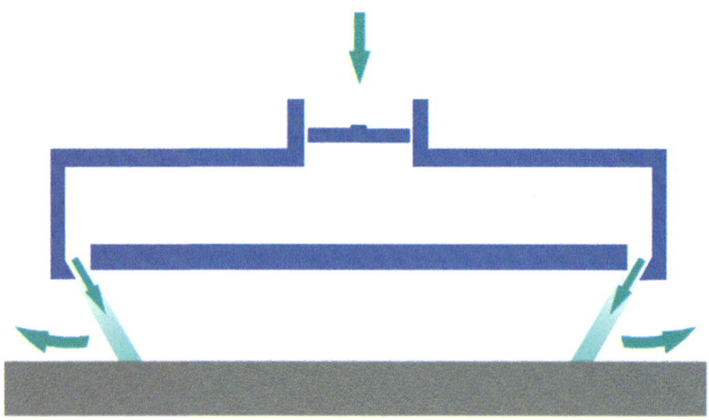

Fig. 7 With this construction, a barrier is generated that generates the peripheral airflow, which is represented as the lower blue rectangle

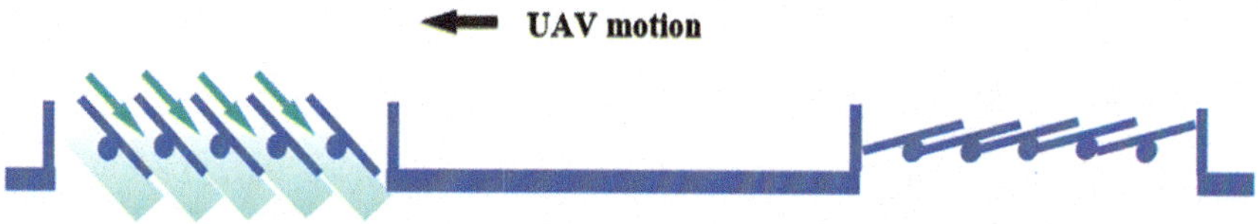

Fig. 8 The valve on the left is open and the air (represented by arrows to indicate its direction), travels through it at a certain angle. On the contrary, the valve on the right is closed, which prevents the transit of air. The result is a force that gives rise to movement

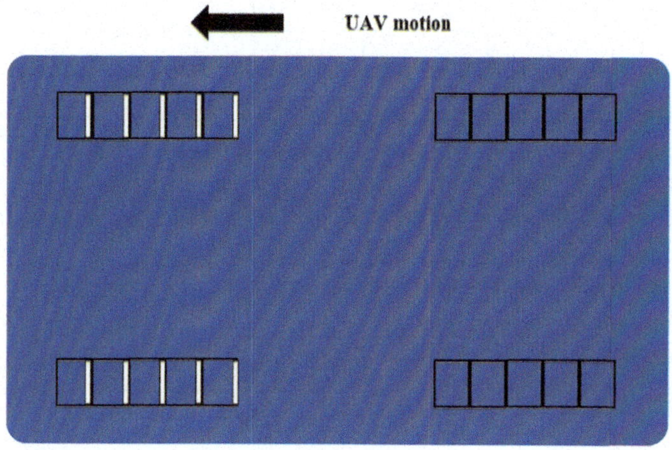

Fig. 9 Hovercraft plant, where air circulates through the valves, causing a force that results in a controlled movement in the indicated direction

As for the calculation of power, flow, and pressure, it is necessary to consider that the flow will remain constant since the increase in exhaust surface when a movement occurs is compensated by the change in speed, as can be seen below. The design requirements are as follows:

- Total mass: m = 80 kg
- Length: l = 1 m
- Width: w = 0.5 m
- Area: A = 0.5 m^2
- Operational altitude: h = 9 mm

First, the pressure to apply on the base of the UAV to support the required weight has to be determined as follows:

$$P = \frac{m \cdot g}{A} = \frac{80 \times 9.81}{0.5} = 1569\,\text{Pa} \tag{16}$$

The air outlet area (A_e) is formed by the plane that surrounds the periphery of the UAV with the operational altitude (applying the principle of continuity, all the air that is pumped will have to leave through this area, disregarding the minimum losses that can be produced by small leaks:

$$A_e = (2 \times l + 2 \times w) \times h = (2 \times 1 + 2 \times 0.5) \times 0.09 = 0.0227\,\text{m}^2 \tag{17}$$

This flow of air will exit through this area with a velocity V_e obtained by applying Bernoulli's law adapted for when there is no difference in height and the air that starts from a stationary state:

$$V_e = \sqrt{\frac{2 \times P}{\rho}} = \sqrt{\frac{2 \times 1569}{1.229}} = 50.53\,\text{m/s} \tag{18}$$

This output speed can only be reduced by reducing the pressure in the system if the requirements of hovercraft dimensions and weight are maintained. Then, the flow rate can be calculated as:

$$Q = V_e \times A_e = 50.5300 \times 0.0227 = 1.147\,\frac{\text{m}^3}{\text{s}} \tag{19}$$

Assuming that pumping losses can be negligible, the pneumatic power required can be calculated as:

$$Pw = \frac{Q \times \rho \times V_e^2}{2} = \frac{1.147 \times 1.229 \times 50.53^2}{2} = 1{,}799\ \text{W} \tag{20}$$

2.6 Motion Control

To control the movement, four valves have been used with whose opening and closing the five basic maneuvers can be implemented: forward and backward, movement to the left and right, and turn (hourly and counterclockwise). The valves will be controlled by two servos placed in the longitudinal middle of the model, one on each side. These servos will move wires of 1 mm in diameter, which will activate the valves according to the type of movement needed in each case. To control the activation and deactivation of the servos, an Arduino microcontroller has been used, in which the PID control of the valves is implemented, receiving the orders from the main system. As motion sensors, it has been chosen to incorporate four Maxbotix MB1040-000 ultrasonic sensors of 42 kHz (powered at 5 V and providing distance readings to the side aisles every 150 ms), and an MPU 6050 (which incorporates gyroscopes and accelerometers), the latter to determine speeds and angles of orientation. Figure 10 displays the connection scheme.

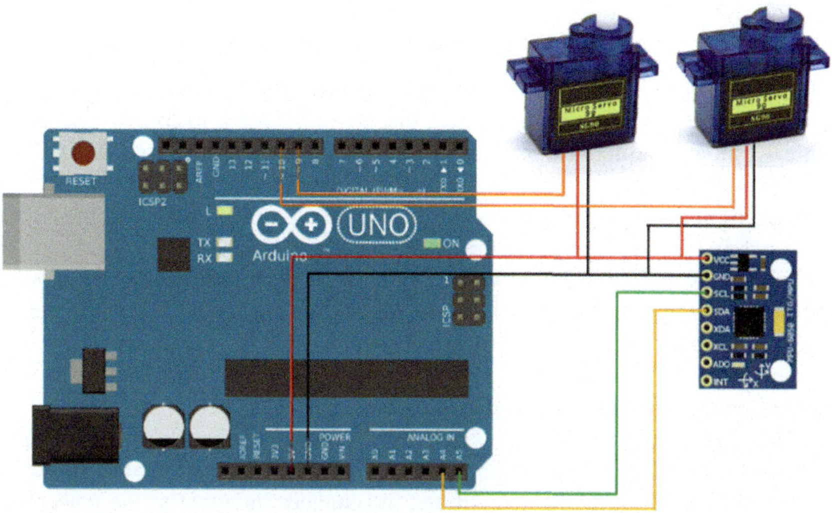

Fig. 10 Schematic of 6050 sensor connections and servos controlling valves

From the mechanical point of view, in Fig. 11, the servos (indicated as A and B) can be seen with the cables that activate the valves, and that will give rise to the movement by the change of airflow through them.

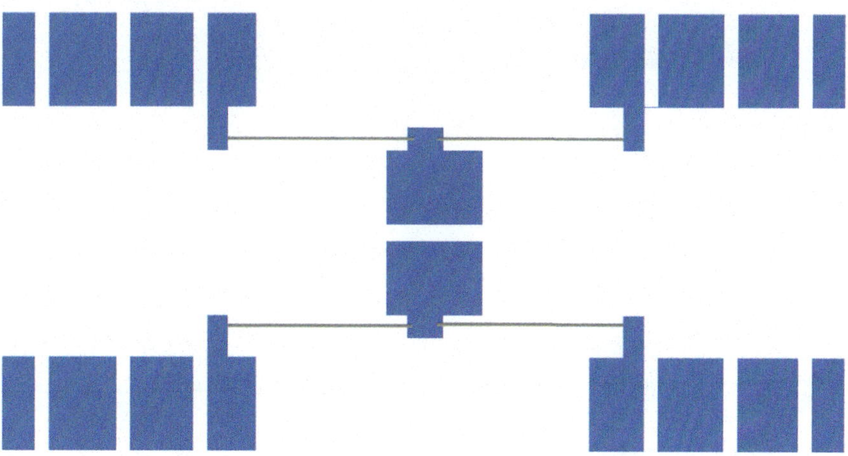

Fig. 11 Diagram of the system of opening and closing the valves with the servos

3 Results

Figure 12 shows the CAD model of the manufactured model.

In the previous figure, an open gate can be seen, for a lateral displacement. Figure 13 shows alternative open gates, to obtain a torque and, consequently, make a turn.

Figure 14 displays the CAD model of the engine supports, which will channel the air through two ducts, to generate the pressure that finally raises the UAV.

This model was manufactured to scale in the laboratory, reducing the dimensions to 50 × 70 cm, and the mass requirements to 4 kg. Using Eqs. (16) to (20) again, a power requirement of 38.5 W has been obtained. Figure 15 depicts the model once the supports of the motors and fans have been installed.

Figure 16 shows the final closed model, with the lid.

In this final model, and to carry out the load tests, lateral supports were implemented to keep the UAV fixed and determine the power required to keep the hovercraft in flight. Figure 17 shows the support and the changes in height as the power was increased from the minimum to the nominal power.

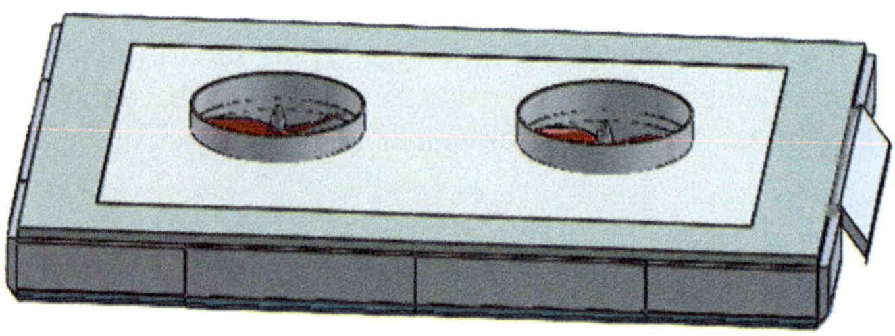

Fig. 12 CAD model of the model

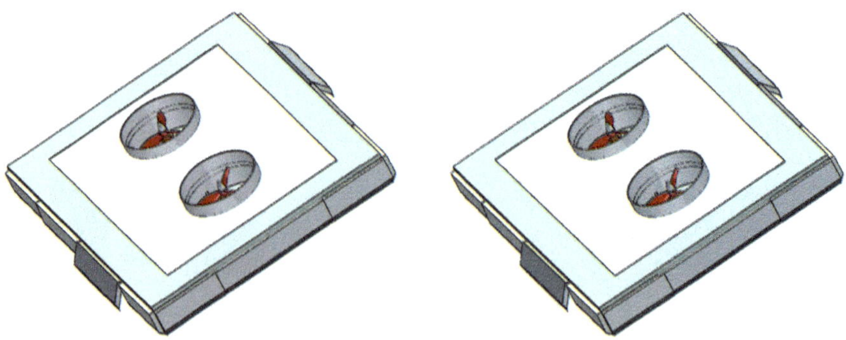

Fig. 13 CAD model with the gates open to get a moment and implement a rotation maneuver concerning the Z axis

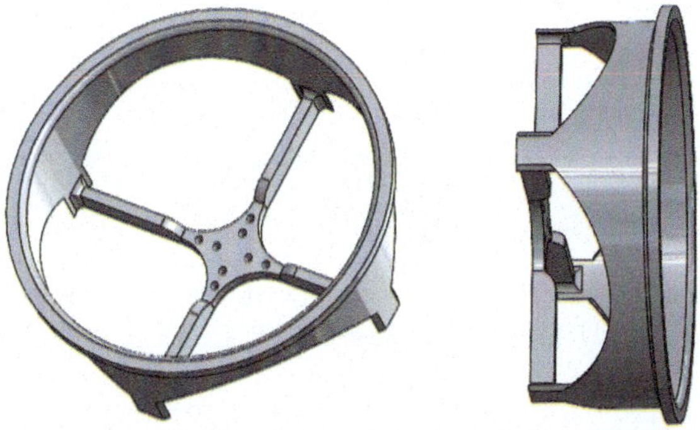

Fig. 14 Front and side views of the motor mount model

Fig. 15 Laboratory model for power and load testing

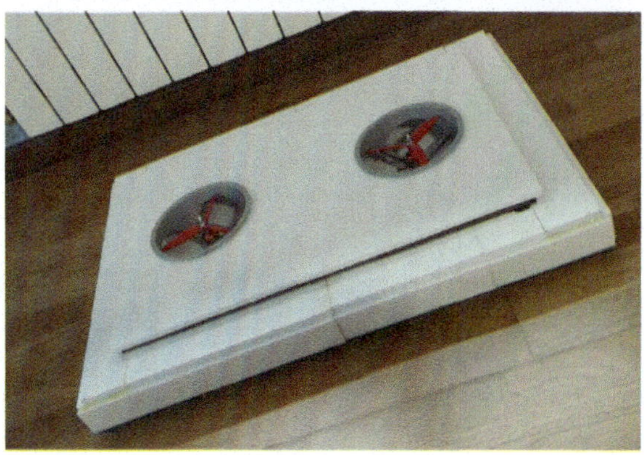

Fig. 16 Final model with the lid

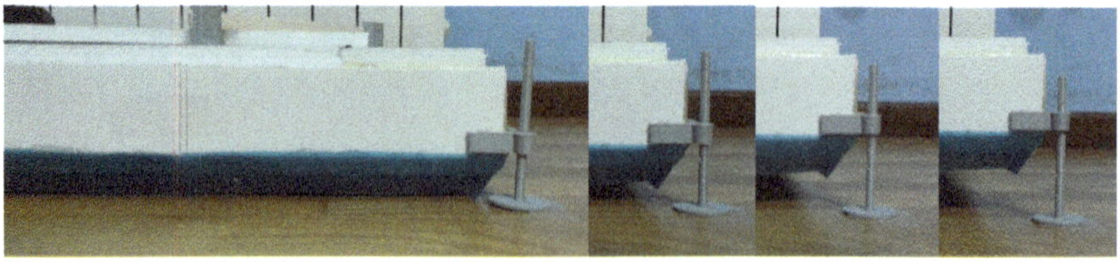

Fig. 17 Four stages of power increase to achieve the takeoff of the UAV, where you can see the added support to perform the tests. From left to right, from minimum power to nominal power

Figure 18 displays the power demand as the load mass increased, from the minimum power of obtaining the flight (the necessary to raise the UAV's mass), to the nominal power for the maximum mass.

The power demanded for the test case, of 4 kg, has been practically 42 W, with a small error concerning the theoretical power of 38.5 W obtained through the calculations made using Eqs. (16) to (20), and that is explained by the logical losses that a laboratory implementation has, as well as the performance of the engines. Figure 19 shows the power meter used connected to the ESC of the UAV.

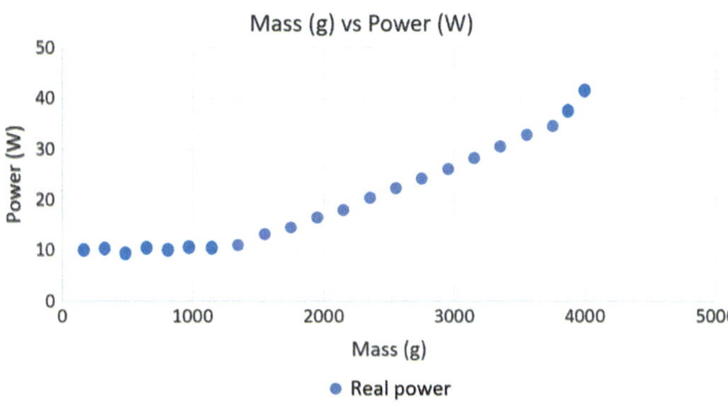

Fig. 18 Power needed to maintain the flight of the UAV as the load increase is made

Fig. 19 ESC power meter

3.1 Motion Control

For the motion control test, an Arduino-based microcontroller has been used, installed both to manage the power of the fans, such as the opening of the valves through the servos, and to obtain the readings of the MPU6050 with which to close the control loop. Figure 20 shows the assembly of the UAV.

Figure 21 exhibits the servo motor that is controlled by the Arduino micro control board, with the cable that activates the air opening and closing valves, which are what enable the movement of the UAV.

Figure 22 displays the end of the connection between the servo and the valve.

The implemented control law is a simple PID, which feeds back the action of the servomotors on the valves to obtain the movement and distance information from the

Fig. 20 UAV controller

Fig. 21 Servo motor that controls the opening and closing of valves

Fig. 22 Valve activated by the servomotor to control the flight of the UAV

Table 1 PID values

Variable	Acronym	Value
Proportional	K_p	4.7
Integral	K_i	0.8
Derivative	K_d	1.2

ultrasonic sensors. The following table shows the values of the controller constants (Table 1).

4 Conclusions

In this study, the design, development, and evaluation of a hovercraft Unmanned Aerial Vehicle (UAV) tailored for efficient light part delivery within a manufacturing plant was presented. The combination of indoor localization using ArUco markers, UWB tags to check position, ultrasonic sensors for obstacle avoidance, and the integration of an MPU6050 for corridor localization has demonstrated significant progress in achieving safe and precise navigation in intricate indoor environments.

The hovercraft UAV system capitalizes on the versatility of ArUco markers, allowing for precise indoor localization and position estimation. The ArUco marker system has proven to be reliable, enabling the vehicle to localize itself with a high degree of accuracy. Furthermore, the use of ultrasonic sensors has greatly enhanced the UAV's capability to ensure the expected position is correct, ensuring that the UAV can navigate through cluttered environments while adhering to stringent safety protocols.

The incorporation of the MPU6050 accelerometer and gyroscope has enabled the UAV to navigate corridors with enhanced stability and maneuverability. The sensor's real-time feedback has been instrumental in maintaining the UAV's orientation and trajectory within narrow spaces, ensuring minimal disruption to ongoing activities within the manufacturing plant.

Extensive experimentation and testing have underscored the feasibility of the hovercraft UAV system for light part delivery in manufacturing plants. The system displayed consistent and accurate performance in indoor localization and corridor navigation, confirming its potential to significantly streamline logistical operations and enhance productivity within industrial settings.

Nevertheless, there remain areas for future improvement and research. The optimization of the hovercraft's propulsion system, battery management, and fine-tuning of the control algorithms could lead to even more efficient and precise movement. Additionally, exploring alternative sensor fusion techniques for even better localization accuracy and robustness is an avenue worth exploring.

In conclusion, the integration of ArUco markers, UWB tags, ultrasonic sensors, and the MPU6050 within the hovercraft UAV system has resulted in a comprehensive and effective solution for light part delivery in manufacturing plants. This research represents a stepping stone toward the realization of safe, efficient, and autonomous UAV-assisted logistics within industrial environments.

References

1. Cook, K. L. (2007). The silent force multiplier: The history and role of UAVs in warfare. In *2007 IEEE Aerospace Conference* (pp. 1–7). IEEE.
2. Fahlstrom, P. G., Gleason, T. J., & Sadraey, M. H. (2022). *Introduction to UAV systems*. Wiley.
3. Samad, A. M., Kamarulzaman, N., Hamdani, M. A., Mastor, T. A., & Hashim, K. A. (2013). The potential of unmanned aerial vehicle (UAV) for civilian and mapping application. In *2013 IEEE 3rd International Conference on System Engineering and Technology* (pp. 313–318). IEEE.
4. Sharma, B., Obaidat, M. S., Sharma, V., & Hsiao, K. F. (2020). Routing and collision avoidance techniques for unmanned aerial vehicles: Analysis, optimal solutions, and future directions. *International Journal of Communication Systems, 33*(18), e4628.
5. Bailon-Ruiz, R., & Lacroix, S. (2020). Wildfire remote sensing with UAVs: A review from the autonomy point of view. In *2020 International Conference on Unmanned Aircraft Systems (ICUAS)* (pp. 412–420). IEEE.
6. Škrinjar, J. P., Škorput, P., & Furdić, M. (2019). Application of unmanned aerial vehicles in logistic processes. In *New Technologies, Development and Application 4* (pp. 359–366). Springer International Publishing.
7. Estrada, M. A. R., & Ndoma, A. (2019). The uses of unmanned aerial vehicles–UAV's-(or drones) in social logistic: Natural disasters response and humanitarian relief aid. *Procedia Computer Science, 149*, 375–383.
8. Gonzalez-R, P. L., Canca, D., Andrade-Pineda, J. L., Calle, M., & Leon-Blanco, J. M. (2020). Truck-drone team logistics: A heuristic approach to multi-drop route planning. *Transportation Research Part C: Emerging Technologies, 114*, 657–680.
9. Boysen, N., Fedtke, S., & Schwerdfeger, S. (2021). Last-mile delivery concepts: A survey from an operational research perspective. *OR Spectrum, 43*, 1–58.
10. Das, D. N., Sewani, R., Wang, J., & Tiwari, M. K. (2020). Synchronized truck and drone routing in package delivery logistics. *IEEE Transactions on Intelligent Transportation Systems, 22*(9), 5772–5782.

11. Sun, Y., Xu, J., Qiang, H., & Lin, G. (2019). Adaptive neural-fuzzy robust position control scheme for maglev train systems with experimental verification. *IEEE Transactions on Industrial Electronics, 66*(11), 8589–8599.
12. Kunze, O. (2016). Replicators, ground drones and crowd logistics a vision of urban logistics in the year 2030. *Transportation Research Procedia, 19*, 286–299.
13. Detweiler, C., Griffin, B., & Roehr, H. (2012). Omni-directional hovercraft design as a foundation for MAV education. In *2012 IEEE/RSJ International Conference on Intelligent Robots and Systems* (pp. 786–792). IEEE.
14. Parker, S. (2007). *The LCAC military hovercraft*. Capstone.
15. Hollebone, A. (2012). *The hovercraft: A history*. The History Press.
16. Toyama, Y., Ono, S., & Nishihara, S. (1992). Design of hovercraft skirt system water scoop by rear skirts and practical solution. *Journal of the Society of Naval Architects of Japan, 1992*(172), 383–391.
17. Pavăl, M. S., Popescu, A., & Zahariea, D. (2019). Numerical analysis of the influence of the lower hull angle of a round skirtless air cushion vehicle. In *IOP Conference Series: Materials Science and Engineering* (Vol. 595, No. 1, p. 012049). IOP Publishing.
18. Rothwell, R., & Gardiner, P. (1985). Invention, innovation, re-innovation and the role of the user: A case study of British hovercraft development. *Technovation, 3*(3), 167–186.
19. Tiemann, J., Schweikowski, F., & Wietfeld, C. (2015). Design of an UWB indoor-positioning system for UAV navigation in GNSS-denied environments. In *2015 International Conference on Indoor Positioning and Indoor Navigation (IPIN)* (pp. 1–7). IEEE.
20. Alarifi, A., Al-Salman, A., Alsaleh, M., Alnafessah, A., Al-Hadhrami, S., Al-Ammar, M. A., & Al-Khalifa, H. S. (2016). Ultra wideband indoor positioning technologies: Analysis and recent advances. *Sensors, 16*(5), 707.
21. Rezwan, S., & Choi, W. (2022). Artificial intelligence approaches for UAV navigation: Recent advances and future challenges. *IEEE Access, 10*, 26320–26339.
22. Cesetti, A., Frontoni, E., Mancini, A., Zingaretti, P., & Longhi, S. (2010). A vision-based guidance system for UAV navigation and safe landing using natural landmarks. *Journal of Intelligent and Robotic Systems, 57*, 233–257.
23. Romero-Ramirez, F. J., Muñoz-Salinas, R., & Medina-Carnicer, R. (2018). Speeded up detection of squared fiducial markers. *Image and Vision Computing, 76*, 38–47.
24. Garrido-Jurado, S., Munoz-Salinas, R., Madrid-Cuevas, F. J., & Medina-Carnicer, R. (2016). Generation of fiducial marker dictionaries using mixed integer linear programming. *Pattern Recognition, 51*, 481–491.

Drones for Space Exploration

Fermin Navarro Medina, Carlos Ulloa Sande, Guillermo Rey, Javier Pampín García, and David Pérez-Tavernier

Abstract The synergy between planetary orbiters and surface vehicles, particularly aerial vehicles in extraterrestrial atmospheres, opens exciting opportunities for exploration and data collection beyond what individual platforms can achieve alone. Orbiters and surface vehicles have already been launched several times to other planets and moons. However, aerial vehicles are incipient agents in space exploration. Aerial vehicles provide an unprecedented capability to traverse difficult terrains and access remote locations, enabling the exploration of regions that were previously inaccessible to surface rovers or landers. This book chapter provides a comprehensive review of the advancements and applications of drones in space exploration, the concept of drones, and their potential benefits in terms of cost-efficiency, versatility, and adaptability to different environments. These applications encompass the exploration of celestial bodies, where drones can assist in mapping, sample collection, and geological and atmospheric analysis. A review of several potential planets and moons to be explored by drones is presented, while two fixed-wing drones are conceptually designed for the most promising targets, one for Mars and the other for Titan. Moreover, the challenges and limitations associated with drone technology in space exploration are addressed. These include issues related to power supply, materials, communication delays, thermal control, and the need for robust navigation systems in unfamiliar terrains. From the challenges, it is concluded that further research and

F. N. Medina (✉) · C. U. Sande
atlanTTic, Universidade de Vigo, Escola de Enxeñaría, Aeronáutica e do Espazo, Aerospace Technology Research Group, 32004 Ourense, Spain
e-mail: fermin.navarro.medina@uvigo.gal

G. Rey
Universidade de Vigo, Escola de Enxeñaría, Aeronáutica e do Espazo, Aerospace Technology Research Group, 32004 Ourense, Spain

J. P. García
Polytechnic University of Valencia, 46022 Valencia, Spain

D. Pérez-Tavernier
Universidad Carlos III, Madrid, Spain

D. Carou et al. (eds.), *Applying Drones to Current Societal and Industrial Challenges*, Management and Industrial Engineering, https://doi.org/10.1007/978-3-031-55571-8_7

development in this field is needed, but also that drones have the potential to revolutionize future space missions, enabling greater understanding and exploration of the universe.

Keywords Space drones · Extraterrestrial atmospheres · UAVs · Mars · Titan

1 Introduction

Space drones have a wide range of applications, including planetary exploration, which has led to research studies focused on their design, optimization, and performance enhancement. Conceptual design is a crucial step in creating space drones as they must be compatible with the flight conditions, which are defined by totally or partially unknown environmental scenarios. They afford non-traditional flight missions with very specific flight objectives. Structural and aerodynamic design and modeling of drones for planetary exploration is a high-priority future technology for space agencies because they open a new point of view for the scientific instruments onboard. They are candidate vehicles, especially on planets or moons where the environmental conditions have already been explored by ground or orbit vehicles, and some data on the surface and the low atmosphere are available. However, designing space drones for planetary exploration presents many challenges to be solved, such as meeting the size, weight, or power constraints and managing risks associated with the flight, the atmospheric conditions, the deployment, or the data return.

Drones, also known as unmanned aerial vehicles (UAVs), have the potential to revolutionize the way humanity explores space, by allowing us to gather data and explore areas that were previously inaccessible. One of the primary advantages of using drones for space exploration is that they can be used to explore regions that are too dangerous or difficult for humans or other autonomous vehicles to reach. For example, drones can be used to explore the surface of Mars from a certain altitude and to gather data in areas where traditional spacecraft or rovers are unable to reach because they are too far or too static on the ground. Overall, drones have the potential to significantly improve humans' ability to explore and study other solar system bodies. As technology continues to improve, more advanced drones are expected to be seen in future space exploration missions.

This chapter is organized as follows. First, a review of the existing projects of space drones is done. Secondly, an analysis of the environmental conditions for flight in several bodies of the solar system is developed. To finish, two study cases of drones on Mars and Titan are presented as examples of potential candidate designs.

2 State of the Art

In this section, a review of the main projects of drones for space exploration is presented. The focus is placed on projects aimed at designing drones for Mars and Titan, which are the main targets of these types of space vehicles. First, a series of UAVs designed for high terrestrial altitudes is presented; later, the main projects of drones for Mars, and finally, some projects focused on Titan are shown.

The Mini-Sniffers were a series of fixed-wing UAVs used for studying the Earth's atmosphere at high altitudes between 1975 and 1982. It comprised three models that were essentially the same aircraft but with modifications and improvements (Mini-Sniffer I, Mini-Sniffer II, and Mini-Sniffer III). The most advanced version was designed to fly at an altitude of 70,000 feet (21 km) carrying a payload of 11.3 kg. For this purpose, it had a hydrazine engine since an internal combustion engine, like the one used in the first model, would not be able to function due to the low density of air at such high altitudes. Considering the similarity between the Earth's atmosphere at the drone's design altitude and that of Mars in the regions near its surface, NASA considered the idea of extrapolating the drone's design to the red planet. This was the first proposal to send an aircraft to the Martian environment. However, the Mini-Sniffer III only managed one flight and failed to reach its target altitude of 70,000 feet. It only flew up to 20,000 feet and did not fly again due to fuel losses.

Considering space drones with flight heritage, the Ingenuity drone was the first, and the unique, in history to achieve sustained controlled and autonomous flight on the Martian terrain. Its successful completion of objectives paved the way for the development of new research projects in this environment using aerial vehicles, particularly those with rotary wings. Considering that Ingenuity was just a technology demonstration, it is expected that more advanced helicopters will be sent in the future to conduct scientific studies that are currently not feasible. The drone has a coaxial configuration with 2 rotors, weighs 1.8 kg, and has a diameter of 1.2 m. The mission plan initially called for 5 flights (which eventually became 13), lasting up to 90 s each, during which the behavior of the aircraft would be demonstrated and analyzed. All this information was transmitted back to Earth through the Base-Station carried by the Perseverance rover, which connected the helicopter and ground operators using radio signals transmitted via antennas.

The Ingenuity's structure is centered around a central mast that supports all other components. At the top of the mast is the electrical interface, which monitors system status, and batteries, and controls and supervises component temperatures before take-off. Just below that, the carbon fiber blades have a diameter of 1.21 m and a chord of 0.1 m, with a shape and twist adapted to low Reynolds (*Re*) number flight. Each of the two-bladed rotors is propelled by a brushless electric motor that induces an angular velocity of about 2800 rpm. Each blade has a collective control (which adjusts the blade angle to provide the necessary thrust and execute yaw maneuvers) and a closed-loop control (which tilts the rotor for forward and lateral movement of the aircraft). The energy required for the system is obtained through a solar panel consisting of high-efficiency inverted metamorphic quadruple-junction cells located on the top of the aircraft, around the electrical interface. The battery, which comprises six lithium-ion cells, is stored inside the fuselage.

One of the most critical aspects of designing the Ingenuity was the fuselage and its thermal protection mechanisms due to the extreme cold conditions on Mars. This is because it houses elements such as batteries, avionics systems, and sensors that require higher temperatures to operate without being damaged. The structure has both passive and active protection methods. The passive methods include CO_2 insulation and a series of thermal barriers to minimize heat loss. Additionally, the outer layer of the fuselage consists of a Kapton film with a dark mirror finish. This material has high absorption and low emissivity, allowing the helicopter to collect solar energy during the day to heat the interior at night, minimizing radiation losses. As for the active systems, only heaters were used, some of which are designed to keep the batteries warm during the night, and others only operate before the drone's flight to ensure that all critical components reach their heat-to-operate temperature range.

Regarding communications, the rover and the Ingenuity are connected through a two-way radio link. The channel used has a frequency of 914 MHz and a data transfer speed of up to 250 kbps. Both vehicles carry an omnidirectional antenna, with the helicopter's mounted on top of the solar panel. The power consumed by the subsystems that make up the aircraft is around 360 W, with the propulsive component being the most power-consuming. The battery capacity reaches 12 Ah [5].

In addition to helicopter-type drones, the ARES project, designed at NASA's Langley Research Center in 2004, is the fixed-wing drone program for Mars that came closest to fruition. It was selected as one of four finalists out of 25 proposals to be funded in the Mars Exploration Program and sent between 2011 and 2013. The aircraft's objective was to complete a controlled flight of approximately 500 km in length at an altitude of less than 2 km in approximately 60 min. During this journey, unique measurements of the atmosphere, surface, and interior of the planet would be taken using scientific instruments (a mass spectrometer, magnetometer, spectrophotometer, and cameras) on board the drone.

The drone has a highly precise inertial navigation system consisting of an Inertial Measurement Unit (IMU), an altimeter radar, and an aerial data system that allows it to always know its position with great accuracy. The UAV will maintain a constant link with the carrier spacecraft through a circularly polarized patch antenna mounted on the top of the fuselage and a diplexer. To mitigate the risk of losing scientific data, the drone has a 40 W RF transceiver operating in the UHF range connected to the antenna. The orbiting spacecraft will then be responsible for sending all information back to Earth through the X-band. The duration of this process could exceed 24 h.

Regarding the propulsion system, the rocket engine was selected because it was safer and more reliable for achieving scientific objectives than the propeller. This is because Mars' low density results in low performance for the propeller method. As a result, a large blade size would be required, which would make packaging difficult given that volume is very limited on this type of mission. Thus, the drone was equipped with a pulsed control thruster based on monomethyl hydrazine (MMH, as fuel) and nitrogen tetroxide with a 3% additive in nitrogen oxides (MON-3) capable of exerting about 62 N of force to achieve a cruise speed of 140 m/s [24].

For thermal control, it only carries a combination of heaters, insulation, coatings, thermostatically controlled local heaters, and a single heat sink for the RF transceiver. Both the fuselage and empennage are constructed of sandwich panels, which combined with the rest of the components and fuel, results in a total mass of 975 kg at liftoff. It should be noted that the structure is divided into three folds, two of which are located between the central area of the fuselage and each wing, and the third at the beginning of the tail section. These will extend once the UAV is ejected from the carrier spacecraft. For the power subsystem, dual 28 V DC buses are used to reduce assembly complexity. Thus, the avionics bus powers the scientific instruments, guidance devices, navigation, computing, data handling, and telecommunications, while the switch bus maintains propulsion, pyrotechnics, altimeter radar, control surfaces, and heaters. Each primary Li-SO_2 battery consists of several chains, with 12 D-cells connected in series [10].

Also proposed for flying on Mars, the MELOS mission (Mars Exploration of Life and Organism Search), operated by JAXA, has the objective of conducting a precision landing and studying the possible existence of life forms on the red planet. The proposed launch year is 2024, with the arrival on the Martian surface expected the following year. When the project was formulated in 2008, despite having the same acronym as now, it had a different meaning and purpose: MELOS (Mars Exploration with a Lander-Orbiter Synergy). At that time, the plan was to send an orbiter and four stationary landing modules that would carry out different measurements. However, in 2015, the concept changed completely, and the approach was reduced to an engineering demonstration and a small investigation through a rover and possibly a UAV. Two variants of drones were proposed for the project, which fundamentally differed in their propulsion configuration (push and pull) [19].

Some other drone concepts for a mission to Mars were proposed by engineering students during a workshop organized by the European Space Agency (ESA) and the European Association of Aerospace Engineers (EUROAVIA) in the summer of 2006. An example of these concepts was the Advanced Reconnaissance Martian Deployable Aircraft (ARMaDA), a long-duration mission plan consisting of three identical drones. The project's objective was to map the surface for the search for landing zones for future human expeditions, analyze soil composition, and study the atmosphere. It was primarily focused on the northern hemisphere. It was proposed that the UAVs remain in operation for days carrying out these tasks. However, the power source (radioisotope) offered the possibility of a lifespan of even years, although, most likely, it would end earlier due to a mechanical failure.

The main characteristics and design properties of interest for drone missions destined for Titan are explained below. The Huygens Probe, launched in 1997, was the precursor to missions to Titan and the first exploration robot to descend and land on the satellite on January 14, 2005, in the Xanadu region. The images and measurements of atmospheric properties were essential for designing future missions as they provided the necessary information to design explorer drones. Several important ideas were drawn from this mission, such as the use of a parachute combined with a thermal shield to slow down atmospheric entry or the possibility of detaching the thermal shield during parachute descent to release the explorer. Apart from useful information about its dynamics and terrain images, Huygens demonstrated the possibility of safely sending a robot to the surface of Titan.

The AVIATR (Aerial Vehicle for In-situ and Airborne Titan Reconnaissance) Project in 2010 consists of a drone with glider morphology. Its design emphasizes the virtues of Titan's atmosphere for continuous flight, with an uninterrupted flight time of around a year. The drone has a mass of approximately 120 kg and is powered by Advanced Stirling Radioisotope Generators (ASRGs). Its equipment is based on terrain imaging, with a very comprehensive set of cameras but without equipment for material composition analysis. It emphasizes information from the study of clouds and terrain to study their evolution. One of the conclusions that can be drawn is energy efficiency. This drone, with a wingspan of just over 6 m, must carry the weight and payload it can afford, as they all consume power. Moreover, it faces the limitation of power provided by its ASRGs, without batteries, limiting its operability and the weight and quantity of equipment that it can carry as it cannot interrupt flight to recharge batteries, which becomes problematic. From a construction standpoint, since its wingspan is too large to handle the dimensions of a launcher and a thermal shield, it had to be folded [6].

The Dragonfly drone, proposed in 2017, was selected by NASA in 2019 to be part of the New Frontiers program. This comprehensive drone, based on the design of a quadcopter with coaxial rotors, presents a robust way of moving around the satellite and its atmosphere and can land in different locations. Its scientific equipment includes a complete set of cameras and equipment for composition measurement and sample extraction. This aircraft proposed by the Johns Hopkins Applied Physics Laboratory launches a series of concepts that are very beneficial for designing this study. The hybrid use of batteries and RTGs is proposed as a viable and rechargeable energy source, which, coupled with the "frog hop" flight strategy—landing, analyzing the terrain, taking off, investigating a potential point of interest, and landing again—make this strategy a solid exploration plan. The importance of having the ability to land and take-off vertically is evident as one of the keys to the feasibility of thorough exploration, as well as being vital for the long duration of the mission, around 2.7 years, allowing battery charging and minimizing risks in landings and take-offs. These functionalities are attractive for this study and must be considered [25].

3 Flight Conditions in Other Atmospheres

According to the space drones described previously, it becomes obvious that the analysis of the flight environment is fundamental to determining the best design options. Various astronomical objects in the solar system, including Venus, Mars, the four outer planets, Pluto, and several moons, have atmospheres. Their characteristics differ depending on many factors like the solar system's body size, gravity, the distance to the sun, gas type, cloud physics, atmospheric chemistry, atmospheric dynamics, pressure, temperature, or wind speed. If the flight operational modes of the drone and the places to take-off or land are considered, such a list of factors can be extended also with the terrain characteristics, the existence of liquid oceans, mountains, valleys, etc. However, of interest for the drone conceptual design during the cruise phase at a given altitude, only the main characteristics of the diverse atmospheres are presented in the following paragraphs.

Comparing extraterrestrial atmospheres to Earth's atmosphere helps in understanding drone sizing. Drones can fly in different solar bodies with atmospheres, but some of them, like Mercury, Jupiter, Saturn, Neptune, Pluto, and other natural satellites, are not suitable for fixed-wing or helicopter-type drones due to high wind speed and low atmospheric temperature. On the other hand, Venus, Mars, Titan, Uranus, and Triton are considered viable targets for drone design. Exploration of the neighboring planets of Earth, such as Mars and Venus, has been a fascinating topic for space scientists and researchers for several decades. With the advancements in technology, scientists have been able to send probes, rovers, and landers to these planets to study their composition, geology, and atmospheric conditions. However, the exploration of these planets is not an easy feat as the harsh environmental conditions pose a significant challenge to the mission's success. Now, each of the candidate planets or moons is described, and later, a comparison based on the characteristics and non-dimensional numbers of the drone flight is presented.

3.1 Venus

One of the most significant challenges in exploring Venus is its thick atmosphere, which is composed primarily of carbon dioxide with trace amounts of nitrogen and other gases. The thick atmosphere of Venus generates an immense pressure that increases with decreasing altitude. The temperature on the planet's surface is hot enough to melt lead, making it impossible for any spacecraft or rover to survive on its surface for an extended period.

However, there is an altitude range where drones can fly without the constraints of excessive pressure and temperature. As mentioned earlier, for an altitude of 45 km on Venus, the temperature is around 353 K, and the pressure is 2.0 bar, while for 60 km, the temperature is 238 K, and the pressure is 0.2 bar. This range of altitude provides an ideal environment for drones to explore Venus's atmosphere and collect data about its composition, temperature, and pressure. However, it remains inconvenient for drones to achieve lift and maintain altitude constantly without the possibility of take-off and landing.

The drones designed for Venus's exploration would need to be built to withstand the harsh environmental conditions of the planet of the acidic atmosphere. Besides, the drone's electronics and sensors would need to be insulated and protected from the high levels of radiation present in Venus's atmosphere. The drone's propulsion system would also need to be powerful enough to navigate through the thick atmosphere and withstand the intense pressure if low altitudes are targeted. Standard resins used to fix carbon fiber materials are not resistant to heat and acid and can simply melt or dissolve in sulphuric acid clouds. Additionally, in such a hot environment, electronics must be cooled to avoid malfunction. Furthermore, stresses and strains associated with wings and rotors must be carefully considered, as materials must not be too flexible at high temperatures.

To address these challenges, the space industry might explore alternative materials like nickel and iron-based alloys renowned for their resistance to high temperatures. Molybdenum, with its very high melting point, low coefficient of thermal expansion, and high thermal conductivity, is frequently used as a coating on the outer parts of spacecraft to shield less heat-tolerant materials against the extremely high temperatures generated during re-entry into the atmosphere. Beryllium copper and similar alloys garner attention from to the space industry due to their high strength, hardness, fatigue resistance, good corrosion resistance, and excellent thermal and electrical conductivities. All these materials, among others, present potential options worthy of further investigation.

3.2 Mars

Mars is the second smallest planet in the Solar System, with approximately one-tenth the mass of Earth. As a result, the planet's gravity is about one-third that of Earth (3.72 m/s^2), which means less lift is required for a UAV to take-off. However, the planet's atmosphere is virtually non-existent, with an average density of 0.016 kg/m^3, making flying on Mars a significant challenge. Moreover, the absence of a gaseous layer that equalizes day and night conditions results in extreme temperature variations. In the summer, daytime temperatures can reach a maximum of 20 °C, while night-time temperatures can drop as low as −80 °C, with an average temperature of −55 °C [18, 28], which makes it impossible to protect all the components of the drone without additional heating power.

On the other hand, another challenge for drones is to survive the dust storms that occur in the southern hemisphere of Mars. These storms can last for months and create hazardous conditions for any equipment on the planet's surface and at low altitudes. Drones, however, can be programmed to fly above the storm and provide crucial data on the weather conditions and any changes in the landscape. However, there is still a lack of information to accurately know the height and the main characteristics of the Martian storms.

A critical aspect of using drones for Mars exploration is their power source. Solar power is a possible option, but it has limitations due to the low amount of sunlight that reaches Mars. Fixed-wing airplanes may provide a more efficient solution to generate energy since they can cover their wings with solar panels. Besides, they can provide better flight performances by covering greater distances and operating for more extended periods than rotorcraft. In terms of energy sources, the use of Radioisotope Thermoelectric Generators (RTGs) has been a popular choice for deep space missions. However, the use of RTGs is not without controversy, and alternatives are being considered.

Nevertheless, one of the key factors in using drones for the exploration of any solar system body is that they must be programmed for automatic control due to the communication delay between Earth and Mars. Drones can operate independently, collecting data and performing tasks without constant human input, but they must be prepared to act against contingencies according to a risk management plan.

3.3 Flyable Atmospheres Further Than Mars

To begin with the analysis of flying drones on Titan, it is important to consider that this is Saturn's largest satellite and has a dense, cold atmosphere rich in hydrocarbon compounds. Titan's environment is suitable for flying, as its low gravity and typical wind range make it an ideal environment. However, the dissociation of methane and nitrogen at high altitudes creates an opaque layer at lower altitudes, which makes it impossible to study Titan from an orbiter. To address this challenge, a mission for flying a drone can be beneficial. This drone would analyze different environments and hidden layers of Titan's atmosphere, measuring atmospheric properties and composition, meteorological phenomena, magnetic fields, relief, and soil layers. Despite the extreme environmental conditions of Titan, including low temperatures, low atmospheric density, and the potential for sudden storms, drone flight on this satellite is an exciting prospect for scientific exploration. To make it possible, innovative solutions and the use of specialized materials and technologies are required. Titan's lower layer density is 5.4 kg/m^3 with a temperature of -180 °C. The atmosphere is rich in complex hydrocarbon compounds, which, combined with the low gravity of Titan (1.35 m/s^2) and typical wind range of 1–5 m/s at low altitudes, makes it an ideal environment for flying [16]. Like Earth, Titan has a water cycle based on methane. Its relief is eroded, with dunes and theories about cryovolcanism and the possibility of an underground sea of ammonia or hydrocarbons. Titan is considered one of the most likely celestial bodies in the solar system to harbor life.

Another challenge is the potential for sudden storms on Titan. These storms can produce high-speed winds and dangerous precipitation, which can pose a threat to drones. The lack of a solar power source also presents a challenge, as well as the absence of fixed-wing airplanes or rotorcraft. RTGs are an alternative power source that can be used in such conditions. The communications delay between Earth and Titan also makes automatic control of the drone necessary. This is because the time delay between sending a command from Earth and receiving the drone's response can be up to 90 min. Moreover, the composition of Titan's atmosphere poses a challenge for materials used in drone construction. Standard resins used to fix carbon fiber materials are not resistant to hydrocarbon solvents and can dissolve. Finally, high-speed winds and potentially hazardous substances raining from a cloud layer made of liquid hydrocarbons make flight operations even more challenging. In conclusion, while drone flight on Titan presents significant challenges, it is also an exciting prospect for scientific exploration. Overcoming these challenges will require innovative solutions and the use of specialized materials and technologies.

Flying drones on Uranus, the seventh planet from the Sun, presents unique challenges due to its extreme environmental conditions. Here are some key factors to consider. Firstly, Uranus is known for its high-speed winds, with gusts of up to 250 m/s at an altitude of 50 m. This makes it difficult for drones to maintain stability and control while flying. Additionally, the slightly lower gravity on Uranus of 8.69 m/s^2 means that drones must be designed to generate enough lift to compensate for the lower gravitational pull. The low temperatures on Uranus also pose a challenge for drone flight. At the surface, temperatures can reach as low as 76 K, and at an altitude of 50 km, temperatures drop to 56 K. The extreme cold can cause electronics to malfunction and materials to become brittle, requiring specialized materials that can withstand the cold environment.

Uranus also lacks a solar power source, fixed-wing airplanes, or rotorcraft. RTGs are an alternative power source that can be used in such conditions. The atmosphere on Uranus has a surface pressure like Earth at 1 bar, but at an altitude of 50 km, the pressure drops to 0.1 bar. This affects the drone's ability to generate lift and maneuver. Communication is also a challenge due to the distance between Uranus and Earth. The communications delay between sending a command from Earth and receiving the drone's response can take several hours, making automatic control of the drone necessary. Finally, the atmosphere's composition on Uranus poses a challenge for materials used in construction. In conclusion, flying drones on this planet presents significant challenges, but the use of specialized materials and technologies can overcome these obstacles. The exploration of this unique planet presents opportunities for scientific research and discovery, making it an exciting prospect for drone flight.

Triton, the largest moon of Neptune, has a thin atmosphere composed mainly of nitrogen, with small amounts of methane and carbon monoxide. Its surface pressure is about 14 microbars, which is about 1/70,000th of Earth's atmospheric pressure at sea level. The surface temperature on Triton is extremely cold, ranging from −235 to −198 °C. The atmosphere is also subject to seasonal changes, with changes in the amount of sunlight affecting the distribution of gases and causing seasonal winds. The thin atmosphere of Triton makes it a challenging environment for flying drones or other aerial vehicles, but it is an intriguing subject for scientific study [8].

3.4 A First Comparison of Flyable Atmospheres

The description of the flight in each planet or moon is quantitatively summarized in this subsection. In Table 1, aimed to do a general comparison, a summary with reference values for the characteristics of the atmospheres is presented. Besides, the dimensions of a reference drone (see Table 2) are used to obtain the mass lifted for each planet. However, when a specific design was afforded much more care should be put in the calculation of each parameter.

The first conclusion from Table 1 is that the same drone can lift more mass on Uranus or Titan than on other planets and that in Triton the mass lifted is very low.

Table 1 Reference values of the main characteristics of the atmospheres and non-dimensionless numbers of a given fixed-wind drone

Reference values	Venus	Mars	Titan	Uranus	Triton
Altitude [km]	45	0	0	0	0
Temperature [K]	353	219	93	76	56
Pressure [bar]	2	6.35E−03	1.45	1	1.50E−05
Main gas component	Carbon dioxide-CO_2 (96.5%)	Carbon dioxide-CO_2 (95%)	Nitrogen-N_2 (95%)	Hydrogen-H_2 (82.5%)	Nitrogen-N_2 (95%)
Secondary gas component	Nitrogen-N_2 (3%)	Nitrogen-N_2 (2.6%)	Methane-CH_4 (5%)	Helium-He (15.2%)	Methane-CH_4 (5%)
Main molar mass	44.01	44.01	28.01	2.016	28.01
Secondary molar mass	28.01	28.01	16.04	4	16.04
Average molar mass	43.31	42.54	27.41	2.27	27.41
Gas constant R [J/kg K]	365.92	365.92	232.89	16.76	232.89
Density [kg/m^3]	1.55E+00	7.92E−03	6.69E+00	7.85E+01	1.15E−04
Gravity [m/s^2]	8.84	3.72	1.35	8.87	0.78
Dynamic viscosity [N s/m^2]	3.10E−05	1.10E−05	6.00E−06	4.00E−06	1.00E06
Cinematic viscosity [m^2/s]	2.00E−05	1.39E−03	8.96E−07	5.10E−08	8.69E−03
Heat capacity ratio	1.282	1.34	1.47	1.6	1.55
Reynolds number	2.50E+05	3.60E+03	5.58E+06	9.81E+07	5.75E+02
Freud number	22.630	53.823	148.012	22.548	256.739
Mach number	0.025	0.031	0.056	0.221	0.070
Drone mass [kg]	9.20E+00	1.12E−01	2.60E+02	4.65E+02	7.75E−03

Table 2 Dimensions of a reference drone used to illustrate the atmosphere comparison

Values of a representative fixed-wing drone	
Lift coefficient	0.3
Wing chord [m]	0.5
Flight speed [m/s]	10
Wind span [m/s]	7
Wing area [m^2]	3.5

This result can be quite confusing because a detailed analysis of the best altitude for flight in each planet and region of the planet should be done. To illustrate the main considerations when designing missions with drones for space exploration, a couple of use cases are presented: one on Mars and one on Titan. The solar system bodies selected are the most likely targets for this type of vehicle, not only because of the existence of an atmosphere, but also because the scientific objectives are better positioned in the list of priorities for the scientific community.

4 Use Case 1: A Fixed-Wing Drone on Mars

4.1 Introduction

Throughout this section, a fixed-wing drone is going to be presented. Despite the attention will be focused on the design process of the UAV, other aspects of the rest of the necessary infrastructures will also be studied. Despite the harsh and cold climate, the interest in Mars exploration has boomed. One of the main reasons for this great interest has been the discovery of resources from which water can be obtained. Currently, it is not confirmed the presence of liquid H_2O on the planet's surface. However, water processes are still present through the existence of evaporites such as halite (sodium chloride), jarosite (potassium and iron sulfate), and plaster (calcium sulfate). Furthermore, the presence of polyhydrated magnesium sulfates (PHMS), a group of evaporites that include epsomite, kieserite, hexahydrite, bloedite, and meridianite is also remarkable. This group of minerals contains between 14 and 62% water and is found in large proportions along the Martian soil. By heating these sediments to temperatures between 100 and 325 °C, it is possible to extract between 5 and 20% of their mass in the form of H_2O [11]. Obtaining water could make the establishment of human civilizations on the planet's surface possible. In addition, the gypsum, once dehydrated, could be used as a building material. Thus, the mission's goal will be to design a drone to analyze the Martian ground in search of these minerals. To survey the largest possible area, a fixed-wing UAV will be employed.

The study zone will be Meridiani Planum, a plain near the Martian equator that extends around the coordinates 0.2°S 357.5°E. This is a safe and accessible landing site for both the probe and the drone which will not experience such extreme temperature variations throughout the year as the high and low latitude regions. Furthermore, within this area, there are some craters such as Endeavour and Iazu which contain gypsum, PHMS, and clay veins on their rims [11]. Given the adverse conditions previously mentioned, along the design process, thermal control and aerodynamic design will be addressed in more depth.

4.2 Mission Description

The mission concept will comprise two parts: the UAV and a rover that will function as a base station, both powered by electrical energy (see Fig. 1). To obtain the energy, the ground vehicle will be equipped with InGaP/GaAs/Ge triple-junction solar panels, a model whose efficiency reaches 25% and has been used on previous missions on this planet such as Mars Exploration Rovers (MER) [13]. It will also have a catapult to launch the drone, a flexible net to catch the aircraft after landing, and a robot arm to pick up the aircraft from the net at the end of each flight and return it to the shuttle for the next one.

While the drone remains on the ground, it will be tethered to the catapult, as shown in Fig. 2. Through an electrical umbilical connected to the lander, it will charge its batteries and transmit the data collected during each flight. This technology has already been used by the Ingenuity drone to power one of its heaters [5].

The duration of the mission will be determined by the degradation of the solar panels (primary energy source). Although the batteries (secondary energy source) will also suffer from this phenomenon, it will not be as significant as in the case of the photovoltaic cells.

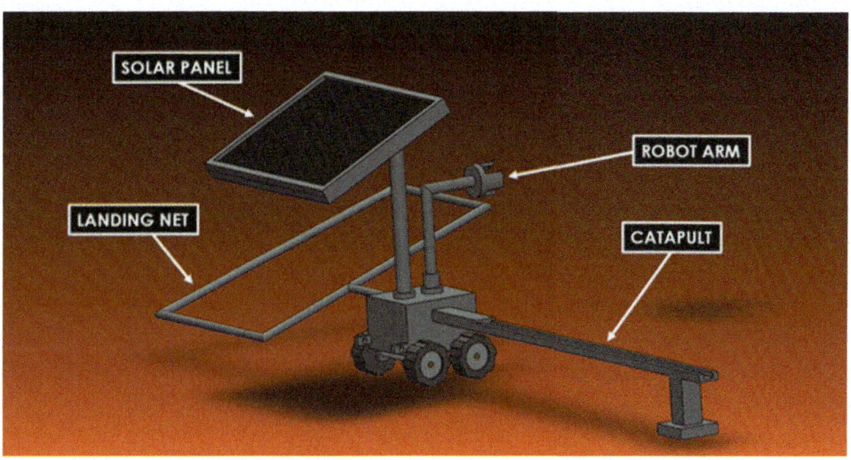

Fig. 1 Conceptual design of the rover

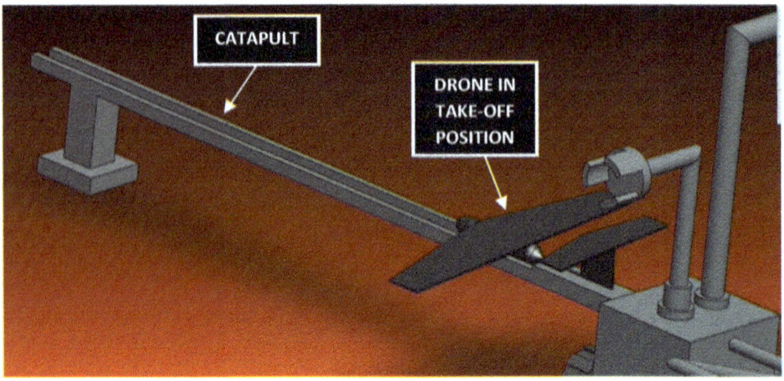

Fig. 2 Conceptual design of the drone located in the shuttle while it remains on the ground

In the Martian environment, several factors will drastically reduce the efficiency of the photovoltaic cells. The main ones are the lower radiation received compared to the Earth due to the planet's greater distance from the Sun; the low operating temperatures; the dust in suspension in the atmosphere (DOD, Dust column visible Optical Depth); and its deposition on the panels (the most important) [20]. Since dust plays a very important role in this area, the first 6 Martian months will be considered the maximum possible duration of the mission as the rest of the year matches the dust storm season [26]. However, considering the degradation of the panels in this environment, the target duration will be reduced to 250 sols. Thus, assuming a maximum time of two Martian days to accomplish the data transferring operations to the lander and the charging of the batteries of the UAV, at least 125 flights are expected to be carried out.

4.3 Science Payload

It is found that all of them have in common the sulfate ion (SO_4) by analyzing the chemical formulation of the gypsum and PHMS. Inspecting the reflection rates of some of these sediments, it is observed that all of them reach their maximum values in the short-wave infrared band (SWIR, 1–2.5 μm), around 1 μm wavelength.

In this way, the UAV will be equipped with the OWL 640 multispectral camera, a light (170 g), and a low energy consumption device (less than 2.5 W) whose operating range encompasses from 0.6 to 1.7 μm wavelength. To maintain the device in a suitable temperature range to protect it from the low Martian temperatures, it will be allocated inside the airframe and will take images through a borosilicate glass [9] assembled on the underside of the drone structure.

4.4 Conceptual Design Process Description

Throughout this subsection, the design process of the drone will be addressed. Firstly, a mass budget will be proposed and the aerodynamic design of the aircraft's wing and the aircraft itself will be carried out, which will provide the power requirements to be met by the propulsion plant. Then, the rest of the subsystems will be described, and commercial model components will be chosen for each one. Although these devices would need to be adapted to operate in the Martian environment, they will serve as a reference to perform a thermal study of the aircraft.

Mass Budget

Considering this value along with considering other UAVs with similar characteristics, the mass fraction that corresponds to each subsystem will be estimated [19, 40]. As a starting point, it will be considered a maximum mass of 4.5 kg for the aircraft, and a mass distribution by subsystems as shown in Table 3.

Airfoil Selection

The study compares different airfoils from the Airfoil Tools database and analyzes their behavior depending on the angle of attack (AoA) for cruise conditions at an altitude of 1000 m on Mars. The selection criteria are based on achieving high-lift coefficients to allow the plane to perform the necessary maneuvers while reaching high-efficiency values, and the momentum coefficient must remain negative throughout the α range. The airfoils that showed the best behavior for these conditions and were selected for a second phase of analysis were the S1091, the AG35, and the SD7037-092-88. The minimum Reynolds number for take-off during the lowest atmospheric density period of the mission is determined to be 16,890, and the S1091 airfoil is selected as the best-performing one. The study allows for the selection of the best-performing airfoil for the different flight phases of the aircraft, considering the evolution of the aerodynamic coefficients versus AoA for an established Reynolds range. The S1091 airfoil meets the selection criteria and is selected as the best option for the drone. The minimum Reynolds number is determined based on the evolution of the variable in the study area, and the velocity at which the UAV is launched is estimated at 30 m/s. The study shows that the selected airfoil is suitable for the cruise and take-off conditions of the drone.

Table 3 UAV mass budget

	Percentage (%)	Mass (g)
Structure	35	1575
Payload	5	225
Electronic devices	10	450
Power subsystem	40	1800
Propulsion system	10	450

Wing Design

Regarding the wing design, an average chord of 0.5 m and an AR (Aspect Ratio) between 5.5 and 6, are chosen as initial parameters based on other UAV prototypes destined to fly on Mars [10, 36]. A cruising speed of 65 m/s and a maximum flight altitude of 1000 m will be established, which leads to a Mach number of around 0.3 (considering the atmospheric features in the mission area). This Mach regime implies that the UAV will always fly in subsonic and incompressible regime, so no shockwaves will be produced. In terms of the Froude number, for cruise flights, this parameter will be 47.7 approximately. Therefore, inertial forces will be more relevant than the gravitational ones. Taking this information and input data into consideration, the aerodynamic studies were performed in the software *XFLR5*.

After analyzing the atmospheric conditions along the Martian sols using the *Mars Climate Database* software [18, 28], it is established that the aircraft will fly at 10 o'clock Martian time each day. At this time, the gaseous layer presents relatively high density due to low nightly temperatures, and there will be enough light to visualize the planet's surface. As explained in the thermal control section, a period of pre-heating is needed before the actuation for the equipment to be inside operational temperature limits.

Taking the wing design inputs previously mentioned as a reference, it has been studied the influence of each wing's geometrical parameter on the aerodynamic performance to create an efficient part. In this way, the following 3 m wingspan (AR = 6) design was proposed. To ease the manufacturing while obtaining high aerodynamic efficiency, a linearly tapered wing was designed. It was found that the highest c_l/c_d coefficient was achieved by implementing a 35 cm chord on the tips and 65 cm on the root after analyzing different distributions. Moreover, it has been observed that it only affects the c_m after performing some simulations varying the sweep angle. By increasing this parameter to the wing geometry, the slope of the moment coefficient becomes more negative, especially in the case of negative sweep angles.

As the payload camera is enclosed within the fuselage under the leading edge of the wing root, a positive sweep of 11.3° will be imposed to reduce the hidden part of the field of view due to the presence of the wingtips during turns. Moreover, the maneuverability loss compared to the straight wing one, may subsequently be compensated by moving forward the center of gravity of the aircraft, if necessary. Finally, a 2° twist will be applied to the root concerning the wing tips. Although the aerodynamic behavior is virtually constant when varying this parameter, a higher AoA in the inner part provides the wing with a smoothed stall peak. This is because the outer sections will continue generating lift while the central area reaches the stall angle.

Plane Design

Once both the wing design and the airframe component layout are completed, the aerodynamic design of the UAV will be completed using XFLR5. For this purpose, a fuselage structure was developed, and the stabilizers were designed using reference

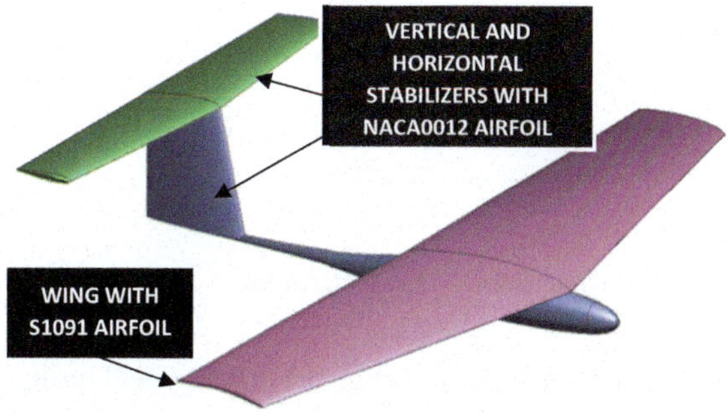

Fig. 3 Complete UAV design

volumetric coefficients for vertical and horizontal tails (see Fig. 3). The fuselage was divided into two sections of the same 95 cm length. The front one with a 17.5 cm diameter for allocating all the electronic components, while the rear one with a 2 cm diameter for supporting the tails. Then the wing was incorporated by placing the central section leading edge at 20 cm from the nose of the UAV.

The stabilizers were designed based on the symmetrical airfoil NACA 0012. Considering the low structural loads foreseen during the flight, a T-tail configuration was employed, because it offers a greater efficiency than other configurations [2]. The vertical stabilizer developed has an area of 0.157 m^2, a chord of 40 cm at the central section, and 29.8 cm at the tip. For the horizontal stabilizer, obtaining an area of 0.438 m^2, a span of 1.7 m, and a chord of 21.8 cm at the root and 13.8 cm at the tips. The weight of the whole aircraft determined for cruise flight is 0.2399. Therefore, an angle of $-3.85°$ (corresponding to this lift coefficient) is imposed for the wing relative to the fuselage. This very low value of the cruising angle of attack corresponds to an airfoil with a considerable curvature that allows for obtaining a big lift coefficient. This also allows enough margin in terms of lift coefficient difference with maximum lift coefficient to provide safe maneuvers at low speeds when taking off or in science mode. The aircraft results in the following UAV design.

Propulsion System Selection

The propulsion system for a drone will be sized based on the maximum thrust required during cruise flight at 65 m/s. The drag generated by the wing will be used as a starting point and the lift coefficient required for cruise flight will be determined by calculating the angle of attack provided by XFLR5. The drag breakdown of the MELOS UAV shows that the wing parasite component accounts for 46% of the total contribution, and extrapolating this data gives a total drag coefficient of 0.09175 for the design. The Scorpion SII-3020-1110 kV (V2) engine and a 9-inch diameter and 4.5-inch pitch APC propeller were selected based on the estimated effective power of 485.7 W, with a safety margin of 25%. To adapt the propeller for flight on Mars, the propeller-UAV mass ratio of the MELOS project drone was taken as a reference, and a mass of 83.57 g and an estimated diameter of 1.13 m were established for the propeller.

UAV Subsystems

Next, the subsystem description and electronic component selection of the UAV will be carried out and presented in Table 4.

4.5 *Component Layout and Thermal Study*

The component layout inside the fuselage will be done in search of a suitable thermal model (keeping all the devices within their respective temperature ranges) and a correct mass distribution (the center of gravity slightly forward of the aerodynamic center to easily achieve a statically stable configuration on the longitudinal axis). In this way, a thermal study is done through SolidWorks 2020, as shown in Fig. 4. For this purpose, a 17 cm diameter fuselage model is created including the electronic components with their real dimensions (provided by the manufacturer) and estimating or extrapolating their thermal properties. For the engine, keeping in mind that it is mainly made of aluminum alloys, the properties of the AW-7075-T6 alloy, widely used in the aerospace field, have been attributed to it. In the case of the devices contained in "Electronic components", estimated values for thermal conductivity and specific heat were established based on printed circuit boards. For the density, an average value for different materials that are commonly used for their manufacture was taken. When flying, the heat power emitted by the engine and batteries is estimated from their efficiencies, which are 90% [12] and 87% [22], respectively. Although the rest of the hardware will also release heat during operation, its consumption will be much lower, so it will be negligible and therefore it will be simplified by assuming that everything they consume is dissipated.

On the other hand, emissivity factors are established for the faces of all the components, imposing 0.95 for the batteries [37], 0.45 for the engine [34], and 0.8 for the rest of the elements. However, these rates could be altered by changing the color of the external surface of the devices. During flight (hot operational case) and before take-off (cold operational case), the estimated component operating temperature ranges will be considered [22]. However, since the aircraft lands and until the moments before a new flight (cold survival case), the devices must be kept within their survival range, and they must be warmed up to their operational lower temperature before operations. In this way, from the hot operational case of study, the airframe distribution will be determined. For these conditions, an ambient temperature of 241 K will be established. It should be noted that for the three analysis results, a safety margin of 10 K will be left to compensate for the limitations of the simulation approach.

As can be seen in the image above, three copper conductive lines (8900 kg/m^3 density, 401 Wm^{-1} K^{-1} thermal conductivity, and 390 Jkg^{-1} K^{-1} specific heat) [17] have been added to conduct heat between the components or to evacuate it to the fuselage [7]. The results achieved by this configuration are shown in Table 5, where it can be observed that all temperatures are inside the allowed limits.

Table 4 Subsystem description and component selection

Subsystem description	Component
Aerodynamic	
It comprises the airframe, wings, stabilizers, control surfaces, and high-lift devices. The material used for its manufacture will be a 0.4 mm thickness CFRP skin made of carbon fiber and epoxy resin with 1540 kg/m^3 density, like the UAV developed by JAXA [19]	
Propulsion	
The propulsion system will be electric, as mentioned earlier. In this way, the engine will be selected according to the aircraft's propulsion requirements, and the rest of the elements of the set must be compatible with it. Selection: a type of scorpion power system with a commander and propeller	
Communications	
It will be based on relay communications. In this way, the UAV will be equipped with a radio to exchange information with the rover, then the rover will establish a proximity link with an orbiter, and this satellite will send the electromagnetic waves to Earth where they will be received by the high-sensitivity receivers of the deep space network stations [15]	
Data handling	
It encompasses collecting, sorting, sending, and presenting procedures of the information collected by the UAV. The large data volumes and the limited data rate make it necessary to resort to these methods. In this way, lossless algorithms (compression ratios around 2:1) and lossy algorithms (higher compression ratios at the cost of some loss of resolution) will be used to reduce the size of the image products [15]	
Control	
A microcontroller will dominate the behavior of the drone. The integration of this device in the UAV control loop provides a complete set of avionics that allows it to complete missions autonomously [23]. However, it is essential to equip the aircraft with positioning systems to enable its operation	
Thermal control	
It will be necessary to equip the UAV with thermal control systems to protect the electronic devices from the extreme temperature variations of the Martian environment. These systems will make it possible to keep all the components in a proper range all the time. The following subsection will address this issue	
Power	
LiPo batteries will be used for energy storage considering their advantages over other types (efficiency of around 90%, no memory effect, high energy, and power density, etc.). The serial inter-cell connection (S) will depend on the supply voltage of the motor and the parallel (P) will determine the flight endurance	

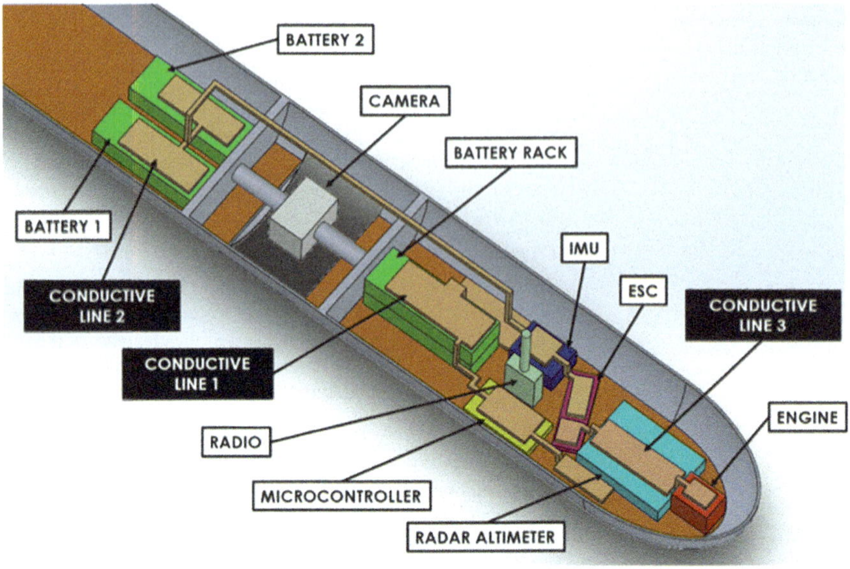

Fig. 4 Airframe layout

Table 5 Hot operational case thermal study results

Component	Operating range (K)	Average temperature (K)	Study result (+10 K)
Battery 1	273–323	301.5	311.5
Battery 2	273–323	309.6	319.6
Battery rack	273–323	311.4	321.4
Engine	253–358	305.8	315.8
ESC	273–318	281.7	291.7
IMU	273–318	290	300
Radar altimeter	273–318	288.1	298.1
Radio	273–318	283.2	293.2
Camera	273–333	281.1	291.1
Microcontroller	273–318	279.3	289.3

The cold survival case is analyzed based on the previous results, followed by the cold operational case. For these two studies, the 10 K safety margin will be subtracted rather than added as they are cold cases. A suitable thermal equilibrium inside the operational temperature limits has been achieved by incorporating electrical heaters on certain aircraft components. For the first case, a power of 117 W will be required and for the second one, it will be 100.3 W. These values would be included in the power budget of the drone.

4.6 *Results and Conclusions*

Along with this study, the development of an electric-powered concept aircraft capable of operating and surviving in the Martian environment has been achieved. Despite the adverse conditions of the Martian atmosphere in terms of flight performance and the low capacity of current batteries, a UAV prototype with good aerodynamic behavior and around 25 min of endurance considering the maximum consumption conditions (cruise flight) has been reached. However, given the mass and volume constraints of existing space probes, it could be complex to deliver the entire required infrastructure to the planet's surface. Even the search for a method to deploy all the rover's accessories could be a complex issue to materialize the project. Nevertheless, considering all the advantages that the use of drones would provide in this field, the conclusions drawn from this study could serve as a reference for the development of future missions.

5 Use Case 2: A Hybrid Drone in Titan: Argonaut

5.1 *Introduction*

As explained before, Titan is Saturn's largest satellite, and it contains a dense and cold atmosphere, rich in hydrocarbons. Its low gravity and wind speed range make this moon ideal for drones to fly. There exist theories about a water cycle based on methane, and the possibility of an underground sea of ammonia or hydrocarbons. Unfortunately, the dissociation of methane and nitrogen at high altitudes creates a dense and opaque layer, making the study from an orbiter complicated. A flying drone equipped with scientific instruments for measuring various Titan characteristics is proposed. The mission is estimated to last 3 years, and the drone will finish by studying the properties of the Kraken Mare.

The mission will use two Multi-Mission Radioisotope Thermoelectric Generators (MMRTG) to operate. The drone will perform experiments at different sites on land and in-flight, through vertical take-off and landing. At the end of the mission, it will float in Kraken Mare to study the properties of the sea. The design illustrated in this chapter was performed iteratively using *Xflr5* accompanied by aerodynamic calculations. Subsequently, wind tunnel tests of a 3D-printed model were performed to verify the validity of the simulations and their results. The non-dimensional similarity parameters in test conditions were fitted as better as possible to Titan's conditions. Additionally, material selection studies, the CAD design of the complete drone, and FEM structural simulations were developed in SolidWorks 2016.

5.2 Mission Description

The scientific objective of this drone mission is the study of the environmental conditions, orography, and chemical composition of Titan's different environments, such as dunes, plains, atmosphere, or hydrocarbon lakes. The aim is to confirm or disprove the numerous theories about the possibility of life, cryovolcanism, surface erosion, and the methane cycle. Therefore, the major phases of the mission that lead to the execution of the proposed objectives will be presented. An overview of the concept of operations (CONOPS) that constitutes this mission, is presented in Fig. 5.

This graphical representation shows the operations that will take place once Argonaut reaches its destination. Starting with atmospheric entry, which is followed by the deployment of the parachutes that will slow the drone during its descent. Once the speed is reduced and the desired altitude is reached, the Argonaut will descend to land for the first time on the plains of Shangri-la. From then on, for 2.5 years, a succession of standard flights (marked in green in the figure) will take Argonaut to explore different sites on Titan, analyzing its atmosphere during the flights and its terrain when landed. Aerial images of the environment will also be taken during the flybys, as well as mapping sites of interest. It will start each flight from a previously confirmed safe location, and with a safe landing destination already determined from the previous flight. During the cruise segments, mapping, imaging, and atmosphere analysis can be performed. Once arrives the next landing point, several imaging and mapping procedures will take place while hovering. This information will be sent

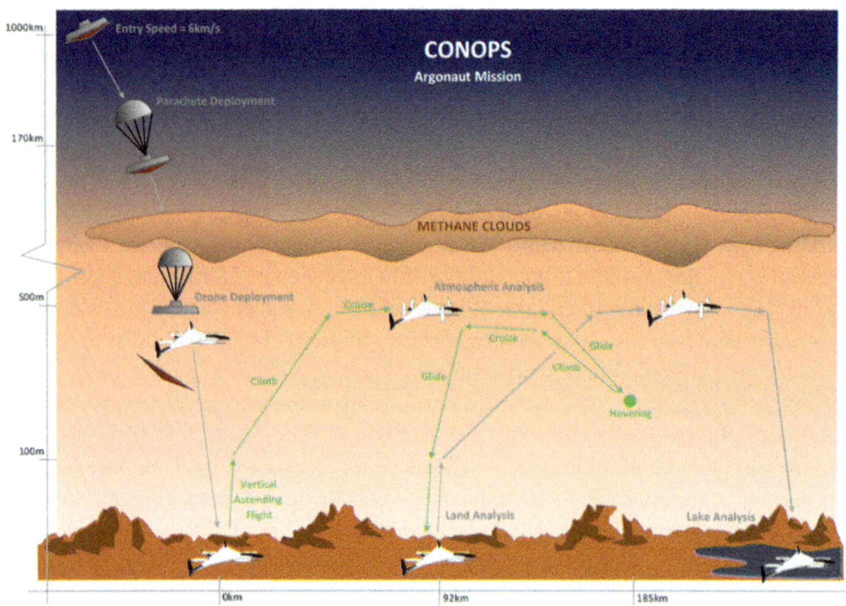

Fig. 5 Argonaut concept of operations

to Earth, once returned to the second landing site, to determine if the new location is safe for landing while recharging batteries and performing soil analysis, usually at night. This cycle was designed with a battery consumption of 75% (7560 Wh fully charged), and an exploration range of 185 km at an altitude of 500 m (without experimentation, only for transporting purposes), much higher than any rover could cover.

At the end of its lifetime, Argonaut will make its final flight to Kraken Mare, where it will analyze the environment and composition of Titan's liquid environment for 6 months, in a floating state on its sea surface. Argonaut will finally have completed its mission and will proceed to shut down its systems. In Table 6, the phases of the entire mission are summarized.

5.3 Argonaut Conceptual Design Process Description

This section will describe the most relevant steps in the Argonaut design process. For the choice of the vehicle configuration, several options were evaluated: first, a conventional quadrotor like Dragonfly due to the vertical take-off and landing capability, and the directional thrust with rotatable propellers; secondly, a Blended Wing Body (BWB) configuration, due to its simplicity without tail and a differentiated fuselage. The BWB has no drag associated with those aircraft components, nor the usual negative lift from the horizontal stabilizer, making it more efficient and with greater lift capacity [27]. One of the drawbacks of the BWB configuration is the limited space for the equipment (only the wing central area). However, the main problem of this configuration is to obtain flight stability in the absence of a tail. There are solutions for this, such as the use of reflex airfoils in combination with airfoils of higher lift capacity and high moment coefficient along the wing, or the use of winglets for lateral-directional stability as vertical surfaces. A scheme of the performance comparison between two types of drones is shown in Fig. 6.

Following the selection of the basic configuration, similar vehicles were studied. In this case, modern BWBs such as MAVERIC and X-48C, Airbus, and Boeing designs respectively, were used [3, 21]. The last study was used as a reference for the planar geometry of the initial vehicle design (see Fig. 7).

Component Selection

The selection of the components of the conceptual design was subdivided into the payload (components with scientific utility and not dependent on the aerodynamic design), as summed up in Table 7, and the service platform (dependent on the design that performs the rest of the functions necessary for the mission), as shown in Table 8. The aim was to select a realistic set of reference components of proven efficiency, mainly from previous missions, that meet the necessary requirements. With the reference list, the total mass, the mass and occupied volume distribution, and the power consumption of the different operational modes were characterized. In Fig. 8, the final

Table 6 Mission phases

Argonaut mission phase	Schedule activities
Pre-operational phase	1.1 **Launch**
	1.2 **Commissioning**
	1.3 **Travel to Titan**
	1.4 **Capture by Saturn and Titan**
	1.5 **Parking orbit**
	1.6 **Heat shield detachment for descent to Shangri-la**
Operational phase	2.1 **Entry**
	2.2 **Parachute deployment**
	2.2.1 First parachute
	2.2.2 Second parachute
	2.2.3 Descent
	2.3 **Argonaut deployment**
	2.3.1 Top heat shield detachment
	2.3.2 Argonaut wings deployment
	2.3.3 Bottom heat shield detachment
	2.3.4 Argonaut vertical descent
	2.3.5 Argonaut glide to Shangri-la landing area
	2.3.6 Argonaut vertical descent for landing
	2.4 **Standard flight exploration**
	2.4.1 Standard flight
	2.4.2 Experimental phase
	2.4.3 Communication
	2.4.4 Battery charge
	2.5 **Last flight to Kraken Mare**
	2.5.1 Glide to Kraken Mare
	2.5.2 Vertical descent for splashdown
	2.6 **Lake exploration**
	2.6.1 Experimental phase
	2.6.2 Communication
	2.6.3 Battery charge
Disposal phase	3.1 **System and subsystems disconnection and passivation**

layout of the components is shown, which considers both component functionality and the proper location of the center of mass for stability.

The design of the thermal subsystem was based on the utilization of the waste thermal energy from the RTG (2000 W) for heating the two-component zones. This selective heat transport is carried out by the heat pipes, which using the designed thermal switches will regulate the heat supply to the component zones separately, as

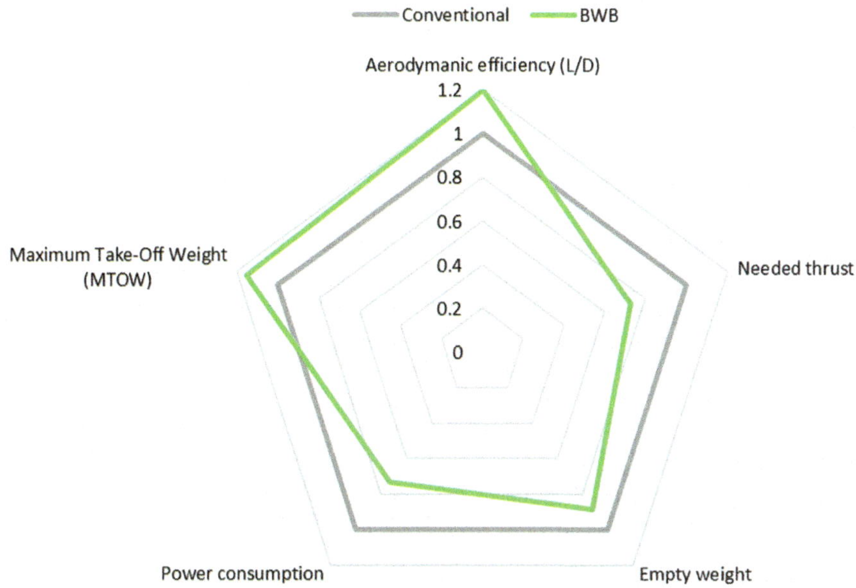

Fig. 6 Performance comparison between the BWB and the conventional configuration (dimensionless values)

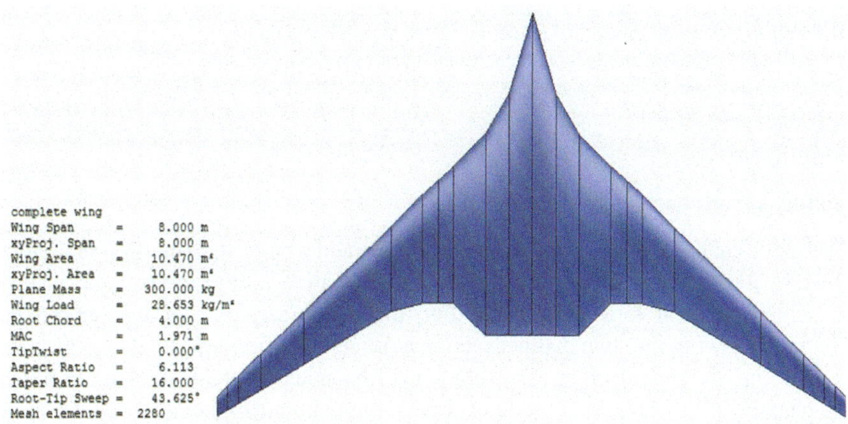

Fig. 7 Initial planar form of the Argonaut (Xflr5)

shown in Fig. 9. For the cooling of these zones, a quantity of Titan's air at −180 °C regulated by the louvers will be allowed to pass through.

Table 7 Components selection for the payload

Component	Function	Reference
Scientific subsystem		
Nano drill	Rotary hammer drill for the extraction of solid samples	Honeybee robotics
MOMA	It combines LDMS (laser desorption mass spectrometry) techniques for solids analysis and GCMS (gas chromatography mass spectrometry) for their volatiles, plus an ion trap	[30]
TLCA	Characterize the atmospheric and lake composition and analyze the presence of any biomarkers	[16]
MAG sensors	Located at one extreme of the Argonaut's structure to detect small variations in Titan's magnetic field	[16]
ASI/MET	System that will characterize the atmosphere and meteorology during entry and flights	[16]
SPP	Sonar with acoustic sensors and accelerometers to characterize the properties of the liquid	[16]
Optic subsystem		
TRS	Obtain the surface profile of Titan's relief in flight, and results from lower layers of the terrain thanks to the penetration of the waves	[16]
VISTA-B	Identification of the orography, study of meteorological phenomena and study of potential landing sites or details of interest	[16]
BIS optics	Study the optical properties of the atmosphere and its meteorological phenomena and analyze the composition of the satellite's fogs and clouds, as well as the composition of the ground	[16]

Airfoil Aerodynamics Design

The aerodynamics design step was crucial for addressing the stability challenges faced by flying wings. Due to the generation of negative C_m by conventional airfoils, to achieve natural longitudinal stability, the aerodynamic twist was employed, dividing the wing into two distinct sections: the central part, dedicated to generating most of the lift, and the outer part, with lower lift capacity but intended to counteract the negative C_m. The central sections of the wing feature a higher chord, with a relative maximum thickness exceeding 10% to accommodate the necessary equipment on the inner side. Furthermore, airfoils with the highest possible C_m values were selected by employing the Airfoil Tools search tools, and other reference sources [1, 4, 35, 39].

The airfoil selection process was conducted in *Xflr5*, consisting of two stages with increasing accuracy. The initial stage involved performing a Batch Analysis of the profiles for each specific zone separately, considering three representative Reynolds numbers for each zone: 35.5, 24, and 13.5 million for the central section, and 13, 7, and 2 million for the outer section, relative to the initial geometry. From each

Table 8 Component selection for the services platform

Component	Function	Reference
Flight control subsystem		
Propellers	Design using the theory of quantity of movement with typical values and conditions for the cruise flight [38]	B-WWR64W of Whirl Wind Propellor
REG 20	Four motor units, one to power each propeller	Rotex Electric
LN-200	Inertial measurement unit to register axial and angular accelerations	[31, 32]
Speed controllers	One by motor, to regulate the angular rate of each one according to the flight computer instructions	MGM Compro HBC Series TMM10063-3
HAPT	To measure the pressure for flight conditions and altitude indication	Bradford space
ZSH 57/1	Total of 18 servo motors for moving parts and surfaces	McLennan
VECTOR	Autopilot with capabilities for both fixed-wing and rotary-wing flights	UAV navigation
Data handling and communications subsystems		
Sun sensor	To detect day/night cycles, to confirm the attitude of the drone, and to orient the HGA toward the Sun (the Argonaut-Earth line of sight is offset by a maximum of 6° from the Argonaut-Sun line of sight)	[16]
HGA	50 cm diameter high gain antenna to amplify and direct the communications with the Orbiter	[16]
DHS	Data handling system to store the data until it is sent to Earth and to handle and convert power	[16]
X/X Transponder, TWTA X-Band and RFDU	System for operations of reception, data processing and decoding, and transmission of the information	[16]
Thermal subsystem		
Thermal louvers	To insulate from a cold thermal environment	Orbital Company's AMPTE
Heat pipes	To distribute heat from the hot spots, typically the RTG area, to the component cold spots when necessary	Thales Alenia space
Temperature sensors	To monitor the temperature of the different areas of interest of the drone	ISTAG P0K2.232.7W (010.02994)
MLI	Multi-layer insulation against the cold exterior and as insulation between the different component areas	Aerospace fabrication PolaWrap and SpaceWrap
Thermal switches	Simple two-position mechanisms that regulate thermal conduction by rotating an insulating plate with an embedded heat-transmitting material	Own design

(continued)

Table 8 (continued)

Component	Function	Reference
Electric power subsystem		
Battery	A lithium-ion battery adapted to the geometry of the component area	[14]
MMRTGs	Two Multi-Mission Radioisotope Thermoelectric Generators to provide 2 kW of thermal energy and 125 W of electrical energy each (at the beginning of their life, after the journey to Titan they will produce 110 W each)	[16]
Structure		
Fuselage	Made of 60% ultra-high strength carbon fiber with epoxy resin, selected as the best after a study with CES Edupack using material indices. Using the law of mixtures for the composite, the yield strength is $4.574 \cdot 10^{11}$ N/m^2, and the theoretical thickness required in the fuselage walls is 1 mm. These results were verified using a FEM study with SolidWorks2016 applying the wing load to one half of the drone verifying a low deformation in cruise flight and that the equivalent Von Mises stress is much below the elastic limit. Material values used in these simulations are at room temperature, although they should be even better at the low temperatures of Titan according to Okayasu and Tsuchiya [33]	Own design
Rotor brackets	To support the stresses on the rotors, both the front and rear rotors are made of the same carbon fiber as the fuselage and with elliptical profiles	Own design
Landing skids	Made of aluminum 2024 T4 these structures serve as support for the Argonaut's landing	Own design

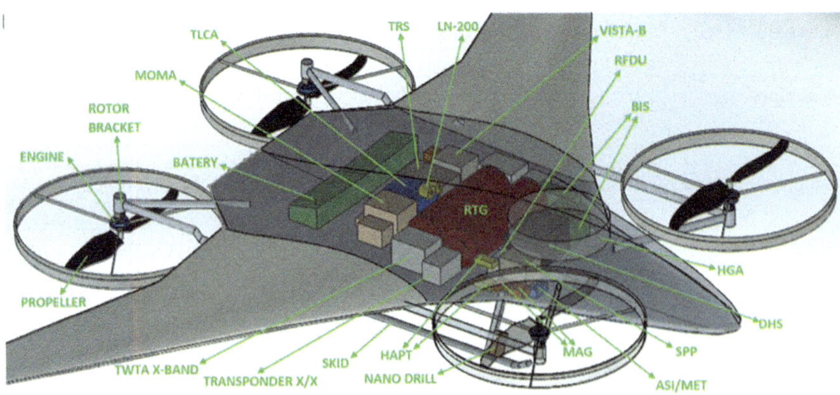

Fig. 8 Argonaut's components' location

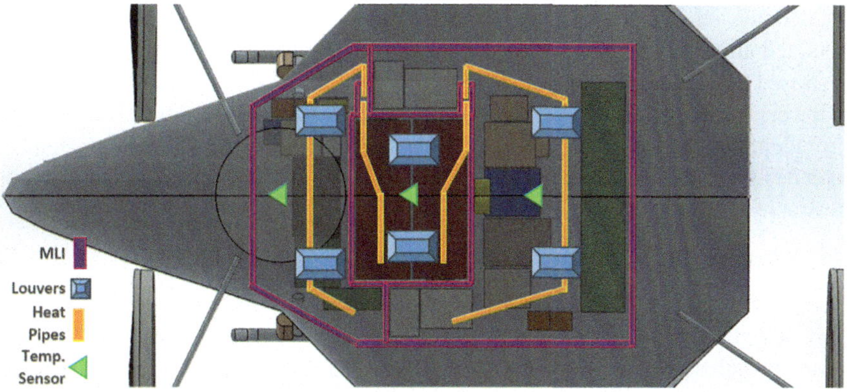

Fig. 9 Thermal subsystem scheme

analysis, the lift coefficient (C_l), drag coefficient (C_d), and moment coefficient (C_m) were extracted for the angle of attack (AoA) that yielded maximum efficiency for each profile. This enabled a rapid comparison of all profiles, facilitating the selection of three candidates from each wing area to proceed to the subsequent phase. The study conditions were established with a density of 5.4 kg/m^3, viscosity of 1.111×10^{-6} m^2/s, and a nominal cruise speed of 10 m/s. These conditions were based on those of Dragonfly, which were estimated to be suitable for the Titan flight study [25]. The selection criteria for the central section favored airfoils with a C_m closer to 0 rather than emphasizing aerodynamic efficiency. This was because aerodynamic efficiency could be compromised if a too-negative C_m had to be corrected in later design phases. Based on this principle, the selection process narrowed down the number of airfoil options to three candidates: Eppler 343, Prandtl D-Root, and NACA 25111. Regarding the exterior area of the wing, since the list of airfoils includes several reflex profiles, positive moment coefficients are observed. It is important to note that when both geometric twist and aerodynamic twist are required to counteract the negative moment in the central area, it results in a loss of aerodynamic efficiency. To minimize the need for excessive geometric twists in the flying wing, the selection criteria remained unchanged, prioritizing airfoils with the highest natural C_m, and among them, the most efficient options. Consequently, three airfoils were selected: LA 2573A, MH 78, and FAUVEL 14.

In the second selection stage, the airfoils were applied to the initial geometry in all nine possible combinations and analyzed to determine the most suitable one. For this purpose, a VLM2 (Vortex Lattice Method) analysis was conducted at a speed of 10 m/s for the nine combinations of central and outer profiles. The results were evaluated at the angle of attack (AoA) that provided a lift of 411 N, equivalent to 300 kg in Titan. At this AoA, the following characteristics were compared: aerodynamic efficiencies, the slope of the C_m curve concerning the AoA (with more negative values being more desirable), and the location of the center of gravity relative to the nose to achieve $C_m = 0$. This location favored the fit of the equipment and the flight dynamics as a quadcopter, requiring it to be positioned around 2.75 m from the nose. The results obtained after the VLM2 analysis are shown in Table 9.

Based on the above results, the combination of the NASA Prandtl D-Root airfoil in the center sections and LA 2573A in the outer sections was selected. These airfoils provide the center of gravity furthest away from the nose, which gave freedom for later design steps, and the second highest aerodynamic efficiency, making it the most desirable combination.

Wing Aerodynamic Design

Starting from the modified initial design geometry with the airfoils selected in the previous subsection, an iterative cycle was initiated, systematically varying the main geometric characteristics of the wing. In each cycle, the first characteristic to be varied was the zonal twist between the central and outer wing areas, allowing the airfoils to operate at different AoA. The twist was adjusted in a range of −1.5° to 2°. A VLM2 analysis was performed for an airspeed of 10 m/s and AoA ranging from −5° to 20°. The sweep was also investigated in the outer wing area while keeping the wingspan, mean aerodynamic chord (MAC), and projected wing area constant for each variation of the design. Sweep angles from 35.1° to 48.1° were tested. The third feature analyzed in each cycle was the winglets, which would function as vertical tail rudders to reduce wetted surface and drag on the model. Five types of winglets were designed, all dependent on the design characteristics fixed in the iterative cycle, such as the distance between the wing and winglet quarter chord (L_{vt}), the tail surface (S_{vt}), the wingspan (b_W), and the wing surface (S_W). The design of the winglets utilized the volumetric coefficient of the vertical tail, which was obtained based on the MAVERIC aircraft, with a value of 0.02. This led to a surface area of 0.55 m^2 between the two vertical tails, and an area of 0.275 m^2 per winglet, resulting in the five designs shown in Fig. 10.

Table 9 List of airfoils and setups comparison

Central section	Outer section	AoA [°]	C_l/C_d	Xcg for $C_{mcg} = 0$ [m]	Slope of C_m versus AoA
NACA 25111	LA 2573A	1.7	21.57	2.68	−0.0365
NACA 25111	MH 78	1.96	21.52	2.58	−0.0368
NACA 25111	FAUVEL 14	2.01	20.96	2.58	−0.0369
Prandtl D-Root	LA 2573A	1.97	22.215	2.80	−0.0368
Prandtl D-Root	MH 78	2.24	22.82	2.70	−0.0369
Prandtl D-Root	FAUVEL 14	2.29	22.21	2.70	−0.0369
EPPLER 343	LA 2573A	1.12	19.068	2.73	−0.0367
EPPLER 343	MH 78	1.38	18.9	2.64	−0.0372
EPPLER 343	FAUVEL 14	1.42	18.512	2.62	−0.0368

Fig. 10 Winglet designs (type 1 to 5 from left to right)

$$C_{vt} = \frac{L_{vt} S_{vt}}{b_w S_w} \tag{1}$$

Within the winglet design iterative process, several VLM2 analyses were conducted to determine the most suitable winglet, providing the geometry for the next iteration of the design cycle. The selection criteria for each variation were based on comparing desirable characteristics for the vehicle, including aerodynamic efficiency, power dissipation due to drag, the appropriate center of mass location for zero moment, and the static margin, all at the angle of attack required for cruise flight. Mission-related constraints were also considered, such as a moderate sweep to facilitate the folding of the drone and its accommodation within the heat shield and launcher. Finally, the dihedral parameter was studied. The variations were made in the exterior wing area, ranging from the existing 0° dihedral to 3°. In addition to the VLM2 analysis, a comprehensive stability analysis was conducted in *Xflr5* for these four dihedral variations. The obtained root locus results enabled the selection of a 3° dihedral as the most stable configuration (indicated in pink in Fig. 11).

Through the analysis of the time response, it was determined that the unitary disturbance values for the different modes were adequately damped, indicating sufficient stability for controllability. The recovery times were as follows: 0.9 s for the short-period mode, 120 s for the phugoid mode, 0.12 s for roll damping, 7 s for the Dutch roll mode (which is naturally stable but often problematic for flying wings), and slightly unstable spiral mode that could still be controlled using automatic flight control. Following this analysis, the iterative design and refinement process of the

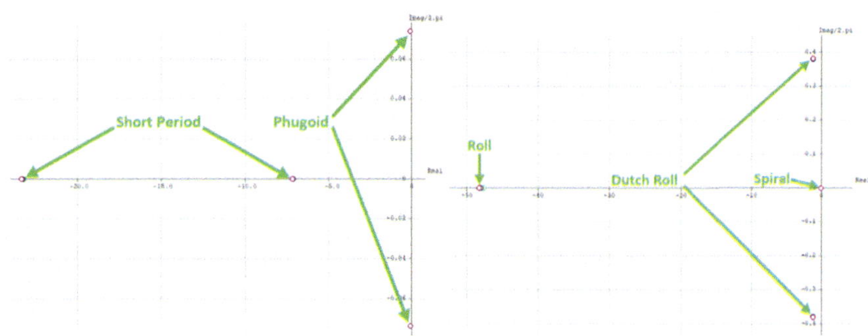

Fig. 11 Longitudinal (left) and Lateral (right) modes root locus

Table 10 Main characteristics of the drone designed for Titan

Lift	479.5 N	AoA	1.6°	X_{cg}	2.53 m
Central section airfoil	Prandt D-Root NASA	Twist	1.5°	Y_{cg}	0 m
		Sweep	37.4°	Z_{cg}	0.07 m
		Dihedral	3°	C_l/C_d	23.1
Outer section airfoil	LA 2573A	Winglet	Tipo 3	X_{NP}	2.67 m
		Wing surface	10.77 m^2	AR	5.9
				Tapper ratio	16
Winglet airfoil	NACA 0009	Winglets surface	0.55 m^2	Dissipated power	207.0 W
		MAC	2.1 m	Speed	10 m/s

Argonaut was completed, resulting in a highly efficient and naturally stable flying wing. The final geometric values and the most significant design parameters are presented in Table 10.

5.4 Experimental Aerodynamic Verification

To validate the use of *Xflr5* as an aerodynamics design tool for the Argonaut, wind tunnel tests were developed (refer to Fig. 12). To accommodate the prototype within the test chamber, the final design geometry of the Argonaut was scaled down to 1/8 of its original dimensions. A PLA model of one of the halves was 3D printed and polished with epoxy resin on its surface. Additionally, a bar was inserted at the calculated center of gravity position in the design, which served as the connection to the Eiffel-type wind tunnel force sensors.

The purpose of the wind tunnel experiment was to compare the drag and lift coefficient results at different angles of attack obtained in the tunnel with those obtained in *Xflr5*. To achieve this, it was necessary to maintain geometric and aerodynamic similarity by scaling the model as mentioned earlier, as well as ensuring equality of the Reynolds number between both cases.

$$Re = \frac{\rho V \cdot MAC}{\mu} \tag{2}$$

where ρ represents density, V represents velocity, μ represents viscosity, and MAC represents the Mean Aerodynamic Chord. A problem arose when considering the Reynolds number for Titan's cruise conditions, which was 18,900,000. This would imply a wind tunnel speed of 1,085 m/s, which was unattainable in the facility and would result in a different Mach regime compared to Titan. The solution to these two problems came from adjusting the velocity to the maximum value that the wind tunnel could offer, 39.8 m/s, which yielded a *Re* of 695,000 under Earth conditions. This, in turn, corresponded to a very low velocity of 0.37 m/s on Titan to maintain

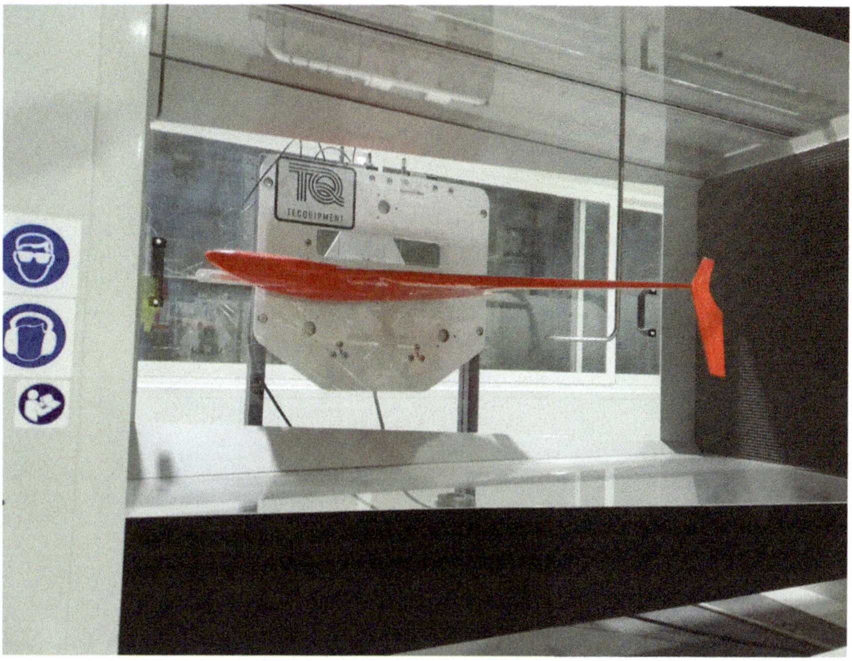

Fig. 12 3D printed wind tunnel model of the Argonaut in position

the same *Re*. As a result, a new analysis at this velocity was conducted in *Xflr5* using Titan conditions, and it would serve for the comparison of aerodynamic coefficients. With this adjustment, both Mach numbers remained below 0.3, ensuring they were in the same incompressible regime. It should be noted that the variation of lift and drag coefficients with Mach number is negligible at low values, so the coefficientts obtained in *Xflr5* for 0.37 m/s and 10 m/s are practically identical, and the velocity change is still valid for verification purposes. Lastly, according to Moreno et al. [29], the Froude number (Fr) should be the same in both cases.

$$Fr = \frac{V}{\sqrt{g \cdot MAC}} \tag{3}$$

However, it is not possible to achieve this given the conditions (where g represents gravitational acceleration). The Froude number in the wind tunnel is 614, while on Titan, it is 0.05. This means that on Titan, the inertial forces are negligible compared to gravitational forces, whereas the opposite is true in the Earth's wind tunnel. As a final note, if it is needed to maintain the *Fr* of Titan on Earth, the velocity would have to be 0.35 m/s in the wind tunnel. Moreover, to maintain the Reynolds number would require an air density of 129.8 kg/m^3, which is impractical. After specifying the test conditions, the comparative results of the lift coefficients and drag coefficients for

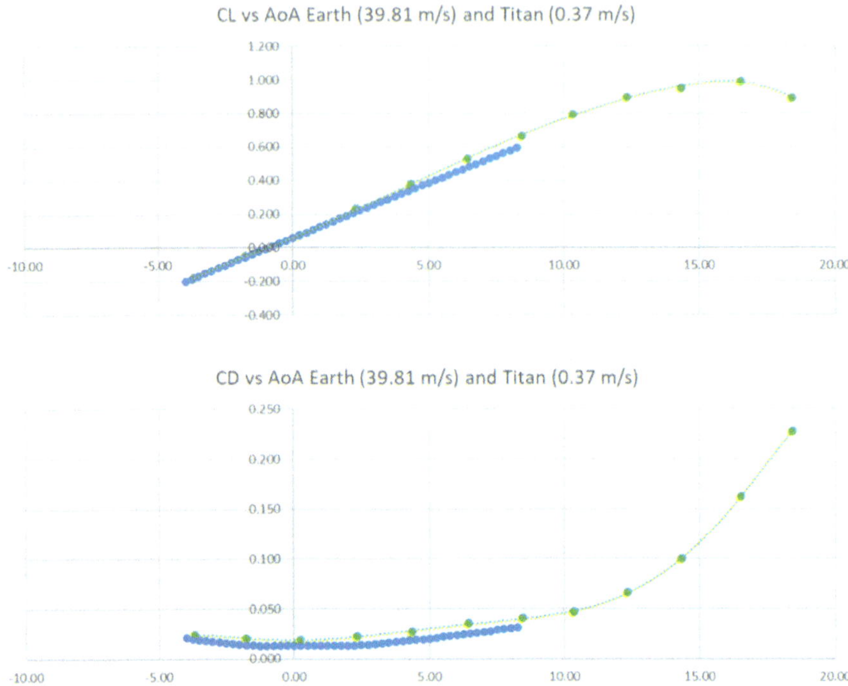

Fig. 13 Comparative results of the aerodynamic coefficients obtained in the wind tunnel (green) and Xflr5 (blue)

different angles of attack (AoA) obtained in the wind tunnel versus those obtained in Xflr5 are presented in Fig. 13.

Based on the comparative graphs, it can be concluded that not only do the *Xflr5* results faithfully represent the trends in the evolution of the coefficients with angle of attack, but their values are also very close to those found in the experimental study. The mean square error was 0.0326 for the lift coefficient and 0.00837 for the drag coefficient. These errors would be even smaller if only the angles of attack used in the Argonaut operation, ranging from -2° to 5°, were evaluated. Therefore, the use of *Xflr5* as a conceptual aerodynamic design tool is considered validated due to the close correlation between its results and those obtained from a real experiment.

5.5 *Results and Conclusions*

After designing the Argonaut and analyzing its performance, several conclusions can be drawn regarding the achievable mission. Firstly, in terms of flight range, the Argonaut surpasses any rover by providing a significantly larger exploring capability in a shorter time frame. Additionally, it offers a safer exploration experience

Fig. 14 Artistic image of Argonaut flying in Titan

by avoiding the hazardous obstacles present on the terrain. The flying wing configuration of the Argonaut, which results in higher aerodynamic efficiency, allows for a flight range over ten times greater than that of a quadcopter drone like the Dragonfly. The advantage of this aerodynamic efficiency also leads to higher lifting capacity at the same energy cost. As a result, the Argonaut can transport a more substantial scientific payload, enabling a greater number of experiments and analysis of Titan's environments. Furthermore, the Argonaut is equipped to study the liquid environment of Titan, adding significant value to the exploration mission. In conclusion, the Argonaut and its mission concept present numerous advantages and are based on existing technologies that can be immediately integrated. This mission can provide a comprehensive analysis of Titan's diverse environments, thus validating the theories currently under evaluation. An artistic image depicting the Argonaut is presented in Fig. 14.

6 Conclusions

In conclusion, numerous projects for space exploration drones have been initiated, but only one has been completed: Ingenuity. The design of space drones is heavily reliant on the requirements set by other carrier vehicles, such as launchers, space vehicles, orbiters, and potentially landers or surface vehicles, to reach distant planets and moons. Extraterrestrial atmospheres possess distinct characteristics, many of which remain unknown at this stage. Mars and Titan emerged as the most promising targets for space drones, and in this chapter, a conceptual design for fixed-wing drones has been proposed for each destination. A thorough understanding of aerodynamics is crucial to comprehend the aircraft's performance in these unique atmospheres, considering key non-dimensional numbers such as Reynolds, Freud, or Mach. However, other subsystems are also under investigation. Thermal control is vital for survival in the extreme conditions of Titan, while propulsion systems on

Mars must compensate for the low thrust resulting from the air-based propellers. Power generation complicates the design because external power sources, such as solar energy, are limited. In contrast, an inner power source like RTG can provide a long lifetime otherwise wouldn't be possible. Communication would need intermediate stages supported by another orbiter of ground vehicles before reaching the Earth. Finally, a huge challenge is for navigation systems that would be completely autonomous in unknown environments. Each destination necessitates a dedicated analysis, but they all share the common challenge of extremely harsh flying conditions. Perhaps it will take a long time before one can see drones flying routinely over Titan, but the journey has already begun through research and development of the technology to make such aerial photographs of the Kraken Sea a reality.

References

1. Alsahlam, A. A., & Rahulan, T. (2017). Aerofoil design for unmanned high-altitude aft-swept flying wings. *Journal of Aerospace Technology and Management*, 1–7.
2. Anderson, J. D. (1999). *Aircraft performance and design.* WCB/McGraw-Hill.
3. Airbus. (2020). *Airbus, media.* Accessed Accessed 23 March 2024 from https://www.airbus.com/en/newsroom/press-releases/2020-02-airbus-reveals-its-blended-wing-aircraft-demonstrator
4. Baig, A., Cheema, T., Aslam, Z., Khan, Y., Sajid Dar, H., & Khaliq, S. (2018). A new methodology for aerodynamic design and analysis of a small scale. *Journal of Aeronautics & Aerospace*, 1–6.
5. Balaram, J., Aung, M., & Golombek, M. P. (2021). The ingenuity helicopter on the perseverance rover. *Space Science Reviews, 217*(4), 56.
6. Barnes, J. W., Lemke, L., Foch, R., McKay, C. P., Beyer, R. A., Radebaugh, J., Atkinson, D. H., Lorenz, R. D., Le Mouélic, S., Rodriguez, S., & Gundlach, J. (2012). AVIATR—Aerial vehicle for in-situ and airborne titan reconnaissance: A Titan airplane mission concept. *Experimental Astronomy, 33*, 55–127.
7. Baturkin, V. (2005). Micro-satellites thermal control—concepts and components. *Acta Astronautica*, 161–170.
8. Benne, B., Dobrijevic, M., Cavalié, T., Loison, J.-C., & Hickson, K. M. (2022). A photochemical model of Triton's atmosphere paired with an uncertainty propagation study. *Astronomy & Astrophysics, 667*, A169.
9. Bernstein, K. S. (2007). Structural design of glass and ceramic components for space system safety. In *Second IAASS Conference: Space Safety in a Global World.* Chicago.
10. Braun, R. D., & H. S. (2006). Design of the ARES mars airplane and mission architecture. *Journal of Spacecraft and Rockets*, 1026–1034.
11. Clarke, J. D. A., & Willson, D. (2017). Southern meridiani planum - A candidate landing site for the first crewed mission to Mars. *Acta Astronautica*, 195–220.
12. Colozza, A. J. (2003). *Comparison of mars aircraft propulsion systems.*
13. Crisp, D., Pathare, A., & Ewell, R. C. (2004). The performance of gallium arsenide/germanium solar cells at the Martian surface. *Acta Astronómica*, 83–101.
14. Damiano, A., Porru, M., Salimbeni, A., Serpi, A., Castiglia, V., Di Tommaso, A. O., & Schettino, G. (2018). Batteries for aerospace: A brief review. 7.
15. Edwards, C. D. (2007). Relay communications for Mars exploration. *International Journal of Satellite Communications and Networking*, 111–145.
16. ESA. (2009). ESA contribution to the Titan Saturn system mission. *ESA*, 1–192.

17. Fierkwap, S. B., & Ilouno, J. (2021). Analysis of the thermal conductivity of locally made cast aluminium pot. *International Journal of Innovative Scientific & Engineering Technologies Research*, 23–32.
18. Forget, F., et al. (1999). Improved general circulation models of the Martian atmosphere from the surface to above 80 km. *Journal of Geophysical Research*, 24155–24175.
19. Fujita, K., Luong, R., Nagai, H., & Asai, K. (2012). Conceptual design of Mars airplane. *Transactions of the Japan Society for Aeronautical and Space Sciences, Space Technology Japan, 10*.
20. Geoffrey, A., & Landis, T. W. (2004). Mars solar power. *National Aeronautics and Space Administration*, 1–10.
21. Gipson, L. (2013). *NASA, aeronautics*. Accessed 23 March 2024 from https://www.nasa.gov/topics/aeronautics/features/X-48_research_ends.html
22. Hanbing Song, C. U. (2007). A mars VTOL aerobot – preliminary design, dynamics and control. *Surrey Space Centre*.
23. Kada, B., & Ghazzawi, Y. G. (2011). Robust PID controller design for an UAV flight control system. *Proceedings of the World Congress on Engineering and Computer Science*, 1–6.
24. Levine, J. S. (2005). The aerial regional-scale environmental survey (ARES) mission to Mars. *Lunar and Planetary Science XXXVI, Part 12*.
25. Lorenz, R. D., Turtle, E. P., Barnes, J. W., Trainer, M. G., Adams, D. S., Hibbard, K. E., Sheldon, C. Z., Zacny, K., Peplowski, P. N., Lawrence, D. J., Ravine, M. A., McGee, T. G., Sotzen, K. S., MacKenzie, S. M., Langelaan, J. W., Schmitz, S., Wolfarth, L. S., & Bedini, P. D. (2018). Dragonfly: A rotorcraft lander concept for scientific exploration at titan. *Johns Hopkins APL Technical Digest (Applied Physics Laboratory), 34*(3), 374–387.
26. Madeleine, J.-B., Forget, F., Montabone, L., & Wolff, M. J. (2011). Revisiting the radiative impact of dust on Mars using the LMD global climate model. *Journal of Geophysical Research*, 13. Retrieved from Martian Seasons and Solar Longitude.
27. Mencos, A., & Krim, L. (2018). Formation of doubly and triply bonded unsaturated compounds HCN, HNC, and CH_2NH via N + CH_4 low-temperature solid state reaction: From molecular clouds to solar system objects. *476*, 5432–5441.
28. Millour, E., et al. (2018). The Mars climate database (version 5.3). *Scientific Workshop: "From Mars Express to ExoMars"*, 2.
29. Moreno, I. A., Jarzabek, A., A González, M., & Perales, J. M. (2016). Optimizing multi-disciplinary scaled tests in terrestrial atmosphere for extraterrestrial unmanned aerial vehicle missions. *Journal of Aerospace Engineering, 230*(1), 77–89.
30. MOMA Science Team. (2017). The Mars organic molecule analyzer (MOMA) instrument: Characterization of organic material in martian sediments. *Astrobiology, 17*(6-7), 655–685.
31. NASA Planetay Data System. (2020). *PDS.NASA.gov*. Accessed 23 March 2024 from https://pds.nasa.gov/datasearch/data-search/
32. Northrop Grumman. (2000). *Caltech.edu*. Accessed 23 March 2024 from http://www.nast-group.caltech.edu/~murray/dgc05/upload/7/7f/Inertial_Measurement_Unit_LN-200.pdf
33. Okayasu, M., & Tsuchiya, Y. (2019). Mechanical and fatigue properties of long carbon fiber reinforced plastics at low temperature. *Journal of Science: Advanced Materials and Devices, 4*(4), 577–583.
34. Pagan, A. S., Massuti-Ballester, B., & Herdrich, G. (2016). Total and spectral emissivities of demising aerospace materials. *Frontier in Applied Plasma Technology, 9*(1), 7–12.
35. Panagiotou, P., Fotiadis-Karras, S., & Yakinthos, K. (2017). Conceptual design of a blended wing body MALE UAV. *Aerospace Science and Technology, 73*, 33–64.
36. Perez Lebbink, L. (2008). *UAV mission design for the exploration of Mars-ESA/EUROAVIA Design Workshop 2006*. Georgia Institute of Technology.
37. Rani, M. F. H., Razlan, Z. M., Shahriman, A. B., Ibrahim, Z., & Wan, W. K. (2020). Comparative study of surface temperature of lithium-ion polymer cells at different discharging rates by infrared thermography and thermocouple. *International Journal of Heat and Mass Transfer, 153*, 119595.
38. Sadraey, M. H. (2017). *Aircraft performance: An engineering approach*. CRC Press.

39. Sanjiv, P., Shailendra, R., Saugat, G., Kshitiz, K. S., & Sudip, B. (2016). Aerodynamic and stability analysis of blended wing body aircraft. *International Journal of Mechanical Engineering and Applications*, *4*(4), 143–151.
40. Tyan, M., & Nguyen, N. V. (2017). Comprehensive preliminary sizing/resizing method for a fixed wing–VTOL electric UAV. *Aerospace Science and Technology*, *71*, 30–41.

Drone Journalism: Where the Human Eye Cannot Reach—Narratives and Journalistic Uses

Ángeles Fernández-Barrero

Abstract One of the most extended uses of drones for journalistic purposes consists of obtaining aerial images that are difficult to cover with the physical presence of journalists. In this chapter, other important narrative utilities, such as their use for investigative and whistle-blower journalism, data journalism, immersive journalism, media coverage of protest sporting and cultural events, urgent media coverage, armed conflict reporting, and live broadcasting, are analyzed. Additionally, a perspective of the obstacles to their incorporation into newsroom routines, despite the existence of an emerging industry is offered.

Keywords Journalism · Drone · Aerial images · Investigative journalism · Inmersive journalism · Media coverage

Introduction

During the last decade, the role of drones in the field of journalism and the role that this technology could play in the production of information has been widely debated by the scientific community. In fact, with incipient users, concepts such as "drone journalism" and "dronalists" have emerged. The wide potential for drones to take images from the sky that are otherwise complex to capture with the subjective perspective of drones has been extensively emphasized by the academic literature. Compared to other types of technologies used by journalism that offer aerial perspectives, drones entail a consequent reduction in costs, as well as greater speed, availability, mobility, and safety for journalists.

However, despite these certain benefits and the media trend toward audio-visual consumption, there has not been a take-off in the journalistic field, proportional to the advantages offered by the technology itself, particularly when compared with other business areas, such as agriculture, surveillance, and security or the delivery of goods. Various factors in different countries have slowed down what was initially seen as an emerging technology with great potential. These include, among others,

Á. Fernández-Barrero (✉)
Universidad de Sevilla, Sevilla, Spain
e-mail: mfernandez10@us.es

D. Carou et al. (eds.), *Applying Drones to Current Societal and Industrial Challenges*, Management and Industrial Engineering, https://doi.org/10.1007/978-3-031-55571-8_8

highly restrictive legislation, some ethical grounds (i.e., privacy protection), and the crisis in the media sector.

An empirical study developed by Adams [1] shows that the use of drones does not appear to be "revolutionizing" newsgathering but changes the viewer's perspective in several distinct ways, using angles or movement that could not have been generated except with an unmanned aerial vehicle (UAV). Adams [1] observes the different purposes of drones for informative uses: "to underline the story content, to add meaning or insight, to provide surprise, novelty or a memorable image, to present a scientific approach, to give the viewer a sense of (unrealistic) power, to provide drama, to suggest surveillance, to add beauty or art, to involve, immerse or distract the viewer, to generate filler shots, to provide unnecessary gloss, or to create a sense of global connection". According to Adams [1], drones have the ability "to make storytelling more engaging, using drama, novelty or beauty but are also doing the opposite, by simply providing padding without new information".

After an initial stage of development, there has been a tendency to prioritize the use of drones in very specific contexts, mainly for photojournalistic purposes. In this sense, drones help in obtaining images from a bird's eye view, which the human eye cannot reach, where the aerial perspective turns small details, which at ground level take a different form, into mosaics, textures, genetic codes, and abstract works.

1 First Images from the Sky: Hot Air Balloons and Carrier Pigeons

The media started using drones in the early years of the twentieth century. At that time, military practice used them for more than seventy years and their experimentation in other areas of knowledge was taking off. Nevertheless, the idea of obtaining aerial images for informative purposes dates back to the mid-nineteenth century, using the technology available in that period.

The photographer Nadar obtained the first aerial photograph from a hot air balloon in 1858. He was able to temporarily immortalize the rooftops of the houses of Petit-Bicêtre, in France, 80 m above the ground, an image that has not been preserved. Two years later, James Wallace Black obtained an aerial photograph of Boston, which is preserved in New York's Metropolitan Museum of Art. It was published in 1860 by the Boston Herald, which promoted it as the first bird's-eye perspective of the earth in the United States. It was obtained at 650 m above sea level by Samuel King from a hot air balloon called Queen of the Air. It is the oldest archived aerial photograph of Boston.

During the American Civil War (1861–1865), correspondents, journalists, and photographers used hot air balloons to take images from the sky, although there is little research on this topic.

Forty years later, in the twentieth century, the photographer George R. Lawrence made photographic flights from 600 m altitude to photograph the bay of the city of San Francisco (California, United States), in ruins as a result of an earthquake [2]. However, it was not easy because the gas from the balloon expanded in the surrounding atmosphere as the balloon rose, inserting itself into the camera and forcing the photographer to take different shots. Fortunately, one of them was useable.

In 1909, Wilbur Wright took the first photograph from an airplane. As technology and aeronautics continued to progress, it became easier to take aerial photographs, while those advances also allowed obtaining more detailed images.

In the last decade of the nineteenth century, lightweight cameras fitted on carrier pigeons were also developed. This method not only reduced the danger for photographers shooting from hot air balloons, but also "this meant that the animal, not the photographer, controlled the apparatus and determined the final image" [3]. According to Hernández [4], in December 1808, the pharmacist Julius Neubronner obtained a patent for the technique of taking images from the sky with carrier pigeons, to which he attached a lightweight aluminum camera and a pneumatic system that controlled the time interval between the automatic shooting of a photograph and the next one. Neubronner was trying to find a formula to detect the route taken by his lost carrier pigeons, which he used to send small medicines answering urgent requests.

Users of aerial photography also resorted to kites, but the carrier pigeons were the ones that were mainly used for espionage tasks, photographing enemy territory during the First World War. Hernandez documents that to remedy the deficiency of these birds returning to their loft, mobile pigeon lofts were used: a trailer of two heights, with a darkroom at the bottom to develop the photos and the loft itself on top of it. The use of carrier pigeons during the Second World War, according to Hernandez, is unknown, as there is limited documentation on the matter, although he concludes that there is enough evidence to believe that they were transported by trained dogs hidden inside baskets to cross enemy lines.

Since the beginning of the century, aerial photography has used airplanes, but pigeons offer more discretion and a higher level of safety for the photographer by not being present when taking the shot.

2 First Cameras on Drones

Wilkinson explains that the contemporary drone could be viewed as the powerfully mechanized equivalent of carrier pigeons [3]. By the time drones reached the media, they had been used by the military industry for more than seventy years. Drones are historically linked to their military use in different ways: as espionage tools and remote-controlled weapons of destruction, and as an informative tool for news coverage of the conflict and prior exploration of the territory. The use of cameras in unmanned aerial vehicles became widespread during the war in Vietnam.

The use of cameras in drones became widespread during the Vietnam War. At that time, the Ryan Firebee model, one of the most common models in the military industry, was used. This model was equipped with a small camera to spy on the adversary in field reconnaissance missions. Using a rocket engine, this drone was launched from the ground. It could also be air-dropped from the air, with a manned airplane. It was then retrieved with the assistance of a parachute that was deployed when it returned to a suitable landing area. In the early 1990s, General Atomics developed the Gnat model, introducing video cameras.

The result was not optimal in the early years, as the images obtained were not of satisfactory quality because of interferences produced by lights or bad weather. Despite this, the cameras captured moving images of pedestrians crossing the Mostar Bridge in Bosnia and Herzegovina. The successor, the Predator model, incorporated technical improvements, such as the possibility to carry and fire two AGM-114 Hellfire missiles or other munitions. It was the beginning of a new era in the development of drones.

The contemporary drone age also coincides with the arrival of civil and commercial drone use. Having become smaller and cheaper, drone models for civil and military use were able to carry lightweight cameras, enabling aerial photography and video to enter into a new dimension.

The cost of the first drones for military use, the complexity of management, and their large dimensions made them prohibitive for use by mass media. However, in the last decade, and despite the distance that still separates the hobby from professional and commercial use, the situation in journalism has changed considerably.

3 Pioneering Journalistic Experiences

The use of drones in journalistic activity differs significantly from one country to another. The level of resourcing available within the industry and the presence of more or less permissive regulations have played a very important role in this matter.

In the United States, since 2016, a new market for journalism has been opened up by allowing the commercial use of drones. However, few companies jumped on this market, but it was already an advantage in comparison to other countries.

CNN (Cable News Network) was the first U.S. media outlet to receive the necessary authorization from the Federal Aviation Administration to operate drones and provide the aerial images obtained to the various stations where it operates. However, many newspapers were reluctant to use drones, questioning the regulation of the Federal Aviation Administration (FAA), as they considered that it resorted to an excessively strict reading of the freedom of the press provided for in the First Amendment to the Constitution. Obtaining the FAA permit was related to passing a test that required knowledge of airspace and the aeronautical navigation map. The speed was limited to 160 km/h, the height to 120 m from the surface of the Earth and the drone had to be within the visual range of the pilot. In addition to these rules, night flights were not allowed.

Since then, CNN has encouraged research into the use of drones for journalistic tasks and has shared the results to help formulate the appropriate framework. For this purpose, it founded a media research enterprise with the Georgia Technology Research Institute. They reached an investigation agreement with the FAA.

The first experiences were highly satisfactory for CNN, and in 2016, it announced the launch of a fleet of drones for news coverage, called CNN Air (CNN Aerial Imagery and Reporting), in collaboration with Turner Broadcasting and Time Warner. In this way, CNN established an aerial imagery and newsgathering department with two full-time aircraft systems operators to distribute images across all CNN networks and platforms. Terence Burke, vice president of National News, explained, "CNN's cutting-edge development of technology to improve the way we tell stories is part of our DNA. We are proud to continue the tradition with CNN AIR and to establish a unit that will expand our technological capabilities for newsgathering" [5].

CNN had already used drones to provide information with greater comprehensibility and context and to enhance storytelling and production value in different events: in the floods in Louisiana in August 2016 and the water crisis in Flint, Michigan in January 2016 (tap water poisoned citizens for two years due to lead contamination), to show the scope of both events. It also used drones in 2015 at conventions to select candidates for both the Democratic and Republican presidential elections; at the primary election debates; for covering the 50th anniversary of the civil rights march on the Edmund Pettus Bridge in Selma, Alabama, with marchers crossing the bridge; and the 10th anniversary Commemorating Acts of Hurricane Katrina.

In earlier years, some other uses were already recorded. The bibliographic review places one of the first experiences of drone journalism or pseudodrone journalism in 2010. The documentary 'Sharks of the French Riviera' used cameras attached to drones to obtain images of Paris Hilton and other celebrities at the Cannes Film Festival and hotels along the coast. Such controversy was generated by the fact that in the 2017 edition of this film festival, organizers banned the use of drones to record or film images.

From 2010 to 2012, there were hardly any recorded experiences of drone journalism, according to the study conducted by researchers Marc Tremayne and Andrew Clarke from the University of Texas at Arlington [6], which focused on natural disasters, protest reporting, and environmental investigative journalism. In 2013, CNN used a camera drone to record the aftermath of Typhoon Haiyan's passage in Tacloban (Philippines). CNN used drones to obtain aerial views of Tacloban ten days after Typhoon Haiyan struck the Philippines. It showed the dramatic scale of devastation. However, it was not until 2014 that the media showed a real interest in putting drone journalism into practice.

4 Contributions to Journalistic Narrative

The great contribution of drones to journalism revolves around obtaining images that are difficult to cover with the physical presence of journalists. Despite the clarity and evidence that images obtained with drones offer them, Gonzalo Prudkin, who focuses on multimedia narratives, argues that there are few signals of the use of such technology as a contextualization complement in the coverage and construction process of the news using a differentiated narrative and language. He thinks drones are used rather for their newness or as a pioneering technical breakthrough to offer to the audience [2].

In this regard, López Hidalgo [7] warns that despite the potential of journalistic applications, for the moment, the images captured by drones are not usually self-sufficient to create autonomous work. Instead, they are shown as pieces to complete journalistic investigations with interviews, images from archives, and other typical resources of television feature stories and reports.

The review of the specialized literature and the vision offered by numerous theorists about the value that drones can add to the information have allowed us to draw up a list of the most important narrative utilities. The following list is a summary of the major contributions of this technology to journalistic narratives highlighted by different researchers specialized in drone journalism, although conditioned by the regulatory framework of each country in which they operate.

Areas with difficult accessibility

Drones offer important advantages for obtaining images of hard-to-reach places due to the natural topographical features of the area (oceans, caves, forests, and deserts), the natural topography of the land (tsunamis, floods, hurricanes, and earthquakes), or intentional disasters (spills, forest, and fires). Remote-controlled flight guarantees access to the area of interest with an added advantage: the safety of the pilot.

These are perspectives that can generate great interest for the viewer and to which audiences would not have access using other means and with the same results. Aerial recordings obtained with helicopters, for example, would be much more expensive. In addition, drones, because of their size, can enter narrow spaces and nooks, between trees, near cliffs, or indoors, and avoid dangers that piloting a helicopter would entail in situations such as the aftermath of natural disasters, fires, or hurricanes. Newsourcing has also seen the development and use of underwater drones [8].

Most of the graphic coverage of the eruption of the La Palma volcano in Spain used drones. During the two months of the volcano activity, until November 2021, different media, and especially television channels, resorted to images obtained by drones to show the advance of the lava, which occupied a total of 258 hectares of surface. Remote piloting ensured safety against the effects of the release of toxic and irritating substances into the atmosphere as a result of the eruption. Radio Television Canaria obtained impressive images with a flight almost at the level of the magma (Illustration 1).

A fact that not only Gynnild [9] but also many other authors point out is the enormous utility of drones to obtain images of geographical areas that are difficult to access due to the natural topography of the site (images of a cave for a documentary, for example) or by the consequences of natural disasters (floods, earthquakes, tsunamis).

Drone journalism could also play an important role in its application to raise awareness about environmental issues, as drone investigation enables long-term environmental monitoring [10]. In the coming decades, a healthier administration of the Earth's resources is considered crucial, and the media has a role to play in keeping

Illustration 1 Image of the La Palma volcano (Spain) obtained by Radio Television Canaria with drones. *Source* RTVC YouTube Channel

people informed and helping them to realize the importance of the situation as well as the engagements that need to be taken.

Investigative and whistle-blower journalism

Some media outlets that have begun to explore the potential of drones are also exploring the advantages of their use for investigative and whistle-blower journalism. This activity, guard and vigilance, has been widely used by police services. Visual documentation of illegal activities strengthens journalistic credibility and can be used to support investigations of criminal activity. Drones have been used, for example, to detect and document spills into rivers, fumigation with toxic substances, or other environmental abuses.

Drone journalism offers considerable documentary possibilities to obtain images without the presence of journalists, which enables them to avoid being observed [11]. Additionally, they are useful in places where access is not easy because of the risk it could involve reporters. Gynnild [9] explains that images could be obtained under safe conditions for journalists without exposing them to unpredictable conditions.

This section also includes the purposes of controlling and monitoring government activities as a counterargument. Concerning investigative use and secret recording, BBC (British Broadcasting Corporation) [12] maintains that "Drones may be used for investigations where there is an evidential purpose in the footage to be obtained and providing there is a strong public interest justification for any intrusion of privacy". For example, drones could be used, according to BBC, to document illegal activities such as fly-tipping, smuggling, or illegal agricultural practices where evidence could not be acquired without the use of an aerial vehicle. Nevertheless, any proposal that risks invasion of privacy by filming with a drone should be considered a proposal for secret filming, which must have BBC preapproval.

Data journalism

A variant of the use of drones for investigative journalism would be for the specific purpose of data journalism, although in this case, highly specialized expertise and large equipment are needed.

Matt Waite, who has been director of the drone journalism lab at the University of Nebraska-Lincoln since 2011, argues in an article by Sengul-Jones [13] for Global Investigative Journalism Network that there are several areas of opportunity for data journalists using drones: for heat mapping, 3D terrain mapping, detecting pollution or creating three-dimensional models of buildings. Waite cites the experience of Radiolab and WNYC's Data News team as representative examples of this trend.

Immersive Journalism

The rise of miniaturization and the high performance of drones and their capabilities, portability, and battery performance have encouraged creativity and enhanced aerial video and photography capabilities, "including capturing news content indoors, as well as in various novel situations such as for documentary production for extreme sports, pointing toward augmented reality, virtual reality, and other immersive applications as well" [8]. This author explains that experiential news and immersive formats utilize media forms that feature six qualities: interaction, immersion (both audio and video envelopment as well as psychological immersion), multisensory content, algorithms and data-driven content, first-person perspective, and natural user interface design (e.g., voice or gesture control). Pavlik [8] identifies at least four implications of drones for virtual reality journalism, mixed augmented reality a mixed: geo-tagged audio and video, volumetric 360°, first-person perspective, video and 3D audio, and the capacity "to generate novel content types or content based on data acquired from a broad range of sensors including 3D photogrammetry as well as electromagnetic spectrum data beyond the standard visible light captured via video cameras".

Media coverage of protests

Drones make it possible to show the size and scale of rallies and protests and to estimate the number of people gathered, thus adding visual context to reports. In 2013, for example, they were used by *TV Folha* for its coverage of the protests in Sao Paulo, Brazil.

Many of the images used by media are provided by the police, which has specialized drone units for vigilance and control of protests.

Aerial images recorded by the media are limited by the legislative framework and authority's control. In many countries, aerial drone photography of densely populated areas with large concentrations of people is prohibited.

Photographer Tomer Appelbaum won a Siena (Drone Photo Award) working in the daily *Haaretz*. It was an aerial image of the protests against Netanyahu in Tel Aviv, Israel, during the COVID pandemic in 2020, when people maintained social distance in crowded places. Images and videos of these silent demonstrations, in which demonstrators wearing face masks protested in long lines in Yitzhak Rabin Square and with social distance to prevent contagion, spread quickly across the world's media. Appelbaum, who frequently uses drones, was arrested, and his photographic equipment and drones were intercepted by police at one of the protests, although the media coverage led to his immediate release. The police claimed a day later that he had been mistaken for one of the activists participating in the protest, even though he identified himself as a reporter.

Journalistic coverage of sporting and cultural events

In the context of live broadcasting of sports and cultural events, drones make it possible to closely follow the trajectory of an athlete in sports that involves movements over long distances. Postema [6] foresees that sports coverage could be more complete in the not-too-distant future, without spectators having to miss crucial moments of the game due to the absence of a camera covering this point. He also recalls that studies are showing that a subjective camera has a significant impact on the visual experience of sports broadcasts.

The media and television channels opt for other quieter and safer systems, such as cablecams, in stadiums and enclosed areas with a high concentration of people. This system, as a counterpoint, requires prior installation work and support points. For the broadcast of the special concert of the Vienna Philharmonic Orchestra in the Basilica of the Sagrada Familia in Barcelona in September 2021, for example, *Radiotelevisión Española* (RTVE) opted for deployment with 13 high-definition cameras, including a cablecam and a telescopic crane, to avoid noise in a cultural event in which the ambient sound was strategic.

Urgent media coverage and live broadcasting

Drones have demonstrated their versatility for urgent coverage, the narration of breaking events, live and broadcasts, unexpected events, and events outside the agenda. The small size of drones ensures their transport and fast mobility, which has a positive impact on immediacy compared to the infrastructure that other technologies require or the time and cost required to deploy them.

To take advantage of the benefits that drone technology can bring to newsrooms for these specific purposes, journalism needs very light drones that facilitate their transport to the new location, as well as sufficient autonomy to obtain images without having to interrupt the flight. There are different aspects in which technological research is advancing at a dizzying pace and where the journalism industry can find an opportunity for a new business model: the dimensions of the aircraft, the quality of the lenses, stabilization systems, and the autonomy of the drones.

Armed reporting

Coverage of armed conflicts with drones provides images of difficult access areas for journalists due to the danger of the event narrated (armed conflicts), enabling informants to disguise their presence in dangerous areas.

In this regard, Postema [6] reflects the intense controversy among authors. Not all of them are in favour of the use of drones in these situations, although these aircraft represent a new working tool that can increase the safety of journalists when capturing images of the conflict. It should be taken into account the number of reporters who have been killed or injured during this coverage. This author especially highlights the advantages of drones in conflicts where reporters are no longer neutral figures but war targets or where journalists are not embedded with troops, having instead a certain margin of freedom to select the images they want to capture to ensure a certain impartiality and objectivity. In these cases, the use of drones can either directly provide images or explore the security and possibilities for journalistic coverage of a particular area.

Critics argue that it is an intimidating element because of its physical similarities to a military spy drone. They also consider that it obtains a mechanistic view that loses human contact. To reduce this mechanistic view, Postema advises combining drone footage with other perspectives, as audiences are accustomed to versatile storytelling.

In 2022, the war between Russia and Ukraine showed the extensive use of military and commercial drones to detect enemy troops and direct attacks. Ukraine came to question the power of the Russian Black Sea fleet using maritime drones. BBC[1] also used drones to show the devastation of large areas in Kyiv, Mariupol, and other locations: buildings and streets destroyed a month after the start of the armed conflict, firefighters and aid workers searching for debris in the rubble, shelling activity, buildings in flames… The bird's eye view images hardly need words to explain the desolation of the war landscape.

Regarding this type of specific application, BBC [12], in its Guidance on the use of drones, in the public states, "Where a drone is used by a BBC remote pilot, it should normally be flown following the country's aviation regulations, unless there is a public interest for not doing so. In a war zone, for example, it may not be possible to seek permission to fly".

5 Pros and Cons

The advantages of drone storytelling, in its different aspects and journalistic uses, are clear. Drones make journalistic work safer, as they make it possible to obtain images and video without risking the integrity of the journalist or the team around him or her. In addition, the images add value to the information itself.

Moreover, the use of drones leads to improvements in innovative models of business since the cost of obtaining these images is much lower than that of obtaining images by manned aerial flights. Using the current technology offered by image stabilizers, it is possible to record with drones and deliver the footage to the client in hand without postproduction work, which also has an impact on the reduction of material delivery times.

[1] https://www.youtube.com/watch?v=1v8OT0rTKI0.

This should be added to the relative ease of operating commercial drones and the Internet's possibilities to seek free advice on how to fly them.

On the other hand, it is important not to lose sight of some obstacles (for example, the impossibility of their use under adverse weather conditions), the dangers of privacy violations, the damage they may cause in the event of an accident, the legislative restrictions, the administrative complexity to obtain legal licenses or permits required to operate and the owners' fear of losing their devices.

Likewise, in daily journalistic practice, the presence of events outside the agenda and with appropriate visibility conditions to be covered with aerial images represent a very small percentage of the total volume of news flow. In these circumstances, it is currently more profitable for medium-sized channels to outsource these services in a context in which there is a growing number of agencies and individuals who offer products and services at competitive prices. In these circumstances, only in the large television stations would it be economically viable to create a drone journalism department to supply images to the different divisions of the channel. If the cost of drones followed this declining trend in prices, the context of the journalistic use of drones could change, but journalists would need additional training and the help of experts in drone technology to tell stories. It is not a simple language.

6 Limitations to the Development of Drone Journalism

Drones are highly versatile gadgets that are widely used in disciplines such as agriculture, engineering, forest fire detection and extinction, assistance in social and humanitarian catastrophes, emergency management, and the delivery of small products. As far as journalism is concerned, there is still a long way to go, and many factors have hampered its development. The literature review identifies different factors:

Legislative restrictions

One of the most commonly argued factors affects legislative restrictions, which vary from one country to another and have limited the use of UAVs due to various issues affecting the safety of citizens. In many countries, legal regulation has been characterized by its limiting nature, both in terms of safety (to reduce the results of any accidents that may occur) and the safeguarding of citizens' intimacy and privacy.

These restrictions confined journalistic drone graphic resources in Spain between 2014 and 2017 to rural or thinly populated areas, which reduced the news material that could be narrated from the perspective of drone journalism. In December 2017, regulations in Spain changed and expanded journalistic uses by allowing drone overflights in the vicinity of buildings, towns, outdoor gatherings of people, and night flights, although they had to submit certain requirements to ensure the safety of operations. Since 31 December 2020, the European UAS (Unmanned Aircraft Systems) regulation, established in the Commission Implementing Regulation 2019/947 and the Commission Delegated Regulation 2019/945, has been applicable, requiring users to register as operators on the AESA (*Agencia Estatal de Seguridad Aérea*) website and obtain a registration number, which must be visibly displayed on the drone. In addition, they must have a minimum level of training following the group in which they are operating: open, for low-risk (no authorization required); specific, for medium-risk flights; or certified, for high-risk flights, which require a regulatory regime similar to manned aviation. Training at nonrisk levels is free and online, and after the successful completion of an exam, the AESA must issue a certificate. An insurance policy must also be taken out to cover civil liability toward third parties in case of any damages that may arise resulting from carrying out each flight performed for both leisure and professional use.

In terms of flight rules, drones are subject to general operating rules which consider factors such as drone weight, proximity to buildings, and professional qualifications. In general, drones may fly up to 120 m, maintaining a direct line of sight with the operator (except in the category under STS BVLOS Operations). According to AESA [14], the operating distance should not exceed 500 m; pilots should avoid flying drones near airports and heliports and stay at least eight km away from airports and three km from heliports for safety. Additionally, they should avoid flying too close to residential areas or populated areas and maintain a safety buffer zone of one km around residential areas. Without consent from the owner or individual, drones must maintain a safety distance of 500 meters from isolated buildings, people, vehicles, animals, and structures. Additionally, flying over military installations, public utilities, archaeological sites, or public or private facilities is strictly prohibited. It is prohibited to fly over national parks, biosphere reserves, wildlife conservation areas, and any type of protected natural space.

UAS Operator's normative indicates that the drone's total take-off mass must be less than 25 kg, and it must be operated within a visual line of sight at a maximum altitude of 120 m. Considering the moderate level of risk involved, flights in this specific category require authorization before the operation.

The future of drones as a tool for journalism is unclear and conditioned by shifting rules and regulations [15], despite their potential benefits. In any case, regulations should not easily excuse the media from the caution they should exercise to avoid collisions and preserve air safety and the safety of citizens.

Protecting the privacy and intimacy of individuals

One of the issues of greatest concern from a legislative point of view is the safeguarding of the intimacy and privacy of individuals. In Europe, regulations make it clear that the right to privacy of the people who may appear in the images captured by the drone must be protected at all times. The media must be particularly careful in publishing images of anonymous people so as not to breach the Data Protection Act.

BBC [12] considers that only public interest justifies filming. There are various potential hazards concerning the use of drones, but the BBC particularly stresses the intrusion into privacy. In this regard, the broadcaster warns, "Drones can be 'highly privacy intrusive' and a privacy impact assessment may be necessary to comply with the Information Commissioner's Office guidance and data protection legislation". The BBC also recalls in its guide that "Drones should not normally be used to identify individuals without their consent or capture close-up images of private areas such as houses, gardens or offices without the consent of the owner unless these areas can be seen from a public vantage point or there is a public interest in showing them".

There are fewer ethical dilemmas for the media in public and semipublic spaces such as railway stations, as it is understood that permissions are not required from people who happen to be captured by a camera by chance, as part of a general shot of the station, unless they are involved in an action where there is a high level of expectation of privacy. Private spaces imply obtaining permission to record, although the invasion of privacy will depend on the area of drone flight and the types of images captured.

In the digital age, people's perspectives on privacy and their zone of immunity have been reduced, while the boundary between public and private has been blurred, as Tremayne and Clark [16] point out, an erosion, they insist, to which drones equipped with video cameras further contribute.

Accident rate

The high rate of drone accidents has been highlighted by Lavín de las Heras and Gallardo Camacho [17]. Although drones have GPS (global positioning system) devices and 3D maps, they can get lost and cause any accident, with the resulting consequences for public safety. In April 2014, for example, the impact of a drone injured an athlete in Australia in the middle of a race. In July 2017, the presence of a drone near Lisbon Airport (Portugal) forced two planes to suspend landing on the runway, and one of the flights had to be diverted to Porto.

The possibility of a collision with a manned aircraft has generated many concerns for aviation agencies because neither the traffic and collision avoidance system (TCAS) nor the controllers detect them, so the possibility of seeing them depends on the visual acuity of the pilot. This danger has led various countries to regulate this aspect and to include in their regulation the safety distance in airports and aerodromes, as well as limitations in areas with high concentrations of people.

Production costs

Another obstacle to the development of drone journalism is related to production costs. Despite being increasingly low-priced, investment costs for the media remain high, considering that drones have journalistic utility in a constant experimentation process, features that advance ahead at a rapid pace, and with a high accident rate.

In a context in which the media are trying to reduce costs to overcome the crisis in devastated newsrooms, they prefer safe investments that guarantee economic viability and survival in a highly competitive sector.

Indeed, the commercial use of drones opened up an interesting path toward lower prices for these products, but the cost of devices for professional use and those that offer appropriate image quality continues to be higher. Bearing in mind that durability may be limited (due to drone accidents, for example) or the constant innovations of a very young technology (new models with higher performance are constantly emerging), investment in drones does not represent a priority for the media.

According to experts, using camera-equipped drones for professional use experiences a continuous development process, with constant improvements that affect resolution, flight range, and dimensions.

Technical and administrative complexity

The process of obtaining licenses for piloting and filming is complex. This fact, linked to regulations, risks to citizens' safety, and aviation involved in their use, has ended up intimidating the major media, which could in any case afford the investment.

The same circumstances could explain the shortage of specialized journalists, who also do not have an adequate academic framework. It would be necessary to provide training in the use of this new information tool to enhance journalistic specialization. Most journalists who have specialized in drone handling have held experimental workshops and multidisciplinary courses in the absence of formal academic training.

University education only adapts subject programs when tendencies and new topics are consolidated, so there is always a certain gap between the training received and professional practice. This gap is often filled by seminars and supplementary training courses.

In the United States, universities tend to sign agreements with research centers and laboratories. These include the agreements reached by the Poynter Institute, the National Press Photographers Association (NPPA), and the Drone Journalism Lab (University of Nebraska). The University of North Dakota has had a degree for drone pilots since 2009. University education has pioneering initiatives, such as the University of Missouri (United States) initiative, which in 2013 launched a course entitled "Science Investigative Reporting: Drone Journalism", to introduce students to investigative journalism with drones.

7 Future Perspectives

All these factors mean that in Spain, in line with the situation in many other countries, the mainstream media make very limited use of drone journalism, with few specialized drone journalism departments in their organizational structures. In television news programs, the images obtained with drones usually come from external purchases (from other pilots or news agencies) or the acquisition of images free from institutions, organizations, and even citizens. Postema [6] recalls, in this regard, the Worcester floods (United Kingdom) in 2014, which were filmed by a drone enthusiast. Alternatively, the views of the Costa Concordia accident in 2012 were obtained by a citizen with a drone, which made it possible to see the enormity of the ship stranded on the Italian coast.

When these services are specifically needed, they are often outsourced. Several industry reports point to the existence of an important industry characterized by the proliferation of small drone-based journalism production companies and freelancers that provide services to others.

As per the initial drone survey conducted by Todrone [18], for the 400 respondents (production companies and freelancers specialized in drone filing, among other actors), broadcasting holds a significant share of 46% of the market. This survey provides an overview of how the industry is currently evolving. The results reveal that the market is dominated by small, very new companies and freelancers (93% of the total), which makes it highly fragmented, volatile, and still very incipient, with substantial growth in the number of operators. Todrone goes on to indicate that production companies using drones in Spain were small companies or freelancers that regard drones as an interesting tool for enhancing their video production or photographic services or for initiating a new business venture. The survey notes that within this industry, innovation is somewhat constrained, and capital expenditure remains minimal.

The Strategic Plan for the development of the civil drone sector in Spain (2018–2021), published by the Spanish Government [19], highlights that the sector has huge potential for expansion. To realize its extraordinary development prospects, the strategic plan highlights two fundamental aspects: innovation and continuous technological progress. The future of drones will depend on technical innovation, but also on the ability to respond to the needs of the media.

References

1. Adams, C. (2019). Tinker, tailor, soldier, thief: An investigation into the role of drones in journalism. *Digital Journalism, 7*(5), 658–677.
2. Prudkin, G. (2016). El Periodismo Drone: Contextualización histórica y posibles usos periodísticos. *Comunicação & Inovação*, 17(33), 7–21.
3. Wilkinson, J. (2013). Animalizing the apparatus: Pigeons, drones and the aerial view. *Shift-Graduate journal of visual and material culture, 6*, 1–21.
4. Hernández, Á. (2014). Palomas fotógrafas, los insólitos drones espía de la Primera Guerra Mundial. Accessed 23 March 2024 from http://www.eldiario.es/hojaderouter/palomas-fotografas-espias-primera_guerra_mundial_0_295020522.html
5. CNN. (2016). CNN launches CNN AIR. Accessed 23 March from http://cnnpressroom.blogs.cnn.com/2016/08/18/cnn-launches-cnn-air/
6. Postema, S. (2015). *News drones: An auxiliary perspective*. Edinburgh Napier University: Edinburgh, UK
7. López Hidalgo, A. (2016). El periodismo que contará el futuro. *Chasqui, Revista Latinoamericana de Comunicación*, 131, abril—julio, 239–256.
8. Pavlik, J. V. (2020). Drones, augmented reality and virtual reality journalism: Mapping their role in immersive news content. *Media and Communication, 8*(3), 137–146.
9. Gynnild, A. (2014). The robot eye witness: Extending visual journalism through drone surveillance. *Digital Journalism, 2*(3), 334–343.
10. Niarchos, M., Stamatiadou, M. E., Dimoulas, C., Veglis, A., & Symeonidis, A. (2022). A semantic preprocessing framework for breaking news detection to support future drone journalism services. *Future Internet, 14*(1), 26.
11. Goldberg, D., Corcoran, M., & Picard, R. G. (2013). Remotely piloted aircraft systems and journalism: Opportunities and challenges of drones in news gathering. Reuters Institute for

the Study of Journalism. Accessed 23 March 2024 from https://ora.ox.ac.uk/objects/uuid:a868f952-814d-4bf3-8cfa-9d58da904ee3
12. BBC. (2022). *Guidance: Use of drones. Editorial Guidelines Issues*. Accessed 23 March 2024 from https://www.bbc.com/editorialguidelines/guidance/drones
13. Sengul-Jones, M. (2021, September 1). Una guía para el futuro del periodismo con drones. *Global Investigative Journalism Network*. Accessed 23 March 2024 from https://gijn.org/2021/09/01/espanol-futuro-periodismo-drones/
14. Agencia Estatal de Seguridad Aérea. (2022). *Normativa europea de UAS/drones*. Accessed 23 March 2024 https://www.seguridadaerea.gob.es/es/ambitos/drones/normativa-europea-de-uas-drones
15. Holton, A. E., Lawson, S., & Love, C. (2015). Unmanned aerial vehicles: Opportunities, barriers, and the future of "drone journalism." *Journalism Practice, 9*(5), 634–650.
16. Tremayne, M., & Clark, A. (2014). New perspectives from the sky: Unmanned aerial vehicles and journalism. *Digital Journalism, 2*(2), 232–246.
17. Lavín de las Heras, E., & Gallardo Camacho, J. (2016). Los drones, ¿una nueva herramienta informativa? In C. Mateos & J. Herrero (Coords.), *La pantalla insomne* (CAC 103, pp. 113–126), Cuadernos artesanos de Comunicación.
18. Fernández Barrero, A. (2018). Journalism and drones. Challenges and opportunities of the use of drones in news production. *Doxa Comunicación*, (26), 35–58.
19. Ministerio de Transportes, Movilidad y Agenda Urbana. (2021). Plan Estratégico para el desarrollo del sector civil de los drones en España 2018–2021. Accessed 23 March 2024 from https://www.transportes.gob.es/recursos_mfom/paginabasica/recursos/plan_estrategico_drones_2018-2021_0.pdf

Challenges of Civil Application of 'Drones' for Commercial Purposes in the Logistics and Transportation Sector

Antonio Martínez Raya and Joaquín Sarrión Esteve

Abstract 'Drone' is a widely used term to refer to any aerial vehicle being operated by using remote commands from a pilot or relying on software for autonomous flight. This aircraft type is more formally known as an Unmanned Aerial Vehicle (UAV), often forming an integral part of an Unmanned Aircraft System (UAS). Introducing such aerial devices in several civil contexts has been considered one of the biggest disruptive technologies of the present decade. Nowadays, such aerial vehicles are being operated for multiple business activities intended for commercial purposes. Shortly, when it comes to the transportation sector, in addition to common use concerning surveillance and security services, it aims to carry goods for logistic services when delivering small parcels and spare parts. It is also expected to make even greater use of commercial drones in the coming years since the massive use of such an aircraft is creating a new market with large economic potential, thus leading the aeronautic industry to point out that the emerging technologies in the drone application have a positive impact on innovation at national levels. Nevertheless, the weakness of the related air operations in terms of safety in crowded urban environments and a lack of a stable global legal framework with adequate guarantees of privacy and other fundamental rights and values complicate the issue still further and constitute an obstacle to the full social acceptance of these flying machines.

Keywords Drone · UAV · UAS · Logistics · Transportation

A. M. Raya (✉)
Department of Organizational Engineering, Business Administration and Statistics, Technical University of Madrid—Universidad Politécnica de Madrid (UPM), 28040 Madrid, Spain
e-mail: antoniomartinez@upm.es

J. S. Esteve
Department of Constitutional Law, National University of Distance Education—Universidad Nacional de Educación a Distancia (UNED), 28040 Madrid, Spain

D. Carou et al. (eds.), *Applying Drones to Current Societal and Industrial Challenges*, Management and Industrial Engineering, https://doi.org/10.1007/978-3-031-55571-8_9

1 Introduction

Over the past years, the use of crewless remote-controlled planes for several civil applications has been intense and fruitful. The term 'drone' is used popularly concerning unmanned aircraft beyond the civil context regardless of whether it is included as an aerial operation in observational use or not. While there are many kinds of drones for civilian uses, either already in service or on an experimental basis, not all of them are necessarily regarded technically as aerial vehicles. Although the present chapter refers exclusively to unmanned artifacts operating in the air mode of transportation, other types of similar vehicles can carry out missions in a wide range of environments, depending on which transportation mode (air, water, and land) such devices can be traveled. Terrestrial drones such as Unmanned Underwater Vehicles (UUV), Unmanned Surface Vehicles (USV), and even the so-called High Altitude Pseudo Satellites (HAPS) are a sort of hybrid air vehicles between drones and satellites.

A huge application of such vehicles, either manned or unmanned, has recently emerged to implement diverse levels of technological solutions for flight controls under both remote and autonomous, and even simultaneous control management capabilities. Although this popular term is a more comprehensive concept to refer to a wide range of these aerial vehicles either in both formal and non-formal flying settings, remotely operated or under navigation autonomously, the huge range of civil missions undertaken by such aircraft has become necessary to update their classification according to new operational roles. This has forced both civil aviation authorities and related government entities to redefine the categories of these aerial systems according to their operational constraints. Major technological breakthroughs have nowadays enabled drone operators to offer more operational capabilities, thus implementing new features for several activities, among others, in the field of logistics and transportation sector.

The administrative organization of the airspace does not necessarily correspond to national borders. Airspace volumes have traditionally been sectorized according to common aviation rules. With that aim, the International Civil Aviation Organization (ICAO) has defined a division of the world's airspace based on flight information regions (FIR). In the case of Spain, there are three regions (FIR Madrid, FIR Barcelona, and FIR Canarias), whose classification of its airspace is also following Sect. 6 of Regulation (EU) 923/2012 establishing the air regulations and common operational provisions for air navigation services and procedures, also known as SERA. The national body for regulating the civil use of UASs in the sectors of Spanish airspace, where it is permitted, either non-controlled or controlled, is the State Aviation Safety Agency, also known as AESA. Besides being duly authorized by such a public body, where flights with drones are conducted inside of controlled air spaces, despite the many constraints facing them, the operation must be monitored by the leading air navigation and aeronautical information service provider in Spain (ENAIRE). In related filed flight plans (hereafter, FPL), excluding those not allowed in prohibited (P), restricted (R), or activated danger (D) areas, and other similar domestic restrictions in the form of FPL buffer zones (FBZ), as well as temporary reserved and temporary segregated areas (respectively, TRA and TSA). Operators may also have to obey state requirements to operate drones or respect the specific measures required by pilots according to safety and security regulations.

Based on the foregoing, drones can also fly by traveling on wind flow, either remote-controlled or in autonomous mode, and have especially been designed with both wings and rotors with blades that create lift for navigation into the airspace following either Instrument Flight Rules (IFR) or Visual Flight Rules (VFR) or even Night Visual Flight Rules (NVFR). The terms Unmanned Aircraft (UA) and Unmanned Aerial Vehicle (UAV) refer to those vehicles flying without a pilot on board, and that is the main element of unmanned aerial systems, which in turn encompass the aircraft, the ground station, and communications link. In that regard, a complete and unambiguous specification of such aerial vehicles has recently been set up to be greatly beneficial for operators and the air industry in general as a basis of definition and scope for its many different civil uses. Five categories can thus be distinguished from other airplanes as follows: Unmanned Aircraft (hereafter, UA); Unmanned Aerial Vehicle (hereafter, UAV); Unmanned Aerial System (hereafter, UAS); Remotely Piloted Aircraft (hereafter, RPA); Remotely Piloted Aircraft System (hereafter, RPAS). Strictly speaking, according to the open category provided by the International Civil Aviation Organization (hereafter, ICAO), it covers those operations with small UA (the so-called 'drones', as stated earlier) weighing less than 25 kg of Maximum Take-Off Weight (MTOW). Under the European rules, recently released by the European Union Aviation Safety Agency (hereafter, EASA), all drones operating in any EU member state shall carry a standardized registration plate, namely a class identification label, depending on a new category based on their weight, capability, and potential risk.

This label distinguishes some drone classes (C0, C1, C2, C3, C4, C5, and C6) depending on their operational constraints and integrated systems. This allows aviation authorities to classify RPAS according to specific characteristics, system components, and equipment for remote control. However, those drones without the class label, before the entry into force of the new EASA regulations, will be considered 'legacy drones' (that is, such new classes provided by the European UAS regulations shall not be applicable).

Nowadays, even though UASs have been designed for the performance of several assignments in a wide variety of operational conditions, the importance of ensuring air safety from flights operating overpopulated settlements, particularly those concerning urban areas within their boundaries, is necessary for all actors involved in the civil use of related aerial vehicles. The issue of urban overflights has been a topic of discussion in the aviation sector for the past years, not only in response to the operational risks of these aircraft in many fields for civil use but also regarding societal challenges, for example, those concerning privacy and data protection policies. In general, overflying urban areas for professional usage is highly restricted not only for privacy reasons but also because of the need to prevent injury to persons and property damage [4]. As the intense use of drones draws near, considerable challenges lie ahead from commercial exploitation for societal and industrial services, particularly those regarding logistics activity and transportation. Hence, standardized licenses and certifications play an essential role here as an element to ensure safety-based initiatives in operating massive drones.

2 Types of 'Drones' According to Their Applications for Transportation and Logistics

As noted earlier, while the concept of 'drone' generally refers to an unmanned aerial vehicle, it is difficult to establish exposure categories and operational limits overall. An approach based on sectoral applications ranging from logistics to transportation is now feasible. Thus, some adequate classifications have been properly proposed below to describe the uniform application of current regulations regarding safety issues and the ongoing adaptation of navigation rules in operational matters.

Firstly, according to the missions entrusted for the transport and logistics industry,[1] seven categories can be distinguished as follows:

- Surveillance, either in monitoring areas or in patrolling worksites.
- Security, either based on zenithal or on panoramic aerial observation.
- Search and rescue, of either people or property in natural and other disasters.
- Maintenance, either for preventive or corrective tasks.
- Shipping, either for parcel or courier service, even food delivery.
- Aerial filming services, either for photography or videography services.
- Data collection, for either tracking or mapping services.

Secondly, according to the operational capabilities,[2] six categories can be categorized as follows:

- Very close range with distance and time not exceeding 5 km and 1 h during the flight, respectively.
- Close range with distance and time not exceeding 50 km and 6 h during the flight, respectively.
- Short range with distance and time not exceeding 150 km and 12 h during the flight, respectively.
- Medium range with distance and time not exceeding 650 km and 24 h during the flight, respectively.
- Long-range with flight distance exceeding 650 km and flight time of up to 24 h.
- Extended range with flight distance exceeding 650 km and flight time of up to 48 h.

[1] This classification distinguishes among drone functions depending on mission features.

[2] This classification distinguishes among drone ranges depending on mission distances.

Thirdly, according to the number of purposes assigned,[3] three categories can be identified as follows:

- A sole purpose, with a single mission.
- Multi-purpose, with various missions.
- Customizable, with interchangeable missions.

Fourthly, according to the operational typologies,[4] three categories can be listed as follows:

- RPA drones (1st generation of UAS technology).
- IoT-enabled drones in autonomous flighting mode (2nd generation of UAS technology).
- AI-based networks in self-organizing flights of swarming drones (3rd generation of UAS technology).

Fifthly, according to the number of horizontal propellers,[5] five categories can be distinguished as follows:

- Monocopter, also called helicopter, is a type of drone that has one rotor.
- Tricopter is a drone that has three rotors.
- Quadcopter is a drone that has four rotors.
- Hexacopter is a drone that has six rotors.
- Octocopter, also called quadcopter, is a drone that has four rotors.
- Coaxial, also called quadcopter, is a drone that has four rotors.

Sixthly, according to the flight line of sight,[6] three categories can be distinguished as follows:

- Visual Line of Sight (VLOS).
- Beyond Visual Line of Sight (BVLOS).
- Extended Visual Line of Sigh (EVLOS).

Finally, according to the base UAV configuration,[7] four categories are differentiated as follows:

- Fixed-wing drone.
- Multi-rotor drone.
- Single-rotor drone.
- VTOL drone, based on either a hybrid or traditional technology.

---

[3] This classification distinguishes among drone features depending on any mission purposes, according to customer needs, except for the limitations established by law.

[4] This classification distinguishes among operations depending on each UAS generation.

[5] This classification distinguishes among UAV types depending on the number of rotor sets, irrespective of their configuration, either coaxial or traditional system, even whether they are mounted with single-blade or multi-blade propellers.

[6] This classification depends basically on whether the drone pilot is losing sight of the UAV during the flight phase.

[7] This classification distinguishes among drone types depending on the rotor mechanism.

- Tilting rotors drone, either with a passive or active mechanism.

Drone applications in the transportation and logistics industry round out their wide range of civil uses. Furthermore, when performing critical mission tasks, drones are calling out to play a key role in avoiding unnecessary risks to workers in dangerous and aggressive environments, in particular those carrying out potentially hazardous work, such as the application of pesticides, electrical practice, and working at heights. Nevertheless, making drone regulations not only at the EU level but also at the international level on the presumable great expansion of drone use can lead to a uniform process in terms of legal certainty across such a range of other sectors beyond transportation and logistics, since the extent of commercial exploitation is growing and is likely to continue to grow in the future.

3 Legal Framework on the 'Drone' Industry

Certainly, in the drone, two different types of technologies can be seen, on one hand, the drone's being (i.e., the drone itself, including the aircraft and its specific systems, mostly them for communications) and, on the other hand, the application that each sector intends to use or develop to develop and optimize the services [21]. In addition, the drone itself is an unmanned aircraft, and therefore, drone operations and activities are developed in the sky (a civil and military sky, see [23]), and the aircraft regulatory regime applies to them largely (1.7 ICAO Cir 328-AN/190 on UAS). However, at the same time, there is a specific developing drone regulatory regimen [29] to answer 'constant issues' as the definition of how to regulate and police the development of drones [18]. Nevertheless, as a disruptive technology, and due to its fast development, drones face important regulatory challenges as the different legal regimes, both internationally and nationally, continue to evolve and follow the drones, in a slower way than the drone technological progress [2].

As ignorance of the law is no excuse—the so-called Selden Law[8]—it is particularly important to correctly identify the current legal framework of drones, particularly concerning the civil industry and the professional use of drones. The main legal tools include, firstly, and considering the international level, the Convention on International Civil Aviation of 7 December 1944 (the so-called Chicago Convention). Although Article 8 of the Chicago Convention refers to 'Pilotless Aircraft', and there is no explicit reference to the category of remotely piloted aircraft (RPA), the consensus is that the definition includes not only autonomous drones but also RPA, making a teleological interpretation.[9] There are several articles applicable to drones, for example, Annex 2 with several rules relating to take-off and landing maneuvers,

[8] A popular amended version of a comment by the English jurist and scholar J. Selden [27].

[9] Including institutions [14, 20] and academia [2, 5, 22, 25, 26].

and other flight operations, Annex 7 on aircraft nationality and registration, and e Annex 16 with the rules on environmental protection.[10]

In addition, within the European Union context,[11] it is important to consider, given the transnational nature of aviation, that the European Union itself, following the competence established in Article 100(2) of the Treaty of the Functioning of the European Union (TFEU), has developed a body of European Air law [26]. This body includes specific legal instruments with basic common rules for drone regulation, with a relevant and progressive development, which, nevertheless, allowed EU Member States to have their own specific rules for military, customs, police, or similar operations, and for small drones.

The first specific drone regulation within the EU was Regulation (EC) No. 1529/2002 on common standards in the field of civil aviation, which established common standards in the field of civil aviation but excluded unmanned civil aircraft with an operating mass not exceeding 150 kg, leaving the regulation of these drones to EU Member States [21]. In the same line, the Regulation (EC) 216/2008 on common rules in the field of civil aviation included a minimum standard regulation applicable to drones with operating mass of more than 150 kg, but excluded droned whose weight was lower or equal to 150 kg. Moreover, the regulation excluded drones, regardless of their mass, for military, customs, police, or similar operations, and research or scientific experimentation, those constructed in their majority, 51%, by an amateur or non-profit amateur association for their purposes and without any commercial purpose, or any aircraft that has been in the service of the military forces [30].

To solve the fragmentation of the legal framework, the risk for the drone market, and the challenges to the protection of fundamental rights, the EU worked for an EU common legal framework for all drones, at least within the civil scope, and regardless of the weighed mass, including small drones [15].[12] It is important to outline the EASA's proposal to create common rules for operating drones in Europe, in September 2015 [11], and the Resolution of the European Parliament of 29 October 2015 on the safe use of remotely piloted aircraft systems (RPAS) in the field of civil aviation.[13]

The EASA launched a collaborative consultation to reach this regulation and prepared and published an Opinion to introduce a regulatory framework for the operation of drones, which would be the basis for the final regulation [12], approved in 2018, the EU Regulation 1139/2018.[14] This regulation establishes, within the

[10] See also, particularly the ICAO Cir. 328 on UAS [20].

[11] It is outside the scope of this chapter other regulations outside the European Union. For the USA, see https://www.faa.gov/uas; For the United Kingdom, see https://www.caa.co.uk/drones/.

[12] Following the Riga Declaration on Remotely Piloted Air Systems: Framing the future of aviation, 6 March 2015.

[13] European Parliament resolution of 29 October 2015 on safe use of remotely piloted aircraft systems (RPAS), commonly known as unmanned aerial vehicles (UAVs), in the field of civil aviation (2014/2243(INI)).

[14] Regulation (EU) 2018/1139 of the European Parliament and of the Council of 4 July 2018 on common rules in the field of civil aviation and establishing a European Union Aviation Safety Agency, and amending Regulations (EC) No 2111/2005, (EC) No 1008/2008, (EU) No 996/2010,

context of the common rules for civil aviation, specific rules for drones, in Articles 55 to 58, (Chap. 3 Substantive Requirements. Section 7 Unmanned Aircraft), in Article 77 (Airworthiness and Environmental Certification), and in Annex IX (Essential requirements for unmanned aircraft).

Certainly, EU Regulation 1139/2018 has been supplemented with other instruments, including Commission Delegated Regulation 945/2019 on unmanned aircraft systems,[15] which develops Articles 58 and 61, and includes the UAS technical design and manufacturing requirements, rules for manufacturing UAS in the open category and electronic ID, and rules for operators from third countries. It is important also to outline that the most important part is the regulation of the characteristics and capabilities of drones to fly safely, including, for example, the requirement for drones to be individually identifiable. In addition, the Commission is implementing Regulation 947/2019 on the rules and procedures for the operation of unmanned aircraft,[16] about Article 57.

This EU legal framework covers practically all types of drones with the main objective of the establishment of a high and uniform level of safety in civil aviation in the Union (Article 1(1) Regulation EU 2018/1139), responding to the challenges of technology with a more homogeneous regulation, regardless of the drone take-off mass, considering the type of operations and activities, and the associated risk. However, the regulation maintains the so-called flexibility notes [31], including the following:

(a) Member States maintain the competence to regulate drones that carry out military, customs, police, search and rescue, firefighting, border control, coastal surveillance, or similar activities or services (Article 2(3a) Regulation 2018/1139), but Member States may choose the application to these activities and operations of the EU regulation (Recital 18, and Article 2(6) Regulation 2018/1139).
(b) Member states may choose to adopt national regulations that introduce additional requirements or conditions for operations with drones for reasons of public safety, protection of privacy and personal data, or others (an open list), following Union Law (Article 56(8) Regulation 2018/1139).

Considering these flexibility notes, one can consider that although the EU regulation is a binding framework directly applicable, in some way, the Member States have to implement it, and there is the possibility of having some variations or additional requirements at the domestic level. Therefore, to access up-to-date and detailed

(EU) No 376/2014 and Directives 2014/30/EU and 2014/53/EU of the European Parliament and of the Council, and repealing Regulations (EC) No 552/2004 and (EC) No 216/2008 of the European Parliament and of the Council and Council Regulation (EEC) No 3922/91. PE/2/2018/REV/1. OJ L 212, 22.8.2018, pp. 1–122.

[15] Commission Delegated Regulation (EU) 2019/945 of 12 March 2019 on unmanned aircraft systems and on third-country operators of unmanned aircraft systems. C/2019/1821.
OJ L 152, 11.6.2019, pp. 1–40.

[16] Commission Implementing Regulation (EU) 2019/947 of 24 May 2019 on the rules and procedures for the operation of unmanned aircraft. C/2019/3824. OJ L 152, 11.6.2019, pp. 45–71.

national information, the best is to consult the web of the national regulatory body of the specific EU member state. In addition, it is also useful to use the website of the Drones Rules project, which informs about the basic requirements at the EU at the national level, including specific up-to-date information.[17]

EASA published on March 2020 a consolidated text containing the regulations and procedures for the flight of drones, in a more readable format, which also included Acceptable Means of Compliance (AMC) and Guidance Material (GM) (EASA, 2020), updated with a revision in September 2022, which included the Commission Implementing Regulation (EU) 2022/425, as regards postponing the transition dates for using certain UAS in the open category and the dates of application for some standard scenarios[18]; and Executive Director Decision 2022/002/R amending Executive Director Decision 2019/021/R on geographical zones, operational authorization forms in the specific category, a procedure and forms for cross-border operations, standard scenarios, training modules syllabus for remote pilots in the specific category, and a new predefined risk assessment.[19]

Nevertheless, this new EU regulation entered force with a long calendar of progressive application, with a staggered efficiency, that has provided the different national legislators with a reasonable time to adapt, and which was delayed due to the coronavirus disease 2019 (COVID-19) until 1st January 2023 [21].

The key elements of the EU Drone legal framework are based on drones (as a specific type of aircraft) categories, with three classes based on their maximum take-off mass (MTOM), linked to the risk, but taking into consideration that they will have access to the same airspace (U-space airspace), including professional and amateur drones. Because of that, Implementing Regulation 947 tries to adopt a series of security measures applicable to both professional and amateur pilots to guarantee the safety in airspace and provides for the adaptation to characteristics of the specific flying area (including population density, presence of buildings, surface characteristics, etc.).[20]

The way to specify the risk is based on identifying these three categories of operations, depending on their risk.

- **Open category**. Includes drones up to 25 kg (drones in the C0, C1, and C2 classes). Operational risks are considered low. Operators need to comply with standardized limitations, such as maximum altitude, distance from people, and visual line of

[17] See https://dronerules.eu/en/.

[18] Commission Implementing Regulation (EU) 2022/425 of 14 March 2022 amending Implementing Regulation (EU) 2019/947 as regards postponing the transition dates for the use of certain unmanned aircraft systems in the 'open' category and the date of application for standard scenarios for operations executed in or beyond the visual line of sight (Text with EEA relevance). C/2022/1445. OJ L 87, 15.3.2022, pp. 20–21.

[19] ED Decision 2022/002/R Regular update of the AMC & GM to Regulation (EU) 2019/947: AMC & GM to Regulation (EU) 2019/947—Issue 1, Amendment 2 | AMC & GM to the Annex to Regulation (EU) 2019/947—Issue 1, Amendment 2.

[20] It is beneficial to use Easy Access Rules for Unmanned Aircraft Systems (Regulations (EU) 2019/948 and 2019/945) [13].

sight requirements. Before starting a flight, no specific authorization is required, but operators must register themselves and their drones.

- **Specific category**. Operational risks are higher. This category is for all operations that are not open or certified. Operators should obtain a Safety Aeronautical Study (SORA), and prior authorization from the competent authority, except in the so-called European Standard Scenario (STS)[21] or when the operator has a UAS operator certificate (LUC) with privileges, i.e., in several cases in which a communication from the operator is sufficient.
- **Certified category**. According to the highest risks (for example, flights over concentrations of people, drones' flights to transport people or dangerous goods, etc.), the requirements are similar for manned aviation, the aircraft will always have to be certified, the activity requires an approved operator by the competent authority and a remote pilot with license.

In addition, on 22 April 2021, the European Commission adopted new rules and conditions for safe, secure, and green drone operations, the so-called U-space package, which included three regulations to create the required conditions for drones—and also manned aircraft—to operate safely in the European Space, the U-Space: the Commission Implementing Regulation (EU) 2021/664 on a regulatory framework for the U-space,[22] the Commission Implementing Regulation (EU) 2021/665 amending Implementing Regulation (EU) 2017/373, on the requirements for providers of air traffic management/air navigation services and other air traffic management network functions in the U-space airspace designated in controlled airspace,[23] and the Commission Implementing Regulation (EU) 2021/666 regarding the requirements for manned aviation operating in U-space airspace.[24] These legal acts introduce new services, particularly for longer-distance and low-level operations, with harmonized conditions, and become applicable as of 26 January 2023.

Finally, it is important to note that drone operations should respect also administrative sectoral-specific legislation, considering the type of operations and activities, in addition to data protection regulation, and environmental legislation. The legal framework affecting the civilian use of drones is not located in a handful of specific

[21] Applicable from 1st January 2024. See Commission Implementing Regulation (EU) 2020/639 of 12 May 2020 amending Implementing Regulation (EU) 2019/947 as regards standard scenarios for operations executed in or beyond the visual line of sight. (Text with EEA relevance). C/2020/2937. OJ L 150, 13.5.2020, pp. 1–31.

[22] Commission Implementing Regulation (EU) 2021/664 of 22 April 2021 on a regulatory framework for the U-space (Text with EEA relevance). C/2021/2671. OJ L 139, 23.4.2021, pp. 161–183.

[23] Commission Implementing Regulation (EU) 2021/665 of 22 April 2021 amending Implementing Regulation (EU) 2017/373 as regards requirements for providers of air traffic management/air navigation services and other air traffic management network functions in the U-space airspace designated in controlled airspace (Text with EEA relevance). C/2021/2672. OJ L 139, 23.4.2021, pp. 184–186.

[24] Commission Implementing Regulation (EU) 2021/666 of 22 April 2021 amending Regulation (EU) No 923/2012 as regards requirements for manned aviation operating in U-space airspace (Text with EEA relevance). C/2021/2673. OJ L 139, 23.4.2021, pp. 187–188.

regulations. Thus, to respond to the legal challenges, one needs to address the entire legal system [6]. Furthermore, over the last few years, fundamental legal changes have occurred in the commercial drone market, and they have had a clear impact on both operators and manufacturers. Nonetheless, regulatory changes have been slow in implementing appropriate global market provisions for parties involved in the commercial exploitation of drones, owing to certain domestic restrictions on transferring knowledge from the defense industry to the civil application. This situation creates numerous opportunities for different local interpretations of operational requirements and different legal levels of enforcement in practice.

4 Technological Challenges for the Sustainable and Efficient Use of 'Drones' in the Transport and Logistics Sector

Nowadays, the challenge faced by the transport and logistics industry is to become more reliable and faster for their customers, either enterprises or individuals, by offering good quality at affordable prices. Hence, the civil use of drones in both sectors will remain massive soon, since the needs. The speed of technological advances in drone engineering and manufacturing is likely so, that it makes lawmakers, regulators, and operators very difficult to update. In addition to that critical issue, other questions remain to be settled, related to the harmonization of pilot licensing and drone certification and to the administrative method of grating them. The question then arises as to whether all existing certificates of airworthiness (hereafter, CoA) should be renewed when the time comes faced with technological changes, or even whether the piloting license's validity should be automatically extended when required by circumstances and in any case subject to recommendations of the corresponding national civil aviation authority (hereafter, CAA). This is particularly the case in respect of future drones carrying passengers on board within the framework of the Advanced Air Mobility (AAM). However, confronted with the challenge of responding to the expected demand levels for such air taxi services into the foreseeable future, more technological efforts are needed in commercially exploiting the electric Vertical Take-Off and Landing (eVTOL). Consistent with this, such alternative transport means expecting to operate under the future urban mobility demand model (hereafter, UMDM) will necessarily entail substantial investments, either from the public or private initiative, in building specific airport infrastructure, so-called 'vertiports'. In addition, air navigation service providers shall be subject to continuous policy improvements for flying drones safely and responsibly in urban areas.

Assuming the fulfillment of operational requirements and strict compliance with drone legislation, the implementation of UAS's forthcoming developments should be targeted toward serious involvement concerning occupational risk prevention and environmental protection, as well as those relating to personal data protection. More specifically concerning the transport and logistics sector, there is great potential for the drone industry in automated business management solutions, particularly those operating micro-unmanned aerial vehicles, as well as lightweight drones for indoor flights. With many interesting possibilities, these drones can carry out several tasks for the transport industry (including shippers, forwarders, and storage firms), such as inventory control, parcel handling, warehouse surveillance, and order preparation. As a result, in future years, the share of machined-aided RPAS will likely increase in the total UAS fleet across the world, paying particular attention to those with information technologies based on machine-to-machine (hereafter, M2M) systems, particularly concerning the Internet of Things (IoT).

5 Regulatory Challenges for the Sustainable and Efficient Use of 'Drones' in the Transport and Logistics Sector

An adequate regulatory framework focused on security and good service for people may boost the drone industry, particularly in the transport and logistics sector. But it is also important to adequately map the threats, from ethical and legal perspectives, of the use of drones [9, 10]. It has been consistently held that it needs to focus on *ethics and law by design*, i.e., in developing drone technology, from the beginning[25] with the introduction of measures designed by default, thus considering not only security issues but also the protection of privacy and other fundamental rights including environmental protection [32].

The EU Regulation 1139/2018 introduces several provisions, including that the drone must be designed and produced in a way that makes it suitable for its function and its operation, adjustment, and maintenance, without endangering people (Annex IX, Article 1(2)). Moreover, it introduces the requirement for the operator and the remote pilot to know the national and EU regulations applicable to operations, in particular regarding the protection of privacy, data protection, liability, insurance, security, and environmental protection (Annex IX, Article 1(1)). If it is necessary to mitigate the risks posed by operations about security, privacy, protection of personal data, and security or environment, the drone must have the required characteristics to consider this, both by design and by default (Annex IX, Article 1(3)). Moreover, the registration requirement is established when operations present risks to privacy, data protection of a personal nature, safety, or the environment (Annex IX, Article 4(2)). Of course, the development of drone technology, focused on *ethics and law by design*, should look for a good balance between technological efficiency and ethical view, the regulation should consider and cover all these issues.[26] Certainly, safety

[25] It is important to focus on ethics and regulation issues, from the beginning, including research and development of projects on drone technologies and operations.

[26] Safety issues, including drones, play a prominent role in regulating the aviation sector. Security concerns include hacking, cyber-attacks, and other vulnerability issues [10]. However, the specific particularities of drones require a broad focus to include other issues.

matters, including drones, play a prominent role in regulating the aviation sector, including concerns about hacking, cyber-attacks, and other vulnerability issues [10]. However, the specific particularities of drones require a broad focus to include other cases.

Moreover, it is important to note that drone operations should also respect sectoral-specific legislation, including particularly environmental legislation [6]. Certainly, drones can generate noise and have ethical, privacy, environmental, and social impacts, and develop an adequate and balanced regulation is the most important regulatory challenge that requires careful consideration.

As the drone industry is global, it needs regulatory consistency in standards of fabrication, development, and improvement of the drone technology, procedures, and services for drone operations and applications, particularly in global and transnational markets such as the European Union, because the differentiation of standards and procedures may limit the growth of the drone industry, particularly in the logistic and transport sectors in the EU Market. EASA and other regulatory bodies and authorities are actively working to address these challenges and adapt regulations to facilitate the sustainable and efficient use of drones, but collaboration between regulators, industry stakeholders, and technology providers is crucial to overcome obstacles and create a conducive regulatory environment. Certainly, collaboration with industry stakeholders is critical. Regulatory bodies must be in contact with drone manufacturers, service providers, and operators. These collaborations help regulators gain insights into technological advancements, operational challenges, and industry needs to develop an adequate legal framework that should be practical and adaptable but with the maximum guarantee of the protection of fundamental rights, including the protection of the environment, for example including social impact for biodiversity, as noted by [28], or for rural social development, as noted by [3].

Moreover, the European Commission conducted an open public consultation in 2021 and used existing consultation platforms for the development of the so-called Drone Strategy 2.0. The Commission Staff Working Document outlines the relevance of mobility and the transport sector to developing "A Drone Strategy 2.0 for a Smart and Sustainable Unmanned Aircraft Eco-System in Europe", which included relevant results regarding the benefits of the use of drones for the eco-system in the future [16]. Moreover, the communication from the Commission on the Drone Strategy 2.0 highlights that the transport sector, including the emerging drone transport sector "should help in achieving this twin green and digital transition" [17].[27]

For this reason, the U-Space integration regulation initiative is a great opportunity to improve the legal landscape with the idea of 'secure and green drone operations' within the U-space packet. The integration of drones in the existing European airspace while ensuring safety and avoiding conflicts with manned aircraft, with clear rules that do not compromise safety and fundamental rights and values, is essential.

---

[27] According to the European Commission, the transport sector, including the emerging drone transport sector "should help in achieving this twin green and digital transition". See European Commission [17].

Another important issue is establishing a regulation that incentivizes the development of adequate infrastructure and charging networks for drones (charging stations, maintenance facilities, and other infrastructure) and adequately regulating liability and insurance issues in the event of accidents, damages, etc.

To study to introduce specific taxes on the use of drones to fulfill the gaps in the legal framework, including direct or indirect taxation, for example, to study the creation of municipality taxes for drone operations in urban space [24]. Although continuous monitoring and legal adaptation of the legal framework according to technological and new advancements and drones' capabilities may help the industry, incorporating lessons learned, it could also create some kind of legal insecurity due to the continuous changes in legislation. Because of that, it is important to look for a more stable regulation for the industry to address legal challenges.

Regarding the challenge as mentioned earlier of the legal and regulatory framework for the civil use of drones, the evolution in the Spanish legislation, in a similar way across the EU, shows that it has been construed by implementing the EASA rules into national law, from Royal Decree 1036/2017 [33] to Regulation (EU) 2018/1139.

6 Discussion on the Issues and Opportunities Facing the Massive Use of Drones in Activities Related to the Transportation and Logistics Industry

Social and economic development depends heavily on the mobility of people and goods to promote affordable passenger and freight transport services, respectively. Two aspects illustrate this point to differing degrees: reliable networks and efficient distribution. Transportation and logistics are a huge industry in which operators can access several different types of technologies. This industry also encompasses both passenger and freight channels, with all their respective business sectors on both transportation and logistics, whether in terms of business-to-business (hereafter, B2B) or business-to-customers (hereafter, B2C). In that regard, a few general considerations that need to be borne in mind are described below.

6.1 Considerations Concerning the Transportation Sector

The expansion of engineering solutions based on practical and versatile drones contributes to becoming a high-mobility multipurpose aerial vehicle. For instance, copter-style uncrewed aerial vehicles, as a subgroup of UAVs, can become a key driver of urban mobility for digital transformation and the ecological transition. Although the UAV technology is seemingly mature enough to introduce scheduled services in operating point-to-point passenger transport, also called drone taxis, its potential operators are still aware of the uncertainties surrounding its future. Besides a few operational issues not yet entirely resolved in carrying passengers, the legal framework for the use of drone taxis is incipient and not yet well defined, in particular concerning civil liability in the event of damage to third parties and setting out rules aimed at preventing accidents in urban environments.

In contrast, drone-related solutions for the transportation of goods through competitive innovation and advanced technologies may make freight costs competitive with cargo shipping and dramatically improve delivery times. Freight transportation is therefore called upon to play a leading role in international business affairs and place itself at the core of formulating and implementing the future norms of international trade. Others are growth niche markets in logistics that hold out the prospect of business opportunities concerning which cargo carriers will be more favorably disposed of.[28]

6.2 Considerations Concerning the Logistic Sector

Although there are not so many studies regarding the social perception of drones, the social view about the civil use of drones seems to be quite positive, including the use of drones in the transportation and logistics industry sector [1, 3]. Drones may have an important impact on the delivery of products and services. One of the legal challenges is to adapt the specific sectorial regulation to the delivery of products by drones. For example, one can think of the current use of drones to solve emergencies, but it is also possible to imagine the use of drones for the regular delivery or dispensation of medical products, with the required adaptation of the specific sectorial legislation [7]. Key factors are driving the industry to the use of drones in the delivery of goods, particularly the development and advances of drone technology and the growth of the e-commerce market. In addition, the logistic sector may include not only the delivery of products by traditional drones—by air—but also the development of marine drones—surface and underwater drones—and drones that both have capabilities to fly and swim [8]. Certainly, drone technology advancement and evolution will be positive for the sector, but beyond concerns and eventual facts, it will be important to assess positive risks [19].[29]

Nevertheless, as pointed out above, drone operations in the logistics sector should not be only aligned with the drone legal framework but also with administrative sectorial-specific legislation according to the various types of drone operations, as well as data protection regulation and environmental legislation beyond the domestic context of the existing national civil authority. For freight transport, despite having

[28] The segment of heavy unmanned cargo drones is nowadays in the second business line behind personal air mobility, particularly those based on solutions for urban drone taxis. Nevertheless, the attention from the logistics sector has recently focused on the future use of long-range UAVs in carrying standardized containers (e.g., the private project of Natilus N3.8 T).

[29] Fox [19] noted the relevance of exploring the risk of the use of drones from a positive potential.

not been widely used so far, drones can carry out various missions, from hazardous tasks to high-risk work, either handling or carrying anything. In addition to mandatory licenses and permits regulated for each flight, proper piloting skills are always valuable for accomplishing complicated actions. It may sometimes be necessary to pilot drones by reinforcing the pilotage with proper communication technologies, particularly those concerning data capture and automation of navigation aids. For instance, the characteristic feature of the so-called First Person View (FPV) is that the pilot flies the device through a single console that enables the flier to remotely control far away from the actual location as if the navigator were in the cockpit. Beyond the immersive experience of flying such remote-controlled unmanned aerial vehicles, that can be seen as supporting technology solutions in improving drone functionalities designed for complex missions, many of which cannot be undertaken autonomously by human beings. One example of that is the drone-based system for navigational aid (hereafter, NAVAID) inspections.[30]

7 Conclusions

Major changes are taking place at present in drone technology since its applications and uses have begun to reflect the interests of the general population. In addition to the legitimate uses of aerial unmanned vehicles for security and defense missions, their inherent capabilities for civilian uses can contribute to enhancing business opportunities in several productive areas. Nowadays, society is witnessing a hesitant global economic recovery with the consolidation of the drone industry within the EU framework, as well as the development of a specific drone legal framework addressed to foster the civil use of drones, as part of the EU strategy on green and digital transformation (Drone Strategy 2.0), as well as the matters relating to the (EU) U-space airspace.

It is important to consolidate a stable legal framework to incentivize the research and development of the drone industry, with the maximum guarantees of ethics and fundamental rights and values, including the protection of the environment and social issues, as a sort of 'ethics and law by design'. In this way, drone operations must comply not only with drone regulations but also with administrative sectorial-specific legislation in functions that correspond to those in the occupational activity under consideration.

Although there are challenges and risks in the use of drones, such unmanned vehicles have come to society to stay. Social and citizens' views on the civil use of drones seem to be quite positive, including the use of drones for the transportation and logistics industry sector. Therefore, drone technology is not part of the problem,

[30] This is related to flight tests-related calibration inspections carried out uniquely with drones, all being necessary for the commissioning and preventive maintenance of instrument landing systems (hereafter, ILS) belonging to airport facilities.

but it seems like part of the solution to the transport and environmental challenges, contributing to a green future eco-system in Europe.

While continuous monitoring and rational adaptation of the legal framework are always valuable for operators and manufacturers to meet technological and new advancements and drones' capabilities, it seems that the global consumer drone market is expecting more stable regulation to address the present challenges. Following the entry into force of legislation that occurred in the last few years, the industry as a whole should be able to adapt quickly to changing regulatory frameworks. It is therefore paramount to discuss the challenges ahead between policy-makers and market players, and for that, it will be vital to meet the needs of the drone industry, which is becoming one of the main driving forces of the global economy. Shortly, regulatory changes due to the intensive use of drones will be likely not only focused on commercial issues (e.g., aerial filming services) but also on key economic sectors (e.g., transportation and logistics). More specifically, those claims relate to technology-related innovations in performing critical mission tasks.

Consequently, one of the drone-related research matters that should be deeply analyzed in further studies consists of the interaction between technology and innovation. There are major societal benefits as the result of applying the most advanced technology along with various factors determining the so-called 'technovation'.[31]

Acknowledgements The authors gratefully acknowledge the development of certain parts of this research within the Chair ISAAC. Individual Rights, Scientific Research and Cooperation, created by a cooperation agreement between UNED and CNR-IFAC, the Research Group Multilevel Fundamental Rights of the UNED, and the Regulation Research Group of the University of Valencia, Spain.

References

1. Bafouni-Kotta, A., Villanueva, J., Kalakou, S., & Dias, J. G. (2023). Upcoming drone delivery services: A perceptions-based clustering of citizens in Lisbon. *Technology in Society, 73,* 102250.
2. Barrio Andrés, M. (2018). Drones. Concepto, clases e historia. Usos actuales. El derecho de los drones. In M. Barrio Andrés (dir.). *Derecho de los drones* (pp. 29–57). La Ley. Wolters Kluwer.
3. Benlloch Domènech, C. (2021). Percepción social acerca del uso de drones. Una aproximación sociológica. In: J. Sarrión Esteve, C. Benlloch Domènech (dir.), & A. Balaguer Pérez (coord.). *Miradas y reflexiones sobre los retos actuales en la regulación de los drones* (pp. 49–60). Centro de Estudios Políticos y Constitucionales.
4. Bergtora Sandvik, K. (2016). The political and moral economics of dual technology transfers: Arming police drones. In Završnik, A. (Eds.), *Drones and unmanned aerial systems* (pp. 45–66). Springer.
5. Bernaw, K. (2016). Drones: The emerging era of unmanned civil aviation. *Zbornik Pravnog fakulteta u Zagrebu, 66*(2–3), 223–248.

[31] This term is widely employed and embraces innovative solutions for emerging society, comprising technical, commercial, and economic or market potential [34].

6. Bombillar Sáenz, F. M. (2021a). Reflexiones jurídicas desde la protección del medio ambiente y la salud en torno a los drones. In: J. Sarrión Esteve, C. Benlloch Domènech (dir.), & A. Balaguer Pérez (coord.), *Miradas y reflexiones sobre los retos actuales en la regulación de los drones* (pp. 155–183). Centro de Estudios Políticos y Constitucionales.
7. Bombillar Sáenz, F. M. (2021b). The legal-administrative regimen of medicinal products in the digital society era: Challenges and opportunities from a Spanish perspective. In: F. E. De la Rosa, J. M. Gil Ruiz, L. M. Hinojosa Martínez, A. Orti Vallejo, S. A. Sánchez Lorenzo, & J. Vals Prieto (Eds.), *Justice, trade, security, and individual freedoms in the digital society* (pp. 217–273). Thomsom Reuters Aranzadi.
8. Burgos Garrido, B. (2021). Los drones acuáticos: un sector carente de regulación necesitado de intervención administrativa. In: J. Sarrión Esteve, C. Benlloch Domènech (dir.), & A. Balaguer Pérez (coord.). *Miradas y reflexiones sobre los retos actuales en la regulación de los drones* (pp. 217–274). Centro de Estudios Políticos y Constitucionales.
9. Custers, B. (2016). Drones here, there and everywhere introduction and overview. In B. Custers (Ed.), *The future of drone use. Information technology and law series* (vol. 27, pp. 3–20). T.M.C. Asser Press.
10. De Miguel Molina, M., Carabal Montagud, M. A. (2018). Legal and ethical recommendations. In: M. de Miguel Molina & V. Santamarina Campos (Eds.), *Ethics and civil drones. European policies and proposals for the industry* (pp. 77–86). Springer Briefs in Law.
11. EASA. (2015). Proposal to establish common rules for the operation of drones. Accessed 23 March 2024 from https://www.easa.europa.eu/sites/default/files/dfu/205933-01-EASA_Summary%20of%20the%20ANPA.pdf
12. EASA. (2018). *Opinion No. 01/2018. Introduction of a regulatory framework for the operation of unmanned aircraft systems in the 'open' and 'specific' categories.* Related NPA/CRD: 2017-05—RMT.0230. Accessed 23 March 2024 from https://www.easa.europa.eu/sites/default/files/dfu/Opinion%20No%2001-2018.pdf
13. EASA. (2022). *Easy access rules for unmanned aircraft systems (Regulations (EU) 2019/948 and 2019/945).* Last revision from September 2022. Accessed 23 March 2024 from https://www.easa.europa.eu/en/document-library/easy-access-rules/easy-access-rules-unmanned-aircraft-systems-regulations-eu
14. European Commission. (2014). *Remotely piloted aviation systems (RPAS): Frequently asked questions.* Memo, Brussels.
15. European Commission. (2015). *Communication from the Commission to the European Parliament, the Council, the European economic and social committee and the committee of the regions.* An Aviation Strategy for Europe. SWD(2015) 261 final.
16. European Commission. (2022a). *Commission staff working document. EU drone sector state of play accompanying the document communication from the commission to the European parliament, the council, the european economic and social committee and the committee of the regions.* 'A Drone Strategy 2.0 for a Smart and Sustainable Unmanned Aircraft Eco-System in Europe'. COM(2022) 652 final. Brussels, 29.11.2022 SWD(2022) 366 final.
17. European Commission. (2022b). *Communication from the commission to the European parliament, the council, the European economic and social committee and the committee of the regions.* 'A Drone Strategy 2.0 for a Smart and Sustainable Unmanned Aircraft Eco-System in Europe'. SWD(2022) 366 final. Brussels, 29.11.2022. COM(2022) 652 final.
18. Fox, S. (2019). Positioning the drone: Policing the "Risky" Skies. *Issues in Aviation Law and Policy, 18*(2), 295–331.
19. Fox, S. (2022). Drones: Foreseeing a 'risky' business? Policing the challenge that flies above. *Technology in Society, 71,* 102089.
20. ICAO. (2011). Cir 328, unmanned aircraft systems. Accessed 23 March 2024 https://www.icao.int/meetings/uas/documents/circular%20328_en.pdf
21. Jiménez-Shaw, C. (2021). La normativa europea y española sobre drones civiles de uso profesional. In J. Sarrión Esteve, C. Benlloch Domènech (dir.), & A. Balaguer Pérez (coord.), *Miradas y reflexiones sobre los retos actuales en la regulación de los drones* (pp. 63–93). Centro de Estudios Políticos y Constitucionales.

22. Konert, A., & Kasprzyk, P. (2021). UAS safety operation—legal issues on reporting UAS incidents. *Journal of Intelligent and Robotic Systems, 103*(51).
23. Lavallée, C. (2017). The single European sky: A window of opportunity for EU-NATO relations. *European Security, 26*(3), 415–434.
24. Martín Rodriguez, J. M. (2021). Retos de la fiscalidad de las actividades con drones en España. Problemas de adaptación en la fiscalidad directa e indirecta y posibilidad de crear tasas locales por sus operaciones en zonas urbanas. In J. Sarrión Esteve, C. Benlloch Domènech (dir.), & A. Balaguer Pérez (coord.), *Miradas y reflexiones sobre los retos actuales en la regulación de los drones* (pp. 185–216). Centro de Estudios Políticos y Constitucionales.
25. Massuti, A. (2012). The regulation of unmanned air system uses in common airspace. *Journal of Law, Information and Science, 21*(2), 212–223.
26. Mendes de Leon, P., & Scott, B. I. (2016). An analysis of UAS aircraft systems under air law. In Završnik, A. (Eds.), *Drones and unmanned aerial systems* (pp. 45–66). Springer.
27. Rawson, H. (2002). *Unwritten laws. The unofficial rules of life.* Castle Books.
28. Sandbrook, C. (2015). The social implications of using drones for biodiversity conservation. *Ambio, 44*(4), 636–647.
29. Sarrión Esteve, J. (2017). El régimen jurídico de la utilización de los drones: Una aproximación multinivel a la legislación europea y española. *Revista de la Escuela Jacobea de Posgrado, 12*, 103–122.
30. Sarrión Esteve, J. (2018). Actual challenges for fundamental rights protection in the use of drone technology (August 27, 2018). Drone law challenges. Working Paper 2. Accessed 23 March 2024 from https://doi.org/10.2139/ssrn.3239562
31. Sarrión Esteve, J. (2021a). Introducción. Aproximación a los retos actuales en la regulación de los drones. In J. Sarrión Esteve, C. Benlloch Domènech (dir.), & A. Balaguer Pérez (coord.), *Miradas y reflexiones sobre los retos actuales en la regulación de los drones* (pp. 17–26). Centro de Estudios Políticos y Constitucionales.
32. Sarrión Esteve, J. (2021b). La regulación de los drones y la protección de derechos fundamentales. In J. Sarrión Esteve, C. Benlloch Domènech (dir.), & A. Balaguer Pérez (coord.), *Miradas y reflexiones sobre los retos actuales en la regulación de los drones* (pp. 95–110). Centro de Estudios Políticos y Constitucionales.
33. The Kingdom of Spain. (2017). Royal Decree 1036/2017 of 15 December regulating the civil use of remotely piloted aircraft and amending Royal Decree 552/2014, of 27 June, implementing the Air Regulations and common operational provisions for air navigation services and procedures and Royal Decree 57/2002 of 18 January approving the Air Traffic Regulations. *Official State Gazette, 316*, 129609–129641. BOE-A-2017-15721.
34. Vasant, P. et al. (2022). Feasibility of drones as the next step in innovative solution for emerging society. In *Artificial Intelligence in Industry 4. 0 and 5G Technology* (pp. 179–208). John Wiley & Sons, Incorporated.

Autonomous Drones and Smart Containers in Medicine

Angela Pirri, Giuseppe Tortora, and Fabrizio Niglio

Abstract In the coming years, the freight transport system will change considerably thanks to the introduction of drones. This low-environmental-impact technology has made considerable progress: to be used exclusively in remote or emergency areas, it is being tested all over the world and in urban areas. The use of artificial intelligence in drones has enhanced their flight safety with the development of drone-embedded software capable of making autonomous decisions in emergency cases such as rotor failure, the sudden appearance of an obstacle, or a cyber-attack. During the most acute period of the pandemic caused by COVID-19, the use of drones has significantly increased in the medical field. Medicines, vaccines, and health equipment have been transported, minimizing the risk of contagion for health workers. Many of the potential uses of this transport system, which is becoming safer every day thanks to the investments of private companies as well as many countries, are unexplored: the transportation of easily perishable organs and personalized medicines (especially expensive oncological medicines) will be one of the next objectives. To date, there are two main obstacles to large-scale deployment: the acceptance of citizens seeing drones rushing over their heads with loads and the regulation on drones, which is not uniform in all countries hindering private investments.

Keywords Smart capsules · Artificial intelligence · Drones · Blood products · Drone transportation

A. Pirri (✉)
Istituto Di Fisica Applicata "Carrara", IFAC, Consiglio Nazionale Delle Ricerche, CNR, Via Madonna del Piano 10C, 50019 Sesto Fiorentino (Fi), Italy
e-mail: a.pirri@ifac.cnr.it

G. Tortora
Smart Medical Theatre Laboratory, ABzero, Polo Tecnologico Navacchio, Via Mario Giuntini 63, 56023 Cascina (PI), Italy

F. Niglio
Azienda Toscana Nord Ovest, Via A. Cocchi 7/9, 56121 Pisa, Italy

D. Carou et al. (eds.), *Applying Drones to Current Societal and Industrial Challenges*, Management and Industrial Engineering, https://doi.org/10.1007/978-3-031-55571-8_10

1 Introduction

Unmanned aircraft systems (UAS) or unmanned aerial vehicles (UAV) are commonly known to the public as "drones" [1]. The European Union Aviation Safety Agency (EASA) [2] and the Federal Aviation Agency [3] define "drone" or "unmanned aircraft" as "any aircraft operating or designed to operate autonomously or to be piloted remotely without a pilot on board". Drones use software-controlled flight plans and navigations in their embedded systems, on-board components such as antenna, receiver, camera, speed and flight controller, altimeter, ultrasonic and collision sensors that avoid their crashes during flights [4], as well as a global positioning system (GPS). Structurally made of lightweight materials to reduce weight and increase their maneuverability, drones are robotic systems constituted by a power source (i.e., a battery [5] or fuel) that determines the level of autonomy, rotors, propellers, and a frame, able to carry payload ranging from a few grams to a few kilograms depending on the model. In essence, drones fall into two main categories: either with a rotor or fixed wing depending on their purpose. The former includes tricopters, quadcopters, hexacopters, and octocopters, which do not require runways. In the latter fall hybrid vertical takeoff and landing (VTOL) aircraft systems. In either case, they do not need a wheeled takeoff which is necessary, for instance, for airplanes or helicopters. Figure 1 reports two typical drones such as a hexacopter (left) and a VTOL (right). Concerning the flight range, drones can cover from a few kilometers for hobbyist drones (around 5 km) to long distances ranging from 50 to 6000 km (professional drones) [6].

Table 1 reports the main technical characteristics of drones.

From the 1930s to nowadays, drone technology [7, 8], their use, and their perception by citizens [9, 10] have changed considerably. The disaster caused by Hurricane Katrina [11] in the USA late in August 2005 was a turning point: drones from objects used exclusively in the field of warfare such as military surveillance and active combat [12–15] were authorized by the Federal Aviation Administration (FAA) to seek survivors during disaster relief missions [16, 17]. In 2010, the United Nations Institute for Training and Research (UNITAR) used drones to estimate the damage caused by the earthquake in Haiti and facilitate aid [18, 19].

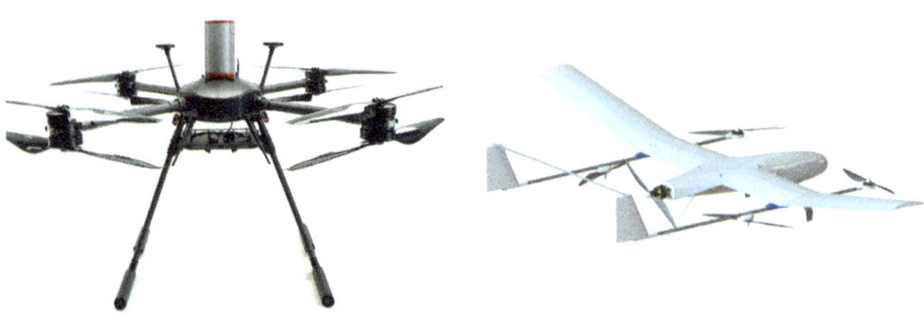

Fig. 1 Structure of a multicopter drone (left) and VTOL (right)

Currently, they are employed in a plethora of civil applications [20–22] ranging from healthcare [23–26] to humanitarian missions, [27, 28] to deforestation [29] and wildlife monitoring, precision agriculture, and natural resources [19, 30, 31], environmental emergencies/disasters, national surveillance, and local security [32], and coverage of sports and music events [33]. Civil protection in countries of the world uses drones for the search and rescue of people [34], covering areas of natural disasters, such as fires [35], typhoons [36], landslides [37], earthquakes [18], or avalanches. Drones have been used in areas affected by nuclear accidents. Thanks to the instrumentation they can carry on board, it is possible to reveal the presence of radionuclides. For example, in the accident at the Fukushima nuclear power plant in 2011 due to a strong earthquake, the use of drones was essential to manage the overheating of the plant, minimizing the risk for the technicians employed in the plant safety operation [38]. They have also been used to map radiation from uranium mines such as the disused mines of Cornwall in the UK [39]. Drones are used to monitor civil engineering infrastructures [40] (such as viaducts, bridges, railways, gas pipelines, mines [41, 42], or high-way infrastructures [43]) located in inaccessible areas [44].

Drones have reached a high degree of technical maturity in just a few years, due to the large economic investments made mainly by the United States (military usage) and, recently, by private companies (commercial usage). It is estimated that in 2026 the entire world supply chain (from drone services to hardware development) linked to drones will be worth over 55 billion dollars (in 2018 it was worth about 14 billion) [44–46] with a compound annual growth rate (CAGR) of 7.3%. To date, drones used by hobbyists account for most civilian drones [33].

According to the European Drone Strategy 2.0 [47], the drone services market in Europe could be worth €14.5 billion by 2030 creating 145,000 jobs. The market for drone-based delivery is expected to grow significantly from 2.3 billion dollars in 2023 and up to 27.4 billion dollars in 2030. In particular, the segment of longer-range delivery (over 25 km) is expected to grow faster [48]. Their use will revolutionize many economic sectors; they are part of the Internet of Things (IoT) to create smart cities [49], and their regulation becomes more and more stringent and fundamental every day.

There are several problems associated with this technology that need to be solved before it can spread. First, they could be a problem for civil aviation since a collision between an airliner and a drone would have disastrous consequences. Secondly, they can fall in populated locations or human-frequented areas. Thirdly, drones can be hacked through targeted cyber-attacks resulting in the loss of transported goods. Finally, they can create problems for wildlife such as birds.

Whenever an operation is not considered a "standard operation"—a payload ranging from 10 to 25 kg is transported—deliveries with drones can be performed in an urban environment only if a specific risk assessment is performed. This analysis allows minimizing risks and identifying mitigations to meet regulatory and safety objectives. Ultimately, the risk assessment methodology is developed to reduce the risks associated with drone operations. The process is known as Specific Operations Risk Assessment (SORA). The SORA is essentially a qualitative and quantitative methodology for risk classifications of drone operations and correct mitigations of such risks. In the SORA process, all the personnel involved in the process are properly trained for performing operations, while the technical requirements of the drone and related systems are analyzed to develop an appropriate procedure that will be part of the operation manual. During the assessment, the ground and air risks are evaluated. The ground risk refers to the risk for people, properties, or infrastructure that could be negatively affected by a potential drone accident. This is directly related to the population density, the type of operation (Visual Line of Sight—VLOS or Beyond Visual Line of Sight—BVLOS), and the drone size; mitigations to the ground risks can be applied by restricting particular areas. The air risks aim to consider the probability of encountering manned aircraft in the airspace. This is derived from the density of manned air traffic in the airspace, such as the presence of other air vehicles. UTM (Unmanned Traffic Management) box, i.e., devices for individuating the presence of air vehicles, can be used for better management of unmanned air traffic. Drone regulation is managed by EASA in Europe and by FAA in the USA [50], and local flight authorities. So far, all flight authorities have collaborated with industry and communities to advance drone operations and integrate them into the national airspace.

Artificial intelligence could provide a way to mitigate ground and air risk: the training of "machine learning" algorithms would ensure the development of drone-embedded software capable of making autonomous decisions in emergency cases such as rotor failure, the sudden appearance of an obstacle, or a cyber-attack.

With the advent of artificial intelligence and machine learning, the market for UAM (Urban Air Mobility) is growing in all sectors. This revolution is enabled by powerful microprocessors and controllers that have enabled massive automation and reliability of the automated processes. AI developments allow the management and integration of novel concepts into the algorithms by optimizing autonomous processes, allowing an unprecedented level of accuracy and precision. One of the applications of AI is related to Artificial Vision, for the recognition of images such as other aerial vehicles or obstacles in real-time. The integration of such technologies will allow the drone delivery market, especially for medical applications, to grow with minimal costs.

The main features that make this new type of mobility extremely attractive can be summarized as follows:

- Self-driving systems: they preserve the safety of the operators in risky scenarios [51].
- Reduction of the workload of healthcare workers: drones can fly 24 h a day, every day.
- Access to inaccessible areas: in some cases, landslides or poor road infrastructure make the road network impassable, so that some places can be reached only by air.
- Low economic cost: they are cheaper than road and manned air vehicles (helicopters and small planes).

- Reduction of delivery times: they follow a straight route instead of a road which can sometimes be tortuous.
- Drones can transport containers capable of carrying heavy payloads.
- They are unaffected by weather conditions like helicopters and planes.

Among the fields of application, the medical field is certainly one of the most important, whose innovations can change the face of traditional medicine by playing an important role in telemedicine and emergency medicine. With C and the wi-fi system, they will be able to visit a patient at home and show, for example, the details of a wound to the medical staff of a hospital. High-resolution cameras (a few millimeters compared to satellite images which have a resolution of tens of centimeters) can play an important role in medical surveillance.

Due to their ability to transport containers in which medicines, vaccines [52], small medical equipment, or health products [53] can be placed, their use will increase in national health systems. The advent of SARS-COVID-19 has accelerated the spread of drones in the medical field since during the pandemic it was essential to avoid contact between people to minimize the risk of contagion. Vaccines and personal protective equipment were transported [54].

As a matter of fact, the market size for medical drones is expected to exceed 399 million by 2025 [55] (in 2018 it was 88.2 million) with a 24.7% CAGR between 2019 and 2025.

2 Drones in Medicine

The possibility of flying autonomously is fundamental in the field of emergency medicine [56] such as road accidents, in war scenarios where it is necessary to set up field hospitals, or in natural disasters. In urban areas, drones reduce delivery times with cost savings [57] by overcoming car traffic while in rural areas they bypass the scarcity of communication routes that sometimes prevent the use of ambulances. It is known that in some emergencies, such as a heart attack or stroke [58], the timely intervention of health professionals is essential to avoid the death of the patient [59–62]. Research carried out in Sweden in the cities of Gothenburg and Kungälv in 2021 [63] has demonstrated how the use of drones to transport defibrillators to places of accident or medical intervention reduces intervention times compared to traditional ambulances. Consequently, the survival rate of people suffering from cardiac arrest occurring out of the hospital is higher. Using a defibrillator in the early stages of cardiac arrest can increase survival rates by as much as 50–70% [64, 65]. It is worth noting that ambulance arrival times have increased over the years in high-income countries due to road congestion and have not improved in low-income countries.

In 2019 [66], a study was carried out to understand whether it was possible to transport insulin-based medicines, which are the most important life-saving drugs for patients with diabetes. It is, however, also a drug particularly sensitive to environmental stress [67] or to thermo-mechanical variations (temperatures, agitations, and vibrations) since the breakage of its structure alters its functionality [68]. The results demonstrated the integrity of the specific drug tested after a flight on both Vertical TakeOff and Landing (VTOL) drones.

Anticancer drugs have also been tested [69] since they are very expensive drugs, perishable in a few hours, and customized according to the physical characteristics of the patient (weight, height, age [70]). Being able to transfer from the place of production to the medical facility in an extremely short time is fundamental for both the health of the patient and the economic sustainability of these treatments by the national health systems [71]. Drugs based on monoclonal antibodies (mAbs) are particularly affected by vibration and shock [72] due to the formation of aggregations that are not visible to the naked eye or chemical alterations of the drug.

The ability to acquire spatial and temporal information in real-time and high resolutions makes drones particularly effective for epidemiological research such as the zoonotic spread of malarial parasites [29] through the acquisition of data on deforestation and agricultural expansion that alter natural habitats influencing its spread. In Spain, researchers traced the spatial distribution and abundance of ungulate mammals carrying tuberculosis and collected data that are essential for managing diseases in multi-host systems [73]. Drones have proven to be very effective tools for medical surveillance. They have been used to detect health risks due to the presence of heavy metals such as copper [74], to measure aerosols in soil, and to track gas levels within soils by predicting and preventing the spread of harmful pathogens to health [75].

In 2007, a system was tested to transport microbiological samples from rural clinics to specialized centers to perform ultra-rapid tests for human immunodeficiency virus (HIV) [76]. In 2016, UNICEF together with the Government of Malawi proposed the idea of reducing waiting times for HIV testing of infants. It was confirmed that transport times, and therefore knowledge of the medical response, have dramatically decreased [77]. In 2014 in Papua New Guinea, the organization Médecins Sans Frontières (MSF) used drones to transport blood tubes to assess for the presence of the tuberculosis virus: there was a 75% reduction in the time spent using an ambulance [78]. Tuberculosis is the second leading cause of death in Papua New Guinea, and it displays an increase in drug-resistant patients (including many multidrug-resistant tuberculosis patients). In 2015, the US government authorized for the first time the flight of a drone to bring medicines from the Wise County regional airport to a clinic located in a rural area (Remote Area Medical clinic) of Virginia [53, 79] with a reduction of the delivery time increased from 90 to 30 min on a total journey of about 63 km.

The demographic increase of the population in various regions of the world, which is often not accompanied by an economic capacity to support families, has triggered studies on the distribution of contraceptives in particularly hard-to-reach areas. The United Nations Population Fund and the Dutch government funded a project to bring contraceptives to women in remote areas of Ghana [80]. In the same country, drones have been used to transport over 1 million doses of targeted deliveries of SARS-Covid2 vaccines during the pandemic [81].

3 Blood Transportation by Drones

The enormous flexibility of drone usage in medicine has prompted the scientific community to explore the possibility of shipping highly perishable biological materials [52, 82] such as organs for transplants or blood products [83]. Biological specimens require stringent environmental controls.

Blood, plasma, red blood cells, white blood cells, and platelets are vital health resources used in a wide range of clinical services and are included in the World Health Organization (WHO) Lists of Essential Medicines [84]. Blood is an essential element for life. It flows in a dense distribution network made up of arteries, veins, and capillaries and performs various tasks depending on the part of the body where it is located (respiratory, excretory, or transport). Red blood cells give blood its typical red color and have the task of transporting oxygen; white blood cells help the body defend itself against infections; platelets are small fragments made by larger cells that help blood clot. The plasma which constitutes the liquid part of the blood has many functions: it transports the cells, transports the nutritional components, allows coagulation, and helps to defend the body from external attacks, to mention a few of them. The amount of blood circulating in the blood vessels is proportional to age, it grows from childhood to adulthood and then decreases slightly in the elderly; in adults about 5 l of blood circulate, which is constantly renewed. White blood cells and platelets are completely renewed every day; red blood cells, on the other hand, last longer and are replaced every 4 months. Blood cells are produced by the bone marrow. Red blood cells collect oxygen from the pulmonary alveoli (O_2) and transport it to all the cells that make up the body. Immediately after having given up the oxygen, they collect the carbon dioxide (CO_2) produced by the cells and return to the lungs along the veins with a reverse path compared to that made when they were transporting the oxygen. When they reach the lungs, by osmosis, they release carbon dioxide which will then pass into the pulmonary alveoli and be eliminated outside with respiration. Red blood cells have such a complex structure that, to date, it has not been possible to replicate the laboratory in a functionally adequate way. Transfusion, which involves transplanting red blood cells from a matched donor to the patient, is commonly used for supportive care in cardiovascular and transplant surgery, massive trauma, treatment of burn patients, cancer therapy, anemia, and rhesus disease (or hemolytic disease of the fetus) as well as complications related to pregnancy. In cases of severe anemia, transfusion is the only possible medical response. The lack of blood can result in an increased mortality rate or serious complications for patients.

According to the WHO annually, more than 118 million blood donations are collected worldwide [85] of which 47 million, corresponding to 40%, are collected in high-income countries where only 16% of the world population lives. In Fig. 2, a typical blood bag is shown. The blood donation rate varies greatly from country to country. Considering 1000 possible donors, WHO estimated the blood donation rate is 5.0 in low-income countries increasing to 31.5 donations in high-income countries. Again, according to WHO data in high-income countries, up to 75% of blood transfusions are for people over the age of 60, while in low-income countries

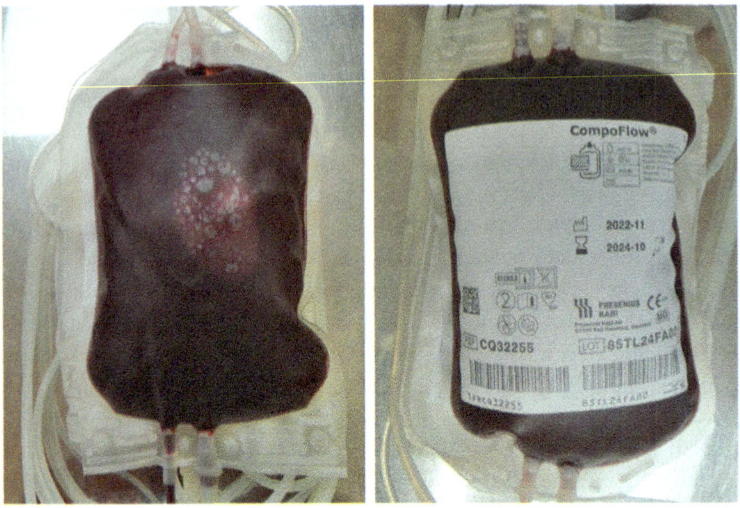

Fig. 2 Typical bags of blood

children under the age of 5 account for 54% of all transfusions. In 79 countries around 90% of blood donations came from unpaid volunteers; out of these, 64 countries achieved 100% of donations from voluntary unpaid blood donors. In 54 countries, 50% came from family/replacements or paid donors. In European countries about 33% of blood supplies are used for surgical purposes while 67% are for medical treatments. However, the demand for blood outstrips supply, and blood transfusion services struggle to meet that demand while ensuring timeliness, quality, and safety.

The WHO classifies the hospitals' need for blood into three possible categories, namely emergency, certain need for blood, and possible need for blood. Emergencies are critical in terms of blood transport for two main reasons: blood must be available within 60 min, and the required blood group is likely the highly demanded group zero, as it is compatible with any other group. Furthermore, urgent transport is very expensive, ranging from €200 to €600 per trip according to public tenders and there is a risk of losing quality (or even wasting blood or organs) due to lack of monitoring and control of transport conditions. In the latest available report prepared by WHO, it is estimated that around half a million blood bags have been wasted due to transportation problems, and almost 6 million blood donations have been wasted due to storage problems [58, 86].

As mentioned before, the structure of blood components is so complex that they cannot be produced in the laboratory but only donated voluntarily, and therefore, their conservation is essential. Furthermore, these resources have a limited duration and must be stored under strictly controlled conditions, or otherwise, they undergo a process of hemolysis [87, 88]. Blood donation is made possible thanks to modern blood collection and storage techniques that allow keeping red blood cells active and transfusable for 42 days. To allow this, the blood must be collected and processed in a very short time, treated with suitable anticoagulants, and, for concentrated red blood cells, stored at a temperature of (4 ± 2) °C. Since the blood collected is so unique and precious, the rules of conservation are strictly preserved even during transport, which may be necessary to get units of concentrated red blood cells where they are most needed based on blood groups (each patient must be transfused with a compatible blood group to avoid serious and often fatal reactions).

Currently, the transport of blood is carried almost all the time on wheeled vehicles, including ambulances or medical cars, capable of transporting tertiary containers storing the blood bags which can maintain the required temperature conditions [89]. Only in particular conditions are aerial means such as helicopters and small transport planes used [90, 91], since they are affected by adverse weather conditions due to the high altitudes necessary for their flight, lower load capacity, and high costs due to the presence of personnel, such as a specialist doctor and pilots. As a result, drone delivery results in a convenient alternative transport system to road transport [92].

Concerning the **common containers**, they are designed in full compliance with the EU Directive 2002/98/EC in Europe. It prescribes quality and safety standards for the entire chain such as collection, testing, processing, storage, and distribution of human blood and blood components [93] and is UN3373 certified (Transport Biological Substance, Category B by Air) [94] (see Fig. 3). In the US, the whole blood chain complies with Full Drug Administration (FDA) regulations and reference documents. The Association for the Advancement of Blood & Biotherapies (AABB) defined, following the transfusion medicine community guidelines, [95] and working directly with the FDA, the rulemaking processes to ensure the safe and reliable collection of blood components, thus guaranteeing an adequate response to the needs of blood.

To meet the regulations, the blood containers should be of enough quality to resist normal transport conditions, including manual and mechanical handling and transfers between different vehicles. Any loss from the packaging should be excluded smoothly avoiding vibrations, variations in temperature, humidity level, or pressure.

Fig. 3 Certified container for blood transportation

The packaging must include at least the following three components: (a) a primary container; (b) a secondary container; and (c) an outer packaging. In addition, the outermost packages should be rigid. The primary container must be packed in the secondary container in such a way as to avoid breaking under normal conditions of transport and without compromising the integrity of the whole container.

In this context, alternative transportation systems such as "drone delivery" play an important role in overcoming the problems mentioned. However, the temperature-control requirement for red blood cells (RBC) is one of the main issues in using drones according to international regulations, i.e., 4 ± 2 °C. Among African countries, the Rwanda government has been authorized to deliver blood supplies and vaccines by drone to 20 regional hospitals since 2011 [96]; their blood is mainly used to treat malaria-induced anemia and for post-birth blood loss. 76.8% of deaths among children under the age of 5 years old are addressed to malaria in 2015, according to the WHO [97]. Deaths from postpartum complications represent one in four deaths worldwide (which is the cause of one in four maternal deaths worldwide) [98].

With respect to costs, examining over 12,000 blood product deliveries over approximately 32 months, Nisingizwe et al. [99] demonstrated that lead times decreased significantly (approximately 30%) and the reduction in 12-month blood product expirations was approximately 67%. However, no data are available about the integrity of the specimens. It is worth noting that according to the WHO report [86] 33% (corresponding to around 37 million) of the total bags are wasted due to expiration, consistently. Countries are invited to put more efforts into reducing as low as possible this issue.

Recently, in Ukerewe Island, district of Lake Victoria (Tanzania), a new drone transport system, based on vertical takeoff and landing aircraft, went into operation to bring medical supplies including blood units for transfusions in emergency surgeries [100].

The first significant scientific test was carried out in 2015 in Baltimore in the US. The Institutional Review Board for Human Subjects of Johns Hopkins Medicine approved the procedure and the investigation. 56 volunteers were involved. 168 blood tubes were flown in a small drone while the other 168 obtained in the same phlebology event were left on the ground. Flight times were between 6 and 38 min; 33 chemical tests among the most performed in medical laboratories (for instance, lactate dehydrogenase, aspartate aminotransferase, or potassium) were performed to measure any alterations. The results demonstrated that the parameters were not affected by drone flight within the limits of the experimental conditions [34]. Ambient conditions were not extreme, and no temperature and pressure control were necessary. However, inadequate temperature control can increase the risk of hemolysis [101] of RBC solution.

European countries are also testing this new type of transport service. For example, Germany has used drones for about three months to deliver urgent medicines between two villages in the Bavarian Alps, Reit im Winkl and Winklmoosalm, with a reduction of over a third of delivery times (from 30 to 8 min) [79]. Tests have been carried out in the UK to transport blood and medical supplies between Guy's and St Thomas's hospitals [102].

Fig. 4 Smart capsule for blood transportation (left) and a coaxial quadricopter carrying the smart capsule

An autonomous drone-based delivery system outfitted with a smart capsule for the transfer of blood products in cities has recently undergone validation testing in Italy (see Fig. 4). An analysis was done on how certain thermo-mechanical factors, including altitude, acceleration/deceleration, outside temperature, and humidity, affected the specimens' integrity. The comparison of the findings from hemolytic tests, which were routinely carried out on samples before and after each drone flight, made it abundantly evident that the blood's integrity was retained and that no negative consequences occurred during the transit. These findings were related to the properties of smart capsules, which enable the integration of real-time quality monitoring, control of the temperature that blood products experience, and mechanical vibrations. Despite the use of a drone-based delivery system has been demonstrated from an experimental viewpoint, it can be considered an innovation as a continuous service [103].

4 Conclusions

Thanks to their main characteristics, such as autonomous driving systems, access to inaccessible areas, low cost concerning road and manned air vehicles, reduced delivery times, and the ability to carry heavy payloads, drones will be of most use in civil activities and will be a key element of future smart cities.

Finally, the usage of drones can decrease the climate-altering gases. The most recent studies have shown that the world's climate is undergoing unprecedented changes. Global warming is causing increased changes in rainfall patterns, ocean ecosystems, and winds in all regions of the world. Higher temperatures and more intense meteorological events involve enormous costs for the economies of the countries and affect the food production capacity with devastating effects, especially in low-income countries.

In the EU, for example, since 1980 climate-related events have caused over 487 billion euros in financial losses, more than 138,000 people have lost their lives to extreme weather and climate events, the economic cost of river flooding exceeds an average of 5 billion euros a year, and the annual economic damage due to forest fires is equal to about 2 billion euros [104]. The EU, to lead the global fight against pollution, launched an ambitious plan called the European Green Deal [105] to achieve climate neutrality by 2050. The use of drone technology serves the goal of decreasing CO_2 emissions as they are powered by electric batteries; the latters can undergo to specific recycling program aiming to make all the manufacturing components of drones 100% reusable. Traditional deliveries come with an environmental cost, as they add congestion to cities, CO_2 emissions, and pollutants to the air. Drone delivery, in addition to being more efficient and faster than car transportation, cuts costs and reduces CO_2 emissions by 13 tons per million population compared to current delivery systems. Traveling 10 km with a gas car (13 km with 1 l) is equivalent to emitting 2 kg of carbon dioxide [106]. Energy-efficient, zero-emission electric drones are used.

Although the market for drones for delivery (in terms of technology and services) is growing strongly [48], the lack of adequate regulations for their use remains an open challenge. Only 57 of the 174 recognized nations in the world have publicly available drone legislation (7), and many of these regulations are outright prohibitions rather than a set of standards for the safe and legal use of drones. Furthermore, there is a lack of cross-border homogenization of regulatory standards. Precisely because of the lack of international regulation, delivery services, from commercial products to medical products or biological material, have been provided for a limited time (a few months at most) or in emergency conditions. Regarding safety, so far, drone delivery systems are based on automatic controls obtained through the flight controller, which is usually a microcontroller. The paradigm on which it is moving is that of the redundancy of artificial intelligence toward computationally more demanding devices. For this reason, at present, the intelligence of drones and autonomous vehicles in general remains confined to the device itself or, in some cases, moved to the cloud or a remote-control station. This solution may not be suitable for reducing the risk associated with flight operations. Therefore, soon, the use of an intelligent system directly on board, capable of managing and supervising the autonomous flight and the medical aspects, is plausible.

The challenges related to the new post-pandemic healthcare needs allow rethinking the regulations to allow the technological development of this new market, which cannot be lost by any high-income country.

If the technological growth rate is as expected, people will soon see drones rushing over people's heads capable of delivering, in a short time, products of all kinds (from medicines to pizza, from books to toys). This will have a positive impact not only in large cities due to traffic conditions that slow down deliveries but also in many villages, especially in Europe. After the pandemic that changed one of the paradigms of work, i.e., the presence in the office, many people have decided to live in villages and small villages far from big cities thanks to the introduction of smart working. One of the limitations of the repopulation of historic villages was precisely the difficulty

Table 1 Comparison of technical characteristics of drones

	VTOL	Fixed Wings	Multicopter
Take off and land everywhere	✓	✗	✓
Long flight time	✓	✓	✗
> 4 kg payload	✗	✗	✓
Easy to use	✗	✗	✓

of connections and access to the Internet. The use of drones would help overcome these difficulties by ensuring high standards of living.

Probably, soon, each house will have in its backyard a small landing platform. Hospitals will employ them to transport blood bags or organs between buildings. The on-board high-definition cameras will be used to support the job of local doctors connected to their hospital colleagues. Only human imagination will limit their potential uses.

References

1. Shaw, I. G. (2014). *The rise of the predator empire: Tracing the history of U.S drones*. Understanding Empire. Accessed 23 March 2024 from https://understandingempire.wordpress.com/2-0-a-brief-history-of-U.S-drones
2. EASA. European Union Aviation Safety Agency. Accessed 23 March 2024 from https://www.easa.europa.eu/en
3. National Archives. Part 107 - Small unmanned aircraft systems. Accessed 23 March 2024 from https://www.ecfr.gov/current/title-14/chapter-I/subchapter-F/part-107
4. Colomina, I., & Molina, P. (2014). Unmanned aerial systems for photogrammetry and remote sensing: A review. *ISPRS Journal of Photogrammetry and Remote Sensing, 92*, 79–97.
5. Goodchild, A., & Toy, J. (2018). Delivery by drone: An evaluation of unmanned aerial vehicle technology in reducing CO_2 emissions in the delivery service industry. *Transportation Research Part D, 61*, 58–67.
6. Anderson, K., & Gaston, K. J. (2013). Lightweight unmanned aerial vehicles will revolutionize spatial ecology. *Frontiers in Ecology and the Environment, 11*, 138–146.
7. Keane, J. F., & Carr, S. S. (2013). A brief history of early unmanned aircraft. *Johns Hopkins APL Technical Digest, 32*(3), 558–571.
8. Newcome, L. R. (2004). *Unmanned aviation: A brief history of un-manned aerial vehicles*. American Institute of Aero-nautics and Astronautics Inc.
9. Oudes, C., Zwijnenburg, W. (2011). *Does unmanned make unacceptable? Exploring the debate on using drones and robots in warfare*. IKV Pax Christi.
10. Finn, P. (2011). *Domestic use of aerial drones by law enforcement likely to prompt privacy debate* (p. 22). Washington Post.
11. Murphy, R. (2015). *Drones save lives in disasters, when they're allowed to fly (Op-Ed)*. Accessed 23 March 2024 from https://www.space.com/30555-beginning-with-katrina-drones-save-lives-in-disasters.html

12. Enemark, C. (2017). *Armed drones and the ethics of war: Military virtue in a post-heroic age.* Routledge.
13. Rife, J. P. (2006). *The sound of freedom: Naval Weapons Technology at Dahlgren, Virginia 1918–2006*. Government Printing Office.
14. Keane, J. F., & Carr, S. S. (2013). A brief history of early unmanned aircraft. *Johns Hopkins APL Technical Digest, 32*(3), 558–571.
15. Chen, S. (2017). *The predator RQ-1/MQ-1/MQ-9 reaper UAV*. Air Force Technology. Accessed 23 March 2024 from https://www.airforce-technology.com/projects/predator-uav/.
16. Robinson, A. (2006). *FAA authorizes predators to seek survivors*. US Air Force. Accessed 23 March 2024 from https://www.af.mil/News/Article-Display/Article/130220/faa-authorizes-predators-to-seek-survivors/
17. Craighead, J., Murphy, R., Burke, J., Goldiez, B. (2007). A survey of commercial & open source unmanned vehicle simulators. In *2007 IEEE international conference on robotics and automation* (pp. 852–857). IEEE.
18. Van Berlaer, G., Staes, T., Danschutter, D., et al. (2017). Disaster preparedness and response improvement: Comparison of the 2010 Haiti earthquake-related diagnoses with baseline medical data. *European Journal of Emergency Medicine, 24*(5), 382–388.
19. Bamburry, D. (2015). Drones: Designed for product delivery. *Design Management Review, 26*, 40–48.
20. Alkouz, B., Shahzaad, B., & Bouguettaya, A. (2021). Service-based drone delivery. In *Proceedings of the IEEE 7th international conference on collaboration and internet computing (CIC)*, Atlanta, GA, USA, 13–15 Dec 2021 (pp. 68–76).
21. Hiebert, B., Nouvet, E., Jeyabalan, V., & Donelle, L. (2020). The application of drones in healthcare and health-related services in North America: A scoping review. *Drones, 4(3)*, 30.
22. Kardasz, P., Doskocz, J., Hejduk, M., Wiejkut, P., & Zarzycki, H. (2018). Drones and possibilities of their using. *Journal of Civil and Environmental Engineering, 6(3)*, 1000233.
23. Rosser, J. C., Jr., Vignesh, V., Terwilliger, B. A., & Parker, B. C. (2018). Surgical and medical applications of drones: A comprehensive review. *Journal of the Society of Laparoendoscopic Surgeons, 22*(3), e2018.00018.
24. Priye, A., Wong, S., Bi, Y., et al. (2016). Lab-on-a-drone: Toward pinpoint deployment of smartphone-enabled nucleic acid-based diagnostics for mobile health care. *Analytical Chemistry, 88*(9), 4651–4660.
25. Barasona, J. A., Mulero-Pazmany, M., Acevedo, P., et al. (2014). Unmanned aircraft systems for studying spatial abundance of ungulates: Relevance to spatial epidemiology. *PLoS ONE, 9*(12), e115608.
26. van Wynsberghe, A., & Comes, T. (2020). Drones in humanitarian contexts, robot ethics, and the human–robot interaction. *Ethics and Information Technology, 22*, 43–53.
27. Cawthorne, D., & Cenci, A. (2019). Value sensitive design of a humanitarian cargo drone. In *Proceedings of the 2019 international conference on unmanned aircraft systems (ICUAS)*, Atlanta, GA, USA, 11–14 June 2019, (pp. 1117–1125).
28. Rejeb, A., Rejeb, K., Simske, S., & Treiblmaier, H. (2021). Humanitarian drones: A review and research agenda. *Internet of Things, 16*, 100434.
29. Fornace, K. M., Drakeley, C. J., William, T., Espino, F., & Cox, J. (2014). Mapping infectious disease landscapes: Unmanned aerial vehicles and epidemiology. *Trends in Parasitology, 30*, 514–519.
30. Malveaux, C., Hall, S. G., & Price, R. (2014). Using drones in agriculture: unmanned aerial systems for agricultural remote sensing applications. In *2014 Montreal, American society of agricultural and biological engineers*, Quebec Canada (p. 1).
31. Hogan, S., Kelly, M., Stark, B., & Chen, Y. (2017). Unmanned aerial systems for agriculture and natural resources. *Calif. Agric. Berkel., 71*(1), 5–14.
32. Thompson, R. M. (2012). *Drones in domestic surveillance operations: Fourth amendment implications and legislative responses*. Congressional Research Service, Library of Congress.
33. Altawy, R., & Youssef, A. M. (2016). Security, privacy, and safety aspects of civilian drones: A survey. *ACM Transactions on Cyber-Physical Systems, 1*, 1–25.

34. Maher, A., & Inoue, M. (2016) Generating evacuation routes by using drone system and image analysis to track pedestrian and scan the area after disaster occurrence. In *Proceedings of the 10th SEATUC (South East Asian Technical University Consortium) conference*, Tokyo, Japan, 22–24 Feb 2016 (pp. 22–24).
35. Cohen, J. (2007). Drone spy plane helps fight California fires. *Science, 318*, 727.
36. Drones, H. J. (2015). *A force for good when flying in the face of disaster*. The Guardian. July 08, 2015. Accessed 23 March 2024 from https://www.theguardian.com/global-development/2015/jul/28/drones-flying-in-the-face-of-disaster-humanitarian-response/.
37. Al-Rawabdeh, A., Moussa, A., Foroutan, M., El-Sheimy, N., & Habib, A. (2017). Time series UAV image-based point clouds for landslide progression evaluation applications. *Sensors, 17*, 2378.
38. Tang, X.-B., et al. (2016). Efficiency calibration and minimum detectable activity concentration of a real-time UAV airborne sensor system with two gamma spectrometers. *Applied Radiation and Isotopes, 110*, 100–108.
39. Martin, P. G. et al. (2015). The use of unmanned aerial systems for the mapping of legacy uranium mines. *Journal of environmental radioactivity, 143*, 135–140.
40. Tkáč, M., & Mésároš, P. (2019). Utilizing drone technology in the civil engineering. *Selected Scientific Papers-Journal of Civil Engineering, 14*, 27–37.
41. Lee, S., & Choi, Y. (2016). Reviews of unmanned aerial vehicle (drone) technology trends and its applications in the mining industry. *Geosystem Engineering, 19*, 197–204.
42. Shahmoradi, J., Talebi, E., Roghanchi, P., & Hassanalian, M. (2020). A comprehensive review of applications of drone technology in the mining industry. *Drones, 4*, 34.
43. Jiang, T., Geller, J., Ni, D., & Collura, J. (2016). Unmanned aircraft system traffic management: Concept of operation and system architecture. *International Journal of Transportation Science and Technology, 5*, 123–135.
44. Finn, R. L., & Wright, D. (2012). Unmanned aircraft systems: Surveillance, ethics and privacy in civil applications. *Computer Law and Security Review, 28*(2), 184–194.
45. Fortune Business Insights. Accessed 23 March 2024 from https://www.fortunebusinessinsights.com/aboutus
46. Drone Industry Insights. Accessed 23 March 2024 from https://droneii.com/product/drone-market-report
47. European Commission. A Drone Strategy 2.0 for a Smart and Sustainable Unmanned Aircraft Eco-System in Europe. Accessed 23 March 2024 from https://transport.ec.europa.eu/document/download/1cb5fb4f-4252-4f97-abf4-c4a167b1c7d2_en?filename=COM_2022_652_drone_strategy_2.0.pdf
48. IEEE Future Directions. Megatrends for this decade. Accessed 23 March 2024 from https://cmte.ieee.org/futuredirections/2020/12/11/megatrends-for-this-decade-xix/
49. Khan, M. A., Alvi, B. A., Safi, A., & Khan, I. U. (2018). Drones for good in smart cities: A review. In *Proceedings of the 2018 international conference on electrical, electronics, computers, communication, mechanical and computing (EECCMC)*, Vellore, India, 28–29 Jany 2018 (pp. 1–6).
50. Rupprecht Law P.A. Drone Law and Drone Attorney Assistance. Accessed 23 March 2024 from https://jrupprechtlaw.com
51. Abrahamsen, H. B. (2015). A remotely piloted aircraft system in major incident management: Concept and pilot, feasibility study. *BMC Emergency Medicine, 15*, 12.
52. Amukele, T. K., Street, J., Carroll, K., Miller, H., & Zhang, S. X. (2016). Drone transport of microbes in blood and sputum laboratory specimens. *Journal of Clinical Microbiology, 54*, 2622–2625.
53. Scott, J. E., & Scott, C. H. (2018). Models for drone delivery of medications and other healthcare items. *International Journal of Healthcare Informations Systems and Informatics (IJHISI), 13*, 20–34.
54. Bright, J. (2020). *Zipline begins US medical delivery with drone program honed in Africa.* Accessed 23 March 2024 from https://techcrunch.com/2020/05/26/zipline-begins-us-medical-delivery-with-uav-program-honed-in-africa/

55. Swain, R. (2023). *Medical drone delivery services market share report, 2023–2032*. Accessed 23 March 2024 from https://www.gminsights.com/industry-analysis/medical-drones-market
56. Cawthorne, D., & Robbins-van Wynsberghe, A. (2020). An ethical framework for the design, development, implementation, and assessment of drones used in public healthcare. *Science and Engineering Ethics, 26*, 2867–2891.
57. Amukele, T. (2020). The economics of medical drones. *The Lancet Global Health, 8*, e22.
58. Mell, H. K., Mumma, S. N., Hiestand, B., Carr, B. G., Holland, T., & Stopyra, J. (2017). Emergency medical services response times in rural, suburban, and urban areas. *JAMA Surgery, 152*(10), 983–984.
59. Arcury, T. A., Preisser, J. S., Gesler, W. M., & Powers, J. M. (2005). Access to transportation and health care utilization in a rural region. *The Journal of Rural Health, 2*, 31–38.
60. Tanser, F., Gijsbertsen, B., & Herbst, K. (2006). Modeling and understanding primary health care accessibility and utilization in rural South Africa: An exploration using a geographical information system. *Social Science and Medicine, 63*, 691–705.
61. Laksham, K. B. (2019). Unmanned aerial vehicle (drones) in public health: A SWOT analysis. *Journal of Family Medicine and Primary Care, 8*, 342–346.
62. Thiels, C. A., Aho, J. M., Zietlow, S. P., & Jenkins, D. H. (2015). Use of unmanned aerial vehicles for medical product transport. *Air Medical Journal, 34*(2), 104–108.
63. Schierbeck, S., et al. (2022). Automated external defibrillators delivered by drones to patients with suspected out-of-hospital cardiac arrest. *European Heart Journal, 43*(15), 1478–1487.
64. Medical Advisory Secretariat. (2005). use of automated external defibrillators in cardiac arrest. An evidence-based analysis. *Ontario Health Technology Assessment Series, 5*, 1–29.
65. Andres, J (2021). *Wytyczne Resuscytacji 2021. Polska Rada Resuscytacji*. Kraków.
66. Hii, M. S. Y., Courtney, P., & Royall, P. G. (2019). *Droni, 3*(3), 52.
67. Akbarian, M., Ghasemi, Y., Uversky, V. N., & Yousefi, R. (2018). Chemical modifications of insulin: Finding a compromise between stability and pharmaceutical performance. *International Journal of Pharmaceutics, 547*, 450–468.
68. Alemu, S., Dessie, A., Seid, E., Bard, E., Lee, P. T., Trimble, E. R., Phillips, D. I., & Parry, E. H. (2009). Insulin-requiring diabetes in rural Ethiopia: Should we reopen the case for malnutrition-related diabetes? *Diabetologia, 52*, 1842–1845.
69. Zhu, W., Oakey, A., Royall, P. G., Waters, T. P., Cherrett, T., Theobald, K., & Lucas, R. (2023). Investigating the influence of drone flight on the stability of cancer medicines. *PLoS ONE, 18*(1), e0278873.
70. Chatelut, E., White-Koning, M. L., Mathijssen, R. H., Puisset, F., Baker, S. D., Sparreboom, A. (2012). Dose banding as an alternative to body surface area-based dosing of chemotherapeutic agents. *British Journal of Cancer, 107*, 1100–1106.
71. Cortes, J., Perez-García, J. M., Llombart-Cussac, A., Curigliano, G., El Saghir, N. S., & Cardosok F., et al. (2020). Enhancing global access to cancer medicines. *CA Cancer Journal of Clincians, 70*, 105–124.
72. Le Basle, Y., Chennell, P., Tokhadze, N., Astier, A., & Sautou, V. (2020). Physicochemical stability of monoclonal antibodies: A review. *Journal of Pharmaceutical Sciences, 109*, 169–190.
73. Barasona, J. A., et al. (2014). Unmanned aircraft systems for studying spatial abundance of ungulates: Relevance to spatial epidemiology. *PLoS One, 9*(12), e115608.
74. Capolupo, A., Pindozzi, S., Okello, C., Fiorentino, N., & Boccia, L. (2015). Photogrammetry for environmental monitoring: The use of drones and hydrological models for detection of soil contaminated by copper. *Science of the Total Environment, 514*, 298–306.
75. Brady, J. M., Stokes, M. D., Bonnardel, J., & Bertram, T. H. (2016). Characterization of a quadrotor unmanned aircraft system for aerosol–particle-concentration measurements. *Environmental Science and Technology, 50*, 1376–1383.
76. Mendelow, B., Muir, P., Boshielo, T., & Robertson, J. (2007). Development of e-Juba, a preliminary proof of concept UAV (Unmanned Aerial Vehicle) designed to facilitate the transportation of micro-biological test samples from remote rural clinics to NHLS laboratories. *South African Medical Journal, 97*, 1215–1218.

77. UNICEF. Drone Testing for HIV Early Infant Diagnosis. Accessed 23 March 2024 from https://www.unicef.org/innovation/stories/drone-testing-hiv-early-infant-diagnosis
78. Medecins Sans Frontieres. (2014). Papua New Guinea: Innovating to reach remote TB patients and improve access to treatment Geneva: Medecins Sans Frontieres (MSF) International. Accessed 23 March 2024 from http://www.msf.org/article/papua-new-guinea-innovating-reach-remote-tb-patients-and-improve-access-treatment/
79. Hackman, M., & Nicas, J. (2015). Drone delivers medicine to rural Virginia clinic. *Wall Street Journal*. Accessed 23 March 2024 from https://www.wsj.com/articles/drone-delivers-medicine-to-rural-virginia-clinic-1437155114.
80. Cousins, S. (2016). *Condoms by drone: A new way to get birth control to remote areas*. National Public Radio. Accessed 23 March 2024 from https://www.npr.org/sections/goatsandsoda/2016/05/19/478411186/condoms-by-drone-a-new-way-to-get-birth-control-to-remote-areas
81. DroneDJ. Zipline drones deliver one million COVID vaccinations in Ghana. Accessed 23 March 2024 fromhttps://dronedj.com/2022/03/30/zipline-drones-deliver-one-million-covid-vaccinations-in-ghana/
82. Lippi, G., & Mattiuzzi, C. (2016). Biological samples transportation by drones: Ready for prime time? *Annals of Translational Medicine, 4*, 92.
83. Glauser, W. (2018). Blood-delivering drones saving lives in Africa and maybe soon in Canada. *Canadian Medical Association Journal, 190*, E88–E89.
84. World Health Organization. WHO Model Lists of Essential Medicines. Accessed 23 March 2024 from https://www.who.int/groups/expert-committee-on-selection-and-use-of-essential-medicines/essential-medicines-lists
85. World Health Organization. Blood safety and availability. Accessed 23 March 2024 from https://www.who.int/news-room/fact-sheets/detail/blood-safety-and-availability
86. World Health Organization. Global Status Report on Blood Safety and Availability. Accessed 23 March 2024 from https://apps.who.int/iris/bitstream/handle/10665/254987/9789241565431-eng.pdf
87. Ando, H., Nieminen, K., Toppila, E., Starck, J., & Ishitake, T. (2005). Effect of impulse vibration on red blood cells in vitro. *Scandinavian Journal of Work, Environment and Health, 31*, 286–290.
88. Kitchen, S., Adcock, D. M., Dauer, R., Kristoffersen, A.-H., Lippi, G., & Mackie, I., et al. (2021). International Council for Standardisation in Haematology (ICSH) recommendations for collection of blood samples for coagulation testing. *International Journal of Laboratory Hematology*, 43, 571–580.
89. Fujita, H., Tojo, Y., Mine, T., & Tanaka, A. (2020). Temperature management of red blood cell solution cell solution transported by car for transfusion at home. *Open Journal of Blood Diseases, 10*, 37–40.
90. Otani, T., Oki, K. I., Akino, M., Tamura, S., Naito, Y., Homma, C., Ikeda, H., & Sumita, S. (2012). Effects of helicopter transport on red blood cell comonenets. *Blood Transfusion, 10*, 78–86.
91. Stein, C., & Caetano, E. (2016). Transportation of blood in a helicopter emergency medical service: The importance of specialized equipment. *South African Journal of Critical Care, 32*, 62–63.
92. Amukele, T. K., Sokoll, L. J., Pepper, D., Howard, D. P., & Street, J. (2015). Can unmanned aerial systems (Drones) be used for the routine transport of chemistry, hematology, and coagulation laboratory specimens? *PLoS ONE, 10*, e0134020.
93. EUR-Lex. Accessed 23 March 2024 from https://eur-lex.europa.eu/eli/dir/2002/98/oj
94. UNECE. About the ADR. Agreement concerning the International Carriage of Dangerous Goods by Road. Accessed 23 March 2024 from https://unece.org/about-adr
95. AABB. Accessed 23 March 2024 from https://www.aabb.org/standards-accreditation/accreditation/become-accredited
96. Balasingam, M. (2017). Drones in medicine—The rise of the machines. *Internationa Journal of Clinical Practice, 71*, e12989.

97. World Health Organization. World malaria report 2022. Accessed 23 March 2024 from https://cdn.who.int/media/docs/default-source/malaria/world-malaria-reports/world-malaria-report-2022.pdf
98. Zailani, M. A. H., Sabudin, R., Rahman, R. A., Saiboon, I. M., Ismail, A., & Mahdy, Z. A. (2020). Drone for medical products transportation in maternal healthcare: A systematic review and framework for future research. *Medicine, 99*(36), e21967.
99. Nisingizwe, M. P., Ndishimye, P., Swaibu, K., Nshimiyimana, L., Karame, P., Dushimiyimana, V., Musabyimana, J. P., Musanabaganwa, C., Nsanzimana, S., & Law, M. R. (2022). Effect of unmanned aerial vehicle (drone) delivery on blood product delivery time and wastage in Rwanda: A retrospective, cross-sectional study and time series analysis. *The Lancet Global Health, 10*, e564–e569.
100. AA. Medical drones help save lives in Tanzania. Accessed 23 March 2024 from https://www.aa.com.tr/en/africa/medical-drones-help-save-lives-in-tanzania/2213162
101. Faghih, M. M., & Sharp, M. K. (2019). Modeling and prediction of flow-induced hemolysis: A Review. *Biomechanics and Modeling in Mechanobiology, 18*, 845–881.
102. Nesta. Flying high. Shaping the future of drones in UK cities. Accessed 23 March 2024 from https://media.nesta.org.uk/documents/Flying-High-full-report-and-appendices.pdf
103. Niglio, F., Comite, P., Cannas, A., Pirri, A., & Tortora, G. (2022). Preliminary clinical validation of a drone-based delivery system in urban scenarios using a smart capsule for blood. *Drones, 6*, 195.
104. European Council. Climate change: what the EU is doing. Accessed 23 March 2024 from https://www.consilium.europa.eu/en/policies/climate-change/
105. European Commission. The European Green Deal. Accessed 23 March 2024 from https://commission.europa.eu/strategy-and-policy/priorities-2019-2024/european-green-deal_en
106. European Environment Agency. Average CO_2 emissions from new cars and new vans increased again in 2019. Accessed 23 March 2024 from https://www.eea.europa.eu/highlights/average-co2-emissions-from-new-cars-vans-2019

Drones in Aerospace Engineering Education: The Case of UVigo Aerotech

Pablo Magariños Docampo, Marco Guiotto Sánchez, Marcos Quiroga Rodríguez, Modesto José Alaez Liñares, José Gómez Fernández, and Pablo Domínguez Estévez

Abstract This chapter plumbs into the challenges of aerospace engineering education, spotlighting the inherent gap between theoretical knowledge and practical application. The section advocates for innovative approaches to address this divide within the constraints of traditional academia. Central to this is the role of student associations, exemplified by UVigo Aerotech, in providing hands-on experiences such as designing and manufacturing fixed-wing drones for students. Throughout it emphasized how these associations not only improve academic learning but also provide essential skills to its members. It also highlights the symbiotic relationship between academic studies and practical experiences as facilitated by student associations, urging for an innovative approach to build a link between theory and application.

Keywords Aerospace engineering · Student associations · Soft skills development · University education · Teamwork in engineering

1 Introduction

The study focused on the aerospace sector is complex, involving numerous and diverse branches of both science and engineering. Subjects such as thermodynamics, mechanics (classical, fluid, flight, or orbital), calculus, and engineering applications in engines and systems are necessary, among others. However, the condensation of this theoretical knowledge in the limited duration of the aerospace engineering degree forces to reduce the time dedicated to practical lessons, which often allow students to directly understand and appreciate real-world applications of the theory they have learned. This leads to a gap between theoretical knowledge and practical experience in the aerospace industry, which can be supplemented through various extracurricular activities that students can engage in during their university education, such as participation in a student association.

P. M. Docampo (✉) · M. G. Sánchez · M. Q. Rodríguez · M. J. A. Liñares · J. G. Fernández · P. D. Estévez
Escola de Enxeñaría, Aeronáutica E Do Espazo, Universidade de Vigo, 32004 Ourense, Spain
e-mail: Pablo.magarinos@outlook.com

D. Carou et al. (eds.), *Applying Drones to Current Societal and Industrial Challenges*, Management and Industrial Engineering, https://doi.org/10.1007/978-3-031-55571-8_11

Focusing on the case of the University of Vigo, the Aerospace Engineering degree lasts for 4 years and includes between 36 and 37 different subjects depending on the chosen specialization. The enrollment of 50 new students each year requires dividing the students into groups for practical lessons. The limited available time and the high cost of materials and infrastructure, which must be shared among the students during the lessons, contribute to the gap between theoretical learning and practical experience that students may face upon completing their degree and entering the job market.

To address this problem, student associations like UVigo Aerotech emerged to bridge the gap between students and the aerospace sector. The association achieves this by designing and manufacturing fixed-wing drones, using cutting-edge materials found in current commercial aircraft, and employing industry-leading software for Computer-Aided Design (CAD) design, Computational Fluids Dynamics (CFD), and Finite Element Method (FEM). Working with an association like UVigo Aerotech while studying for the degree allows students to compare, understand, and observe the application of the theory they learn during their studies without compromising the time dedicated to academic learning.

Engaging in extracurricular activities, whether through internships or involvement in student associations, also equips students with additional skills and knowledge beyond what they acquire during their degree. This differentiation becomes valuable in the job market, as companies do not need to provide them with basic training.

Furthermore, among the different extracurricular activities, the work of student associations encourages students to investigate various phenomena and aspects of the aerospace world, depending on the purpose of the association. This way students can stay at the forefront of the industry and learn from the latest and most modern studies in the aerospace sector.

Due to the rapid progress and research in this industry, the teachings in certain subjects offered by the university may become outdated relatively quickly. Therefore, the coordinated work between associations and universities has positive consequences in educating students with the most modern and up-to-date knowledge.

2 Benefits of Engineering Associations

There are several advantages of this type of university association. Among them, some of the main ones are shown below.

2.1 Teamwork

In engineering, industry leaders consider teamwork to be one of the fundamental soft skills. Experience gained through working in a multidisciplinary organization like UVigo Aerotech provides its members with this invaluable skill.

The team comprises six departments, ranging from aerodynamics to electronics and marketing. Therefore, effective communication becomes pivotal to the overall success of the group. Those Aerospace Engineering students who do not engage in these associations miss out on valuable learning opportunities and personal growth. Transmitting ideas between departments with diverse technical backgrounds fosters a nuanced approach, requiring members to proficiently communicate engineering concepts to individuals with varying expertise levels in the field.

Successfully coordinating a team of nearly thirty members requires strong leadership. Consequently, UVigo Aerotech implements a straightforward command chain while ensuring the optimal utilization of each member's abilities. At the time of writing, the Team Leader assumes the role of ensuring adequate functioning and optimization of the team and its member's capabilities. Meanwhile, the Technical Leader and department heads handle task micromanagement. Although this organizational scheme was used before shifting toward a scrum approach, it still allows for different layers of teamwork and coordination to manifest. Regular board of directors' in-person gatherings serve the purpose of aligning everyone and maintaining shared understanding. Additionally, the team's size permits meetings among all members to occur once or twice monthly, offering everyone an opportunity to voice their opinions. These gatherings serve as crucial forums for debating and further emphasize the necessity of exceptional communication.

To sum up, UVigo Aerotech serves as an example of the paramount importance of teamwork in engineering. The organization's multidisciplinary nature nurtures essential soft skills, with communication emerging as the cornerstone of team success. As the team continues to evolve, its relentless focus on outstanding communication and collaborative problem-solving maintains a culture of excellence in teamwork (Fig. 1).

2.2 *Creativity Freedom*

Another of the virtues of UVigo Aerotech is providing its members a platform to test their ideas. In short, freedom to learn via trial and error. The design process allows testing several iterations of mechanisms, structures, and aerodynamical solutions. Accordingly, everyone can learn from the mistakes they made in their designs. Remember that this is a collaborative effort, and at this stage, no individual carries a disproportionate workload. This freedom to apply ideas and learn from mistakes not only fosters innovation but also strengthens the sense of ownership and commitment among team members.

Often students have ideas that cannot be implemented due to financial or technical constraints. However, thanks to its member's skills and previous experience in design and manufacturing, associations prove a valuable platform for students to bring their ideas to life. Thus, UVigo Aerotech closes the loop between academia and real-world application, allowing students to apply their theoretical knowledge in a practical setting and gain valuable hands-on experience. To achieve this, the teams often have

Fig. 1 UVigo Aerotech at the XtraChallenge 2023 in Valencia

a sponsorship model, where different companies and entities put their trust in the projects and support them financially or with different resources.

In summary, UVigo Aerotech's commitment to providing its members with the freedom to learn, innovate, and apply their ideas cements its position as an exceptional organization that enriches the academic journey but also prepares students for the challenges and opportunities ahead in their careers.

2.3 *Manufacture Learning*

Manufacturing is a key step in the engineering process since it is the point where the product leaves the CAD, and is materialized in the world. This turning point comes with a set of challenges that may be well known and can be explained in the classroom but seeing them happen first-hand allows a more in-depth learning and a lifetime impression of these set problems.

The most important benefit of being able to manufacture during university is the enrichment that takes place from experiencing these challenges, which combined with the knowledge acquired in the classroom allows the students to make what they learned a significant tool for the future.

This learning is not only in the field of manufacturing, but the most interesting part is how it influences the understanding of all the processes before and after manufacturing. For instance, the design of a product can kill its viability even before it is tested because it can't be manufactured, or the manufacturing process is too expensive. These real problems appear when participating in activities such as UVigo Aerotech, where the engineering process is followed from an initial design to its final testing in a real drone. The manufacturing that takes place in between allows students to see the associated technical problems (like mold design, how materials react to different processes or tools, etc.) and non-technical problems (like the cost problem or the availability of the manufacturing procedures).

For example, one of the most interesting and up-to-date processes in engineering is carbon fiber composite manufacturing. This process is not individual and unique, but a set of viable processes to choose from depending on the component, and even within a single process many variations take place to obtain the exact part needed for the job. Being able to manufacture something in carbon fiber helps in understanding how the materials react and how this affects the final product, and a better understanding of how the initial design conditioned the manufacturing and how it can be improved to make manufacturing easier, cheaper, and faster. In UVigo Aerotech, for instance, carbon fiber manufacturing is one of the key learning objectives, and the opportunity to work with these materials during the formation process helps students interact with the materials and processes that they studied and improves not only the understanding of said processes, but it has proven to improve the quality of the designs made by the same students on subsequent years, meaning manufacturing can be more efficient and more precise every year that it passes in a process that is done entirely by students who learned from the experience and their own mistakes, working as a real-life engineering department, carrying these skills into the world as future engineers.

2.4 Networking

Knowing and getting to know each other is one of the greatest benefits of being part of university projects. From the first moment the student is immersed in the discipline of their section of the project, it leads them to ask themselves a series of questions about what they have learned in class, and they can seek answers from the teaching staff, who are very open to help and guide them. It is at these moments that the students can learn more about the teacher's department while increasing their knowledge and discovering related academic or research opportunities.

On the other hand, interacting with companies seeking sponsorship can lead the student to meet workers in the sector who are delighted with the student's drive and interest. This, in turn, can be a way for the student to get an internship, by having contact with a specific company and later being hired.

Finally, it is important to assess what the aim or objective of the project entails. This may involve trips to other universities, where you can get to know the perspectives of students from other centers, get to know their facilities, and finally enjoy a rewarding experience, in which the efforts of a whole year are generally reflected, far removed from a numerical value such as grades in studies.

3 Simultaneity Between Degree and Student Team

One of the most important aspects to consider when contemplating membership in a university team is the effect it may have on the academic aspects of the degree being pursued. Despite the significant benefits that come with being part of university associations, the degree takes the forefront, and other types of activities are relegated to a secondary level. Therefore, it is crucial to emphasize that team activities should be developed in a way that does not pose any inconvenience to academic commitments. This can be achieved through the following methods:

- Team expansion: by including more members, the distribution of tasks improves, reducing individual workloads and bringing in more ideas and points of view.
- Implementation of reduced work periods: selecting specific intervals during which the university workload will be higher than usual, such as times when assignments or exams are due.

To achieve good results for the team, it is essential to maintain members' motivation, which requires a careful balance between academic pursuits and team involvement. If the association demands too much work from its members, they may feel pressured and not perform at their best, while an excessively low workload could lead to disengagement due to a lack of interest. Paradoxically, a slight excess of workload can contribute to improved performance by habituating individuals to work under stress, thereby enhancing their ability to perform under pressure—a valuable skill in many situations.

Nevertheless, through careful organizational efforts by the association's leadership, effective planning can be achieved, allowing the team to reach its seasonal objectives while maximizing the potential of each member and enabling them to pursue other activities. Maintaining this balance reveals an extremely positive experience of belonging to the team, facilitating not only the application of knowledge acquired throughout the academic journey but also fostering the learning of new theoretical concepts and soft skills. Examples of these skills include the organization of temporal and human resources, as well as collaborative teamwork.

As a result of the aforementioned factors, a feedback loop exists between university studies and student associations. Knowledge acquired during the degree is put into practice and expanded while developing the team's activities. Simultaneously, improving the understanding of theoretical concepts enhances academic performance in the degree, and the organizational skills acquired can be applied in numerous situations, both within and beyond the university context.

In summary, the simultaneity between pursuing a degree and being involved in a university association is not only entirely achievable but also extremely beneficial for its members.

4 Innovation in Associations

So far, the chapter highlighted the ability of associations to introduce students to real projects and thus develop the skills and knowledge acquired during their studies, but there is another opportunity open to be discovered, and that is research and development. There is a big opportunity for the students to learn, discover new fields, and go a step further in aerospace engineering knowledge.

UVigo *Aerotech* has recently created a new project division called *UVigo Aerotech: Research & Development*, where the students can work on some of the most advanced fields in the aerospace industry, such as autonomous flying, solar-powered aircraft, computer vision, the application of machine learning on aircraft design, or the development of morphing wing, among others. To achieve this, modern methodologies are being used, applying the principles of the agile methodology to improve the results obtained. In this way, all the members of this project can maximize the development achieved in each of the fields and can result in graduates of the aerospace engineering degree specializing in various subjects of great value, both now and in the future.

As *UVigo Aerotech* does, every student-led project possesses the opportunity to take the next step into state-of-the-art technology, creating an ecosystem of research that benefits not only the teams but also all the students and the entire school to learn about new and emerging fields.

5 Conclusion

This chapter has systematically explored the critical intersection between aerospace engineering education and its practical application, underscoring the invaluable role of hands-on experience in cultivating essential skills for future engineers. Through the lens of student-driven initiatives such as *UVigo Aerotech*, the narrative has compellingly demonstrated the enhancement of academic learning through real-world project involvement, such as the design and construction of fixed-wing drones.

Acknowledgements The authors wish to express their profound gratitude to the sponsors of *UVigo Aerotech*. The work and advancements of the team would not have been possible without the unwavering support and commitment of the various entities and enterprises collaborating with the project. A significant number of these sponsors hail from the industrial sector of Galicia, complemented by a substantial contribution from both national and international corporations. Collectively, these entities contribute to the project's success by providing state-of-the-art resources and technologies, which are crucial for the advancements achieved.

Main sponsors: *Escola de Enxeñaría Aeronáutica e do Espazo, Universidade de Vigo—Coasa—Mecadis—Gurit—Europrecis—Utingal—Marine Instruments—Altair—Centro de Investigación, Transferencia e Innovación—Symula—Civil UAVs Initiative*

Conclusions

Pedro Orgeira-Crespo

This book aims to introduce the ever-expanding frontier of Unmanned Aerial Vehicle (UAV) and Unmanned Aerial Systems (UAS), where innovation intersects with diverse industries, promising transformative solutions, and unparalleled capabilities. The book commences with an introductory chapter that lays out the foundational concepts pertinent to the topic. It offers readers a historical overview and a classification system, providing essential context for understanding drones. Subsequent chapters delve into a myriad of UAV applications, illustrating their pivotal role in molding the modern technological landscape. From autonomous drones navigating the intricacies of medicine to cutting-edge Counter-Unmanned Aircraft Systems (C-UAS) safeguarding critical infrastructures, each chapter encapsulates the breadth and depth of UAV capabilities.

"*Autonomous Drones and Smart Containers in Medicine*" unravels the transformative impact of UAVs in healthcare, illustrating how they enhance medical logistics, facilitate rapid response, and redefine the delivery of critical supplies.

The imminent future of freight transport is poised for significant transformation with the widespread adoption of drones. Originally designed for use in remote or emergency areas, drones are now being extensively tested in urban environments worldwide. Notably, the integration of artificial intelligence (AI) in drones has augmented their safety measures, enabling autonomous decision-making during emergency scenarios such as rotor failure, sudden obstacles, or cyber-attacks. The COVID-19 pandemic witnessed a substantial increase in drone use in the medical field, facilitating the transportation of medicines, vaccines, and healthcare equipment while minimizing the risk of contagion for frontline workers.

P. Orgeira-Crespo (✉)
Aerospace Area, Department of Mechanical Engineering, Heat Engines and Machines, and Fluids, Aerospace Engineering School, University of Vigo, Campus Orense, 32004 Orense, Spain
e-mail: porgeira@uvigo.es

D. Carou et al. (eds.), *Applying Drones to Current Societal and Industrial Challenges*, Management and Industrial Engineering, https://doi.org/10.1007/978-3-031-55571-8_12

Despite the progress made, large-scale deployment of drones faces two main hurdles: public acceptance of drones in their airspace and inconsistent drone regulations across countries, impeding private investments. The untapped potential of drone applications includes the transportation of perishable organs and personalized medicines, particularly expensive oncological drugs. Drones' unique features, such as autonomous driving systems, accessibility to inaccessible areas, cost-effectiveness compared to traditional transport, reduced delivery times, and the ability to carry heavy payloads, position them as pivotal elements in future smart cities and civil activities.

Crucially, the adoption of drone technology holds promise in mitigating climate change by reducing climate-altering gases. Traditional delivery methods contribute to environmental costs, including congestion, CO_2 emissions, and pollutants. In contrast, drone delivery, powered by energy-efficient, zero-emission electric batteries, emerges as a more efficient and environmentally friendly alternative. Although the drone delivery market is experiencing robust growth, the lack of standardized international regulations remains a significant challenge. As the world grapples with the challenges posed by the COVID-19 pandemic, rethinking regulations is imperative to foster the technological development of this evolving market, ensuring it becomes an invaluable asset for high-income countries and beyond.

Moving from healthcare to security, "*C-UAS in the Protection of Critical Infrastructures*" explores the vital role UAVs play in safeguarding essential facilities, underscoring their significance in proactive threat mitigation. The text highlights the critical vulnerability of democratic societies' socio-economic activities, which rely on critical infrastructures located in remote areas. Drones, originally developed for civilian applications, now pose a potential threat as technological options for both armed forces and malicious purposes such as terrorism and smuggling. The danger lies in the ability of low-cost drones to carry significant explosive payloads, enabling terrorist organizations to target critical infrastructure easily. Recognizing this risk, the chapter emphasizes the need for the development of Counter-Unmanned Aircraft Systems (C-UAS) as defense mechanisms capable of detecting, identifying, and intercepting aerial threats.

The researchers provide a comprehensive review of challenges related to protecting critical infrastructures from unmanned aircraft acting as loitering munitions. The analysis considers military and non-military threats from various drone systems, highlighting the heightened risk posed by rotary-wing drones with high payload capacity. The study advocates for the adoption of multi-sensor solutions, combining RF sensors with RADAR, acoustic, and image systems across visible and infrared spectrums. The integration of LiDAR, with its affordability, robustness, and compatibility with other sensing systems, is proposed as a promising avenue for further exploration. The accessibility of drone technology, coupled with its relatively low cost, underscores the urgency of advancing C-UAS technology to mitigate potential misuse in terrorist activities.

The logistics and transportation sector takes center stage in "*Challenges of Civil Application of 'Drones' for Commercial Purposes in the Logistics and Transportation Sector*" offering insights into the hurdles and triumphs of integrating UAVs into the logistics landscape. The chapter introduces UAVs in this arena, emphasizing their disruptive impact on various civil contexts and the significant economic potential in the transportation sector. Drones are anticipated to play a pivotal role in commercial activities, including logistics services for delivering small parcels and spare parts. However, challenges such as safety concerns in crowded urban environments and the absence of a stable global legal framework pose obstacles to the full social acceptance of drones. The lack of clarity in regulations and guarantees of privacy and fundamental rights complicates the integration of drones into societal norms.

The study highlights the transformative changes occurring in drone technology, aligning with the interests of the general population and reflecting the potential for enhancing business opportunities in various sectors. The consolidation of a stable legal framework is deemed crucial to incentivize research and development in the drone industry while ensuring ethical considerations, fundamental rights, environmental protection, and adherence to sector-specific legislation. Despite challenges and risks, the overall societal view of the civil use of drones, particularly in transportation and logistics, appears positive. The technology is seen as a solution to transport and environmental challenges, contributing to a green future ecosystem in Europe. The discussion emphasizes the need for ongoing dialogue between policymakers and market players to address evolving challenges and meet the needs of a drone industry poised to become a driving force in the global economy. The focus on regulatory changes anticipates shifts not only in commercial issues but also in key economic sectors, emphasizing the relationship between technology and innovation in the drone landscape.

Shifting the focus to industrial applications, "*Transport of Light Parts for Logistics Supply in Industrial Manufacturing Plants by Means of UAV*" delves into the precision and efficiency that UAVs bring to industrial logistics, streamlining processes and optimizing workflows. The study introduces a hovercraft Unmanned Aerial Vehicle designed for efficient light part delivery within manufacturing plants.

The researchers highlight the successful design, development, and evaluation of the hovercraft UAV system. The combination of technologies, including ArUco markers and UWB tags for indoor localization, ultrasonic sensors for obstacle avoidance, and the MPU6050 for corridor navigation, demonstrates significant advancements in safe and accurate navigation in complex indoor environments. The ArUco marker system proves reliable for precise indoor localization, and ultrasonic sensors enhance the UAV's capability to navigate cluttered environments safely. The MPU6050 accelerometer and gyroscope contribute to superior stability and agility during corridor navigation, minimizing disruption in manufacturing plants.

Extensive experimentation confirms the hovercraft UAV system's viability for light part delivery in manufacturing plants, showcasing consistent and accurate performance in indoor localization and corridor navigation. The research represents a significant step toward realizing safe, efficient, and autonomous UAV-assisted logistics in industrial environments, contributing to streamlined logistical operations and enhanced productivity.

The battle against natural disasters unfolds both in "*Monitoring of the Evolution of the Firefront of Forest Fires Using Unmanned Aerial Vehicles (UAVs) in Night*

Conditions" and "*Unmanned Aerial Vehicles (UAVs) in Firefighting*" where UAVs emerge as indispensable tools in disaster response and fire management.

In "*Monitoring of the Evolution of the Firefront of Forest Fires Using Unmanned Aerial Vehicles (UAVs) in Night Conditions*", the researchers discuss the transformative role of unmanned aerial vehicles in air power projection, emphasizing their significance in surveillance and tracking missions during wildfires to reduce the risk of personal injury. The chapter focuses on the Firepoctep Interreg Spain Portugal cross-border cooperation project, detailing the contributions in both hardware and software. The hardware involves the design and construction of a new drone suitable for hostile environments, such as forest fires. The software development aims to improve piloting conditions in low visibility, including night flights.

The chapter highlights the comprehensive approach to using UAVs for monitoring and analyzing wildfire evolution. The integration of cutting-edge technologies results in new UAV hardware and a sophisticated visualization system, offering valuable insights into fire surveillance operations. The software enhancements, including digital cartography and improved viewer functionality, provide substantial benefits. The integration of multiple drones within the same system presents a versatile solution for managing and visualizing multiple UAV operations. Through rigorous testing, the system's performance is examined under various conditions, ensuring robustness and adaptability to unpredictable scenarios crucial for effective decision-making during firefighting missions. The exploration of integrating the visualization system with real UAV equipment reveals the potential for enhanced accuracy and relevance by incorporating external sensors like smartphones. The new UAV hardware design and the developed assistance tool have the potential to significantly enhance the efficiency of wildfire surveillance operations, providing real-time insights and informed decision-making for pilots. The project establishes a robust foundation for future advancements in UAV-assisted wildfire monitoring, aligning with efforts to mitigate the catastrophic impacts of wildfires on the environment and communities.

On the other hand, "*Unmanned Aerial Vehicles (UAVs) in Firefighting*", focusing on the FIREPOCTEP project, details the analysis, design, and testing of a specific UAV tailored for night observation of forest fires. Forest fires present significant challenges, and UAVs can offer a fast, cost-effective solution without endangering pilots' lives. The designed UAV aims to provide safe, reliable, and economical monitoring of the firefront in forest fires occurring during night conditions.

The project's goal was to design, manufacture, and test a UAV capable of monitoring fire fronts in forest fires, particularly in the border area between Spain and Portugal, such as Galicia. The initial phase involved reviewing the characteristics of forest fires in the project's areas of interest. Subsequently, a comprehensive study based on wing airfoils and 3D wings was conducted to optimize autonomy, and simulations were performed to achieve the best results. The design process covered various elements, including the fuselage, empennage, engines, propellers, and on-board electronics, all tailored to fulfill the observation mission. Special attention was given to the structural aspect to ensure the UAV's lightweight yet robust design, maintaining necessary drag limits and structural integrity during flight. The text highlights the emphasis on autonomy evaluation through software simulation and real-world tests. The UAV's autonomy not only met but exceeded expectations, surpassing the values predicted by simulation and initial objectives. This positive outcome was attributed to the careful study of forest fire shapes and sizes in the project's study area. Overall, the designed UAV proves to be an effective solution for monitoring forest fires during night conditions, showcasing the successful application of UAV technology in addressing critical challenges.

The exploration of new frontiers in journalism is unveiled in "*Drone Journalism: Where the Human Eye Cannot Reach. Narratives and Journalistic Uses*" highlighting the transformative role of UAVs in journalistic storytelling. The chapter discusses the diverse applications of drones in journalism beyond traditional aerial imaging. It emphasizes their role in investigative and whistle-blower journalism, data journalism, immersive journalism, media coverage of protests, sporting and cultural events, urgent news coverage, reporting on armed conflicts, and live broadcasting. The chapter provides insights into the narrative utilities that drones bring to journalistic practices, enabling unique perspectives and coverage that may be challenging with traditional reporting methods. Despite the emerging industry, the text acknowledges obstacles to the widespread adoption of drones in newsrooms, shedding light on the challenges faced in integrating this technology into routine journalistic practices.

"*Drones in Aerospace Engineering Education: The Case of UVigo Aerotech*" discusses the complexity of aerospace engineering studies, encompassing various scientific and engineering branches such as thermodynamics, mechanics, calculus, and applications in engines and systems. While theoretical knowledge is crucial, the limited duration of the aerospace engineering degree often results in reduced time for practical lessons, creating a gap between theory and real-world applications. This gap is addressed by student associations like UVigo Aerotech, specifically at the University of Vigo, which focuses on bridging this divide by engaging students in hands-on projects like designing and manufacturing fixed-wing drones. This approach allows students to apply theoretical concepts in a practical setting, gaining valuable experience and insight into the aerospace sector.

The University of Vigo's Aerospace Engineering degree spans four years, covering numerous subjects and involving group divisions for practical lessons due to time and resource constraints. UVigo Aerotech serves as a platform for students to actively participate in extracurricular activities, contributing to their understanding of aerospace applications. By working on projects like drone manufacturing, students not only apply theoretical knowledge but also gain additional skills and knowledge beyond the standard curriculum, enhancing their competitiveness in the job market. The collaboration between student associations and universities ensures that students stay updated with the latest advancements in the aerospace industry, providing a dynamic and relevant educational experience.

Also covered in the book is the chapter '*Key Technologies for UAVs in Search and Rescue (SAR) and Surveillance Operations*' on the utilization of unmanned aerial vehicles (UAVs) in search and rescue (SAR) missions, which has evolved significantly, driven by advances in video analytics, precision navigation, and communication technologies. This chapter explored several applications that show the transformative impact of UAVs on SAR operations, emphasizing the critical role played by cutting-edge algorithms and technologies. Special emphasis is placed on video analysis algorithms, which allow unmanned aerial vehicles to carry out efficient searches for missing people, taking advantage of image processing and pattern recognition techniques. UAVs intended for these types of applications are equipped with advanced video analysis, which allows them to continuously monitor isolated people, improving situational awareness and response capabilities. It also explains the application of unmanned aerial vehicles in the planning and execution of rescue missions, which has been drastically improved thanks to real-time video analysis, which facilitates rapid decision-making and optimizes resource allocation. In particular, video analysis helps assess the extent of damage caused by natural disasters, enabling rapid and targeted response efforts. Efficient deployment of UAVs ensures optimal utilization of resources, minimizing response time and maximizing the effectiveness of SAR missions. In such missions, seamless communication is facilitated by various protocols such as RF data links, 3gPP networks, SATCOMS, and IEEE 802.XX is crucial for real-time data exchange between UAVs and ground stations. Finally, it describes how simulation is an indispensable component of SAR mission planning, as it helps in movement planning, and route design and ensures the preparation of UAV systems for complex and dynamic environments.

The final frontier beckons in "*Drones for Space Exploration*" where each is beyond Earth, providing a glimpse into their potential role in the exploration of outer space. The text discusses the potential synergy between planetary orbiters and aerial vehicles in extraterrestrial atmospheres, emphasizing the exciting opportunities for exploration and data collection that arise from their collaboration. While orbiters and surface vehicles have been used in space exploration, aerial vehicles are relatively new in this context. These vehicles offer unique capabilities, enabling exploration of difficult terrains and remote locations that were previously inaccessible to traditional surface rovers or landers. The chapter provides an extensive review of drone advancements in space exploration, highlighting their benefits in terms of cost-efficiency, adaptability to different environments, and versatility.

The applications of drones in space exploration encompass mapping, sample collection, and geological and atmospheric analysis of celestial bodies. The text presents a review of potential planets and moons that could be explored using drones, focusing on Mars and Titan. Conceptual designs for fixed-wing drones tailored to these destinations are proposed. The chapter also addresses challenges and limitations associated with drone technology in space, including power supply, materials, communication delays, thermal control, and the need for robust navigation systems in unfamiliar terrains. Despite the challenges, the text concludes that drones have the potential to revolutionize future space missions, providing greater understanding and exploration of the universe.

In conclusion, while numerous space exploration drone projects have been initiated, only one, Ingenuity, has been completed. The design of space drones is heavily influenced by the requirements set by carrier vehicles and the unique characteristics of extraterrestrial atmospheres. Mars and Titan are identified as promising targets, and conceptual designs for fixed-wing drones are proposed for each destination. The text emphasizes the importance of understanding aerodynamics and other subsystems, such as thermal control, propulsion systems, power generation, communication, and navigation. Each destination requires dedicated analysis, but the common challenge is the harsh flying conditions. The text acknowledges that realizing routine drone flights over Titan may take time, but research and development are underway to make such aerial exploration a reality.

Index

D. Carou et al. (eds.), *Applying Drones to Current Societal and Industrial Challenges*, Management and Industrial Engineering, https://doi.org/10.1007/978-3-031-55571-8

The manufacturer's authorised representative in the EU is Springer Nature Customer Service Centre GmbH, Europaplatz 3, 69115 Heidelberg, Germany. If you have any concerns regarding our products, please contact ProductSafety@springernature.com

Printed and bound by CPI Group (UK) Ltd, Croydon, CR0 4YY
30/09/2026
02225435-0005